FIRE AND EMERGENCY SERVICES SAFETY AND SURVIVAL

FIRE AND EMERGENCY SERVICES SAFETY AND SURVIVAL

Travis Ford
Nashville Fire Department
Volunteer State College

Pearson

Boston   Columbus   Indianapolis   New York   San Francisco   Upper Saddle River   Amsterdam
Cape Town   Dubai   London   Madrid   Milan   Munich   Paris   Montreal   Toronto   Delhi
Mexico City   Sao Paulo   Sydney   Hong Kong   Seoul   Singapore   Taipei   Tokyo

Publisher: Julie Levin Alexander
Publisher's Assistant: Regina Bruno
Executive Editor: Marlene McHugh Pratt
Senior Acquisitions Editor: Stephen Smith
Development Editor: Jo Cepeda
Associate Editor: Monica Moosang
Editorial Assistant: Samantha Sheehan
Director of Marketing: David Gesell
Senior Marketing Manager: Brian Hoehl
Marketing Specialist: Michael Sirinides
Marketing Assistant: Crystal Gonzalez
Managing Production Editor: Patrick Walsh
Production Liaison: Julie Boddorf
Production Editor: Karen Fortgang, bookworks publishing services
Senior Media Editor: Amy Peltier
Media Project Manager: Lorena Cerisano
Manufacturing Manager: Alan Fischer
Creative Director: Jayne Conte
Interior Designer: Wanda España
Cover Designer: Karen Salzbach
Cover Photo: Hugh Wingett
Composition: Aptara®, Inc.
Printing and Binding: R.R. Donnelley/Willard
Cover Printer: Lehigh-Phoenix Color/Hagerstown

10 9 8 7 6 5 4 3 2 1

www.pearsonhighered.com

ISBN 13: 978-0-13-701548-1
ISBN 10:   0-13-701548-8

CONTENTS

Chapter 4 Unsafe Practices 76

Chapter 10 Grant Programs 227

The staggering number of firefighter line-of-duty deaths and injuries continues each year in the U.S. fire service. It is not just a local problem. It is not just a state problem. It is a national problem that affects volunteer, career, and combination departments alike. Firefighter safety has been an issue that we talk about and write about often. We even spend a lot of time, energy, and effort on training and educating firefighters about and making plans for catastrophic accidents, natural disasters, terrorist events, and other hazards. Emergency medical services is another area that has taken up a lot of training and education time; yet, nothing that has been done has truly impacted the number of firefighter line-of-duty deaths and injuries. Instead of talking about it and writing about it, something needs to be done about it, and that something refers to you.

In the past, several organizations have challenged the U.S. fire service to reduce firefighter line-of-duty deaths. Almost 40 years ago, the *America Burning* report, issued by the National Commission on Fire Prevention and Control, pointed out that firefighters themselves were a big part of the nation's death and injury problem. It seemed up until that time firefighter deaths and injuries were considered just part of the job, and nobody was paying attention to the overall issue of firefighter safety. And to date it appears that not much has changed!

It is essential to realize how our ways of thinking and acting have brought us to today. It is not that we do not know what prevents us from succeeding; it is that we just do not take action on what we know is our greatest obstacle. If we want to be safer, we must be willing to let go of some old ways of thinking and begin to adopt new ones.

In March 2004, the first ever National Firefighter Life Safety Summit brought more than 200 fire service leaders together to focus on how to prevent line-of-duty deaths and injuries. As a result, the 16 Firefighter Life Safety Initiatives and the Everyone Goes Home® Program were created. Reducing firefighter fatalities by 50% over the next 10 years was the overall major goal. The summit was yet another attempt to get a handle on the issue of firefighter deaths and injuries that continues to plague the U.S. fire service. Unfortunately, the number of firefighter fatalities has not yet been reduced by any notable percentage.

Every day the fire and emergency services finds itself in a more complex and challenging environment. Even though we work in one of nature's most difficult and unforgiving environments, we must ask, "When will the time come when we are ready to step up and resolve the issue of safety and start to reduce the number of firefighter deaths and injuries?" In order to honor those individuals who have made the ultimate sacrifice, we must be willing to tell the whole cold hard truth. Until then, we will continue to dishonor every firefighter in the U.S. fire service.

Maybe our attitude and habits toward safety come from our symbol, the Maltese cross. According to James Parker, the symbol that is shared by all firefighters is an insignia of the fire service tradition:

> The Maltese Cross is a fitting symbol, a badge of honor worn by members of a culture wherein self-preservation can come second to getting the job done. In keeping with this tradition, the fire service through the years has accepted death in the line of duty as part of the job. . . . If fatalities were a cost of doing business, the cost to the fire service has been staggering. (2002, p. 12)[1]

The essence of the badge is filled with our history of tragedy. However, there is no honor in a needless death or injury of any firefighter anywhere at any time.

[1]Parker, J. G. (2002, September/October). "Fire Service: No More Maydays." *Every Second Counts, 4,* 12–16.

We cannot continue to hope that everything will turn out okay and that everyone will go home safe. Somehow pride has distorted our perspective of reality. What we need to realize is that to get what we have never had, we must do what we have never done.

So what would it take to make the problem disappear, or what steps are needed to minimize the number of deaths and injuries? Answers to those questions will affect our profession in the long term. However, we must first all agree that safety is an issue and that there is a need to develop proper attitudes and habits toward it. Attitudes develop when we learn to think correctly without acting. Habits develop when we learn to react correctly without thinking. Right attitudes alone will not reduce the number of line-of-duty deaths and injuries without cultivating the right habits. Right habits will not keep us safe with the wrong frame of mind. To develop a comprehensive set of attitudes and habits with the primary focus of safety will take a commitment from everybody in the fire and emergency services to create a comprehensive set of attitudes and habits.

A lack of knowledge and skills is another area that warrants closer attention. Have you ever asked yourself the question, either on or off the emergency scene: "What the hell were you thinking?" or made a statement such as this: "I knew something like this was going to happen!" We cannot expect firefighters in stressful emergency situations to act in a safe manner and make appropriate decisions if they are not trained and educated to do so. If we teach the right behavior but do not model it, we cannot expect it to be effective. What we teach, say, and do need to match. Sometimes we train and educate firefighters and think that they will automatically make the right decision when the time comes. However, in order to do that we must first be willing to evaluate every emergency run every day and understand that the evaluation process never ends. If we want to improve, we should start training and educating toward goals. Those goals should be the foundation for proper decision making and not just a way to meet more local, state, and federal requirements. Somewhere along the way we have developed the culture of "Don't make me think; just tell me what is on the exam!" Although improvement should occur from our evaluation process, improvement seldom occurs unless someone is pushing us. Therefore, the responsibility for safety is a two-way street, involving everybody from top to bottom in the fire and emergency services.

To what degree do we allow members of our department to give just 90%, 80%, 70%, or less of what is required of them? Somewhere along the way we have come to accept and even celebrate merely trying. To use the word *try* means to quit with respect. If the goal is to try, it is like premeditated failure. We should never mistake good intentions for results. Everybody has heard, "I will try to do better next time." However, the next time comes, and you get the same results followed by the same statement. The difference between success and failure can be in doing something nearly right or doing it exactly right. Even though absolute perfection is impossible, asking firefighters to follow very strict safety policies and procedures in certain areas is not nearly as difficult as some make it out to be. When it comes to safety, we should be striving for consistently near-perfect performance.

It is amazing what we get used to. We have become comfortable with and tolerant of the way things are; and our continued careless behavior has become normal and accepted in everything we do in the fire and emergency services, both on and off the emergency scene. Our culture has allowed us to get and remain out of shape both physically and mentally, while still expecting top-notch performance on the emergency scene. The bottom line is that we cannot continue to make excuses and think we will be successful in reducing the number of line-of-duty deaths and injuries. It is widely agreed that an overwhelming majority of the deaths and injuries are preventable, and the total number every year is unacceptable. The same message every year should be unacceptable to us as a profession. When is "I will try to do better next time" not enough? Our attitudes and habits toward safety are repairable. We continue to hear excuse after excuse and think that we will be successful in conquering the issue of safety. We need to quit believing we are safe and start expecting firefighters to be safe by putting action behind what we believe. We can no longer continue to ignore the problem of firefighter safety by looking the other way and choosing not to say something to those individuals who continually count on luck each shift they work and at every incident to which they respond.

As you prepare to read *Fire and Emergency Services Safety and Survival,* you are making an investment—in yourself, your department, and your future. It is important to learn from the mistakes of others, because you do not have the time to make all of them yourself. My recommendation is that you first read through the entire book to get a big-picture view of safety and an understanding of how the chapters are interrelated. Then go back and read through each chapter separately, taking time to reflect on the subject and focus on the implications to you and your department. Doing so will not only help you understand the principles but also emphasize how it directly applies to you and your department. You should try to read through and process a new chapter each month. However, certain chapters will require more time than others to learn and build upon to get you to the right attitude, habit, and behavior toward safety. Once you gain a firm grasp of the principles contained in each chapter of the book, you will have a good foundation for understanding the need to put safety first and be serious about it; why you should not turn the responsibility of safety over to anybody but yourself; and when it comes to safety, why you should be saying no excuses, no exceptions.

Safety is not going to happen overnight, so do not expect dramatic change right away. When you go back in to work, it will be the same people, the same department, and the same issues. However, you cannot transform your colleagues or department until you know how to transform your own performance. Do not resign yourself to blaming the system; start by taking responsibility for yourself. Just remind yourself that safety can begin with one commitment and one specific action. Eventually, you will create a ripple effect that will start to transform everyone around you and your department.

If we have identified that safety is truly an issue in the fire service—and we know that it is—then having the commitment to change is necessary. Whenever people have the option to avoid change, they will. Preaching change is much different than practicing it. In order to truly reduce the number of firefighter line-of-duty deaths and injuries, the U.S. fire service will need a total commitment from everybody concerned in order to create the type of change that challenges the existing attitudes and habits that have become commonplace. If we are committed to changing our attitudes and habits toward safety, then we will start to see a difference in the number of line-of-duty deaths and injuries. When you are *interested* in doing something, you do it only when it is convenient. But when you are *committed* to doing something, you accept no excuses and no exceptions, only results.

Safety should be a primary concern throughout our entire career. Remember, you can spend your career any way you want, but you can do it only once, so choose to do it safely. At the end of a long career it would be nice to say, "I was never injured myself and no one was ever injured as a result of my errors." The bottom line is to take safety seriously.

Simple Truths About Firefighter Safety

Firefighters given the option to avoid change will.

Initiate a 24/7 commitment to safety.

Remember, do not complain about what you permit.

Every firefighter counts regardless of rank.

Firefighters have an incomplete understanding of their own technology.

Individual firefighters can make a big difference.

Good safety habits are not always addressed by a standard.

How little changes can somehow have big effects.

Train so that you do not confuse output with outcomes.

Expect that it is your responsibility to do the right thing.

Realize that not all knowledge that we accumulate through "experience" is accurate.

Safety depends on many things, but mainly you.

Aim to change a little bit every day and you will start to see a big difference.

Firefighters are not persuaded by what they hear, but by what they understand.

Even though you say something 100 times, it is not the same as seeing it.

Tolerance of unsafe acts should not be accepted.

You rarely make significant improvements accidentally.

Travis Ford

NATIONAL FALLEN FIREFIGHTERS FOUNDATION

Over the last five years, Everyone Goes Home® and the 16 Firefighter Life Safety Initiatives® that support it were developed under the purview of the National Fallen Firefighters Foundation (NFFF). When the NFFF was chartered by Congress in 1992, it was charged with a clear mandate to honor all firefighters throughout the United States killed in the line of duty, and to provide support to the survivors as they rebuild their lives. In addition, through programs such as Everyone Goes Home®, the foundation is working with the national fire service community to reduce the occurrence of firefighter fatalities and embrace a culture of safety throughout.

In order to accomplish this, a commitment to higher education must be made, and we now have the opportunity to lead more in-depth studies on the application of this blueprint for change at the academic level.

Through FESHE, this course has become a core competence for a fire science degree and, in our view, rightfully so. Through partnerships with the learning institutions, publishing houses, and the National Fallen Firefighters Foundation, we can travel together down the path to preventing line-of-duty death.

Ronald J. Siarnicki

A portion of every purchase of this book will be donated to the National Fallen Firefighters Foundation.

ACKNOWLEDGMENTS

This project would not have been possible without the help and guidance of many individuals; I am forever grateful to each one. This book is dedicated to the love of my life, Sheila, for her patience and understanding throughout the project. To my daddy, Ernest A. Pfeiffer, who was the first one that actually exposed me to the fire service and proudly served every day of his 37-year career in the Nashville Fire Department on Truck 19. To my friend forever, Jimmy Dean, who always took the time to show and introduce me to more in the fire service than any body could ask for. For all of those officers and firefighters that I have had the pleasure of working with side by side—thanks for making it enjoyable. To those individuals that I had the pleasure to be taught by and that I conversed with during my training and education at the National Fire Academy and other conferences—thank you.

Special thanks goes to the various contributors whose breadth of fire and emergency services experience added a dimension of geographical influence to the content. This book was possible because of you and I thank you for being a part of the team. I would also like to extend a special thanks to my development editor, Jo Cepeda, and to Monica Moosang and Samantha Sheehan on the Brady team for organizing and preparing the project for publication.

Chapter Contributors

A Review of Fire and Emergency Services Line-of-Duty Deaths: Adam K. Thiel— Mr. Thiel is with the City of Alexandria, VA.

Chapter 1: John B. Tippett, Jr.—Mr. Tippett is the deputy chief of Operations for the City of Charleston (SC) Fire Department. He spent 32 years with the Montgomery County (MD) Fire and Rescue Service before going to Charleston in 2009.

Chapter 2: Nicol P. Juratovac—Ms. Juratovac is acting captain with the San Francisco Fire Department. She holds the executive fire officer (EFO) designation and has earned a BA in English Literature from the University of California at Berkeley and is a Juris Doctorate.

Chapter 3: Marc A. Revere—Chief Revere is the fire chief of the Novato Fire Protection District, an internationally accredited agency in Marin County, CA. He has 34 years in the fire service, is an EFO, a CFO, and a Harvard Fellow. He is the first recipient of the Ronny Jack Coleman Leadership and Legacy Award from the Center for Public Safety Excellence 2010. He holds a bachelor's degree in Management from the University of Redlands.

Chapter 4: John F. Sullivan—Mr. Sullivan is the deputy fire chief of the Worcester (MA) Fire Department. He holds a master's degree in Public Administration and is an executive fire officer (EFO) and chief fire officer (CFO).

Chapter 5: Ronny J. Coleman—Chief Coleman, Fellow of the Institution of Fire Engineers (FIRireE), has earned an associate of arts degree in Fire Science, a bachelor's degree in Political Science, and a master's degree in Vocational Education. He is designated as a chief fire officer (CFO) in California.

Chapter 6: Martha Ellis—Ms. Ellis is a division chief and the fire marshal for the Salt Lake City Fire Department, a certified fitness coordinator, and the fitness editor of *Fire Rescue* magazine. She is also the five-time Women's World Champion in the Firefighter Combat Challenge.

Chapter 7: Charles R. Jennings—Dr. Jennings is a Member of the Institution of Fire Engineers (MIFireE) and a chief fire officer (CFO). He is a faculty member at John Jay College of Criminal Justice of the City University of New York, where he is also director of the Christian Regenhard Center for Emergency Response Studies.

Chapter 8: Martin J. DeLoach—Chief DeLoach earned a bachelor of science degree from Madonna University and a master's degree in Interdisciplinary Technology from Eastern Michigan University. He is currently the fire chief for Palm Beach Shores, FL.

Chapter 9: Dale R. (Rob) Rush—Mr. Rush is the captain of the Palm Beach County Fire Rescue Fire and Arson Investigations Division; a director for the International Association of Arson Investigators; and a past president and life member of the Florida chapter of the International Association of Arson Investigators.

Chapter 10: Brian P. Vickers—Mr. Vickers holds a BS in Computer Information Systems Management and is the chief executive officer of Vickers Consulting Services, Inc. He is also an IFSAC/NFPA Firefighter 1, a TX SFFMA Intermediate Firefighter, a driver/operator, an Emergency Medical Technician–Basic, and a TX SFFMA Level 2 Instructor; as well as a member of the Cy-Fair Volunteer Fire Department, Cypress, TX.

Chapter 11: Wil Dane—Mr. Dane holds a master's degree in Public Administration and the executive fire officer (EFO) designation. He has been the fire chief of the City of Three Rivers and Scio Township (MI) (retired); the battalion chief/shift commander of the City of Victoria (TX); the chief administrative state fire marshal in Louisiana; the assistant director of the LSU Fire and Emergency Training Institute (LA); a training coordinator in Louisiana; and a firefighter-EMT in Michigan.

Chapter 12: Richard Gist, Vickie Taylor, and John F. Neeley—Mr. Gist, Ph.D., is the principal assistant to the director of the Kansas City (MO) Fire Department. Ms. Taylor, licensed clinical social worker (LCSW), is the division manager of Youth, Adult and Family Services of the Prince William County (VA) Community Services Board. Mr. Neeley is the deputy chief of Special Operations for the Kansas City (MO) Fire Department.

Chapter 13: Beverley E. Walker—Ms. Walker is a lieutenant with the Hall County (GA) Fire Services Training Division. She holds a master's degree in Education, a master's degree in Public Administration, and the executive fire officer (EFO) designation. She has over 20 years' experience in the fire service, specifically in fire prevention and safety education.

Chapter 14: A. K. Rosenhan—Mr. Rosenhan is a professional engineer, Fellow of the Institution of Fire Engineers (FIFireE), and chief fire officer (CFO). He is the fire services coordinator in Oktibbeha County (MS)

Chapter Reviewers

I would like to thank many other individuals for taking the time and effort to review the book. Their comments and suggestions were invaluable throughout the process. To those individuals who gave support and information throughout this project that did not want to be mentioned and to those that I have probably missed—thank you.

John E. Barrett
Firefighter
Plano Fire Department/Founder,
    President—HONOR THEIR
    SACRIFICES
Plano, TX

Don Beckering
State Director
System Safety and Emergency Services
    Education
Minnesota State College and
    Universities

James M. Broman, CFO, MIFireE
Fire Chief, Lacey Fire District
Lacey, WA

Jill Napoletan Craig, M.S. Sports
    Medicine
Austin Fire Department
Austin, TX

Jim Crawford (Retired)
Fire Marshal
Portland Bureau of Fire and Rescue
Portland, OR

Leo DeBobes, MA (OS&H), CSP,
    CHCM, CPEA, CSC, EMT
Assistant Professor/Fire Protection
    Technology Program Coordinator
Suffolk Community College
Selden, NY

John DeIorio
Fire Chief, Ocala Fire Rescue
Ocala, FL

Kerri L. Donis, CFO, EFO, MIFireE
Fresno Fire Department
Fresno, CA

Gary W. Edwards
Fire Science Program Director
MSU—Billings, College of
    Technology
Billings, MT

Kim C. Favorite
Occupational Health & Fitness
    Strategic Advisor
Seattle Fire Department
Seattle, WA

Ryan P. Fox
Campbell County Fire Department
Gillette, WY

Larry J. Hill, FF/EMT-P
Georgia

Steve Kerber, P.E.
Underwriters Laboratories
Northbrook, IL

Jeff Kimble, M.S.
Associate Professor
Fire Safety Engineering Technology
The University of North Carolina at
    Charlotte
Charlotte, NC

Daniel T. Kistner
Fire Chief
Lufkin Fire Dept
Lufkin, TX

Robert Lantman
Captain
Clearcreek Fire District
Springboro, OH

Murrey Loflin (Retired)
Battalion Chief
Virginia Beach Fire Department
Virginia Beach, VA

Daniel Madrzykowski, P.E. FSFPE
National Institute of Standards and
    Technology
Gaithersburg, MD

Byron Mathews
Cheyenne Fire and Rescue
Cheyenne, WY

Mike McEvoy, Ph.D., REMT-P,
    RN, CCRN
EMS Coordinator
Saratoga County, NY

Gary P. McGinnis
Firefighter
Maryland State Fireman's Assn.
Sykesville, MD

Don Mehl, Training (Retired)
Captain
El Paso Fire Department
El Paso, TX

James G. Munger, Ph.D., MIFIreE, CFPS
President of James G. Munger and
    Associates, Inc.
Adjunct Faculty of the National
    Fire Academy
Cullman, AL

William Opsitnik, BSAS
Captain, Liberty Township Fire
    Department
Youngstown, OH

Randall F. Parr, EFO, CFO
Tomball Fire Department
Tomball, TX

Mike D. Pritchard (Retired)
Assistant Fire Chief
El Paso, TX

Vickie Pritchett
Project Manager
Fire Team USA

Shane Ray
Fire Chief
Pleasant View Volunteer Fire
  Department
Pleasant View, TN

Scott B. Richardson
Lieutenant
South Metro Fire Rescue Authority
Centennial, CO

Jennifer Schottke
Municipal Fire and Public Safety Policy
ESRI
Vienna, VA

Christopher J. Schutte
Fire Lieutenant
Milwaukee Fire Department
Milwaukee, WI

Robert Scott, Ph.D., CTS
Director, FAFD Behavioral Health and
  Wellness Programs
Los Angeles, CA

Dr. Ray O. Shackelford
Fire Protection and Safety Programs
West Los Angeles/Harbor College
Wilmington, CA

Chris Shimer (Retired)
Deputy Chief
Howard County Department of Fire &
  Rescue Services
Columbia, MD

Heidi Simon
Lieutenant
South Metro Fire Rescue Authority
Centennial, CO

Robert D. Sjolund
Lead Instructor
Central Oregon Community College
Redmond, OR

Michael Stanley, BS, EMT-P
Captain
Aurora Fire Department
Aurora, CO

W. Faron Taylor (Retired)
Office of the Maryland State Fire
  Marshal
Pikesville, MD

George L. Thomas IV
Safety Lieutenant
Frederick County, Maryland Division
  of Fire and Rescue Services
Frederick, MD

Jim Tidwell (Retired)
Fort Worth Fire Department
Fort Worth, TX

LaRon Tolley
Director
Fire Services Administration
Western Oregon University
Monmouth, OR

Brad Van Ert
Fire Captain
Downey Fire Department
Downey, CA

Allen Walls
Division Chief
Training & Safety
Colerain Township Department of
  Fire & EMS
Cincinnati, OH

James E. White, EFO, CFO, MIFireE
Chief of Department
Winter Park Fire Rescue
Winter Park, FL

Battalion Chief Mark J. Wilson
Hallandale Beach Fire Rescue
Coordinator, Continuing Education,
  Broward Fire Academy
Terrace Davie, FL

ABOUT THE EDITOR

Travis Ford is the director of the Fire Science Technology Program at Volunteer State Community College in Gallatin, Tennessee. He additionally serves as an adjunct instructor for the U.S. Department of Homeland Security's National Emergency Training Center in several areas, including the National Incident Management System, Incident Command for High-Rise Operations, and Fire Protection Systems for Emergency Operations. Mr. Ford holds an associate's degree in Fire Science from Tennessee State University, a bachelor's degree in Fire Administration from the University of Memphis, and a master's of science degree in Human Resource Development from the University of Tennessee. He has also completed the Executive Fire Officer Program at the National Fire Academy.

Currently serving as a district chief on shift at Station 9, "The Bottoms," with the Nashville Fire Department in the busiest district in the city with 39 stations, he has ascended the ranks over his 25-year career with Nashville. He has been responsible for developing and presenting programs in leadership for company and chief officer training and high-rise operations around the country, and recently served as vice-chair of the high-rise committee for the International Fire Service Training Association (IFSTA). Additionally, Mr. Ford served as the vice-chair of the National Fire Science Associate's Degree Program Committee and is currently serving on the High School to College Pathways Committee with the Fire and Emergency Services Higher Education committee in the development of model fire science course curriculum.

Fire and Emergency Services Higher Education (FESHE) Grid

The following grid outlines Principles of Fire and Emergency Services Safety and Survival course requirements and where specific content can be located within this text:

Course Requirements	1	2	3	4	5	6	7	8	9	10	11	12	13	14
Define and describe the need for cultural and behavioral change within the emergency services relating to safety, incorporating leadership, supervision, accountability, and personal responsibility.	X	X												
Explain the need for enhancements of personal and organizational accountability for health and safety.		X												
Define how the concepts of risk management affect strategies and tactical decision making.			X											
Describe and evaluate circumstances that might constitute an unsafe act.				X										
Explain the concept of empowering all emergency services personnel to stop unsafe acts.				X										
Validate the need for national training standards as they correlate to professional development inclusive of qualifications, certifications, and recertifications.					X									
Defend the need for annual medical evaluations and the establishment of physical fitness criteria for emergency services personnel throughout their careers.						X								
Explain the vital role of local departments in national research and data collection systems.							X							

Course Requirements	1	2	3	4	5	6	7	8	9	10	11	12	13	14
Illustrate how technological advancements can produce higher levels of emergency services safety and survival.								X						
Explain the importance of investigating all near misses, injuries, and fatalities.									X					
Discuss how incorporating the lessons learned from investigations can support cultural change throughout the emergency services.									X					
Describe how obtaining grants can support safety and survival initiatives.										X				
Formulate an awareness of how adopting standardized policies for responding to emergency scenes can minimize near misses, injuries, and deaths.											X			
Explain how the increase in violent incidents impacts safety for emergency services personnel when responding to emergency scenes.											X			
Recognize the need for counseling and psychological support for emergency services personnel, their families, as well as identify access to local resources and services.												X		
Describe the importance of public education as a critical component of life safety programs.													X	
Discuss the importance of fire sprinklers and code enforcement.													X	
Explain the importance of safety in the design of apparatus and equipment.														X

A Review of Fire and Emergency Services Line-of-Duty Deaths

Adam K. Thiel

What Is a Line-of-Duty Death?

At least four fire and emergency services organizations track the number of firefighters in the United States who die from causes associated with their duties each year: the International Association of Fire Fighters (IAFF), the National Fallen Firefighters Foundation (NFFF), the National Fire Protection Association (NFPA), and the United States Fire Administration (USFA).

Each of those organizations applies detailed criteria to determine whether or not the circumstances of a firefighter's death should be included in its annual listing. Each organization's criteria are slightly different from those of the others. For instance, some organizations count "on-duty" firefighter deaths (the firefighter is included in the count only if he or she is actually on duty at the time of death), whereas other organizations count "line-of-duty" deaths (LODDs). (See Table 1.)

In 2009 the Safety, Health, and Survival Section of the International Association of Fire Chiefs (IAFC) called a meeting of numerous fire and emergency services organizations, including the four just mentioned, to discuss a single national definition for line-of-duty deaths and injuries. As this book goes to press, the final outcome of that effort is still to come.

TABLE 1	Public Safety Officers' Benefits (PSOB) Programs

- Since 1976 the Public Safety Officers' Benefits (PSOB) Programs have been providing benefits to survivors of public safety officers who have died or been permanently disabled in the line of duty.

- The Hometown Heroes Survivors Benefits Act of 2003 expanded federal benefits to include public safety officers who die from a heart attack or stroke within 24 hours of a nonroutine stressful or strenuous physical public safety activity or training.

A Long and Tragic History

The United States has a long and tragic history of fire and emergency services line-of-duty deaths. While the courage displayed by firefighters in extreme situations cannot be denied, many are killed each year under circumstances that too often are preventable.

Over the past 30 years, the number of U.S. firefighters killed in the line of duty steadily decreased, as displayed in Figure 1, using data from the U.S. Fire Administration (USFA, 2009c).

Despite the overall 30-year downward trend in firefighter line-of-duty deaths, there are still several causes for alarm. For the last two decades, each year an average of 112 firefighters were killed while responding to, on the scene of, or returning from emergency incidents. That statistic does not include the 343 New York City firefighters killed on September 11, 2001, which would increase those numbers even further.

The deaths of more than 100 firefighters each year is even more shocking considering the overall decline of fire incidents in the United States. According to the USFA, since 1978 the rate of firefighter fatalities per 100,000 fire incidents actually peaked in 1999 (USFA, 2002)! As seen in Figure 2, over the 10 years from 1999 to 2008, the annual rate of firefighter fatalities per 100,000 fire incidents averaged 3.5, ranging from a low of 2.81 in 2005 to a high of 4.5 in 1999 (USFA, 2009c). (See Figure 3.) The latest statistics in 2008 separated 118 firefighter line-of-duty deaths into three categories: 66 volunteer firefighter, 34 career firefighter, and 18 wildland agency responder deaths (USFA, 2009a).

A 2002 study performed by National Fire Protection Association (NFPA) researchers on firefighter fatalities at structural fires revealed another troubling finding: "The rate of firefighter deaths at structure fires in the late 1990s was roughly the same as the rate in the late 1970s" (Fahy, LeBlanc, and Molis, 2007).

Discussions about firefighter line-of-duty deaths often are confined to those occurring immediately before, during, or after an incident. Yet the effects of chronic exposure to a hazardous work environment can have long-term implications on firefighter survival.

FIGURE 1 Firefighter line-of-duty deaths, 1977–2008. Note: This analysis does not include the unforgettable and tragic loss of 343 firefighters in New York City on September 11, 2001. *Graph created from data published by USFA, October 14, 2009*

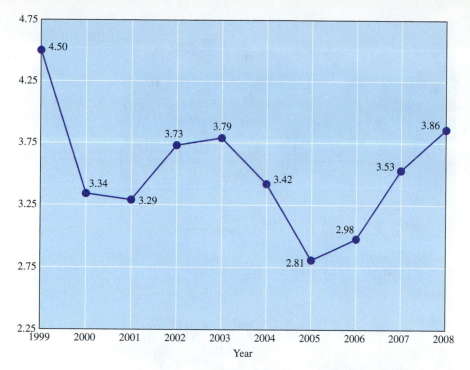

FIGURE 2 Firefighter line-of-duty deaths per 100,000 fire incidents, 1999–2008. Note: This analysis does not include the unforgettable and tragic loss of 343 fire-fighters in New York City on September 11, 2001. *Graph created from data published by USFA, October 14, 2009*

FIGURE 3 This memorial commemorates the tragic deaths of nine Boston firefighters from a build-ing collapse at the Hotel Vendome fire on June 17, 1972. *Courtesy of Adam K. Thiel*

Medical research suggests that firefighters are at greater risk than the general population of experiencing certain cancers, heart and lung disease, and other chronic illnesses.

Line-of-duty deaths are not the only threat to firefighter safety and survival. Industry estimates place the annual number of firefighter injuries in the United States at approximately 80,000 (NFPA, 2009g). The actual number of U.S. firefighter injuries occurring each year is unknown, but could be significantly higher than estimated.

Firefighter injuries place a great burden on fire departments, on community institutions, on families, and most importantly, on the injured responders themselves. According

FIGURE 4 Thirty-five years after the Hotel Vendome collapse, and almost to the day, nine firefighters died in Charleston, South Carolina, at the Sofa Super Store fire on June 18, 2007. *Courtesy of Adam K. Thiel*

to a study performed by the National Institute of Standards and Technology (NIST), the total economic cost of U.S. firefighter injuries and prevention efforts ranges from $2.8 billion to $7.8 billion per year (NIST, 2004, p. ii).

Despite meaningful advances in technology, personal protective equipment, industry standards, emergency scene operating practices, and the concerted efforts of many fire and emergency services organizations, death and injury statistics reflect a clear and compelling need for a greater emphasis on all aspects of firefighter safety, health, and survival. The challenge presented by firefighter deaths and injuries has not diminished in proportion to the overall national decline in structural fires over the past few decades. (See Figure 4.) Fire and emergency services leaders at all levels—from the jump seat to the chief's office—must further commit to preventing the needless loss of life and livelihood suffered by career and volunteer responders each year.

References

Bureau of Justice Assistance. (n.d.). *Public Safety Officers' Benefits (PSOB) Program: Hometown Heroes Survivors Benefits Act of 2003*. Retrieved November 17, 2009, from http://www.ojp.usdoj.gov/BJA/grant/psob/psob_heroes.html

Bureau of Justice Assistance. (n.d.). *Public Safety Officers' Benefits (PSOB) Programs*. Retrieved November 19, 2009, from http://www.ojp.usdoj.gov/BJA/grant/psob/psob_main.html

Fahy, R. F. (2002, July). *U.S. Fire Service Fatalities in Structure Fires, 1977–2000*. Quincy, MA: National Fire Protection Association. Retrieved November 18, 2009, from http://www.nfpa.org/assets/files/pdf/FFFStructure.PDF

Fahy, R. F., LeBlanc, P. R., and Molis, J. L. (2007, June). *What's Changed over the Past 30 Years?* Quincy, MA: National Fire Protection Association, p. 1. Retrieved November 18, 2009, from http://www.nfpa.org/assets/files/PDF/whatschanged.pdf

Karter, M. J. (2007, February). *Patterns of Firefighter Fireground Injuries*. Quincy, MA: National Fire Protection Association. Retrieved November 18, 2009, from http://www.nfpa.org/assets/files/PDF/OS.PatternsSummary.pdf

National Fire Protection Association. (2009a). *Deadliest/Large-Loss Fires: Deadliest Fires in the U.S. with 5 or More Firefighter Deaths at the Fire Grounds, 1977–2007*. Retrieved April 2, 2009, from http://www.nfpa.org/itemDetail.asp?categoryID=954&itemID=35210&URL=Research/Fire%20statistics/Deadliest/largeloss%20fires

National Fire Protection Association. (2009b). *Deadliest/Large-Loss Fires: Deadliest Incidents in the U.S. Resulting in the Deaths of 8 or More Firefighters*. Retrieved April 2, 2009, from http://www.nfpa.org/itemDetail.asp?categoryID=954&itemID=40025&URL=Research/Fire%20statistics/Deadliest/large-loss%20fires

National Fire Protection Association. (2009c). *The U.S. Fire Service: Firefighter Activities, Injuries, and Deaths*. Retrieved April 2, 2009, from http://www.nfpa.org/itemDetail.asp?categoryID=955&itemID=23605&URL=Research/Fire%20statistics/The%20U.S.%20fire%20service

National Fire Protection Association. (2009d). *The U.S. Fire Service: Firefighter Deaths (U.S. Fire Service)*. Retrieved April 2, 2009, from http://www.nfpa.org/itemDetail.asp?categoryID=955&itemID=23674&URL=Research/Fire%20statistics/The%20U.S.%20fire%20service

National Fire Protection Association. (2009e). *The U.S. Fire Service: Firefighter Deaths by Type of Duty*. Retrieved April 2, 2009, from http://www.nfpa.org/itemDetail.asp?categoryID=955&itemID=23471&URL=Research/Fire%20statistics/The%20U.S.%20fire%20service

National Fire Protection Association. (2009f). *The U.S. Fire Service: Firefighter Fireground Injuries by Nature of Injury*. Retrieved April 2, 2009, from http://www.nfpa.org/itemDetail.asp?categoryID=955&itemID=23635&URL=Research/Fire%20statistics/The%20U.S.%20fire%20service

National Fire Protection Association. (2009g). *The U.S. Fire Service: Firefighter Injuries by Type of Duty*. Retrieved April 2, 2009, from http://www.nfpa.org/itemDetail.asp?categoryID=955&itemID=23466&URL=Research/Fire%20statistics/The%20U.S.%20fire%20service

National Institute for Occupational Safety and Health. (2008, November). *Fire Fighter Fatality Investigation and Prevention Program: Leading Recommendations for Preventing Fire Fighter Fatalities, 1998–2005*. Morgantown, WV: Department of Health and Human Services, Centers for Disease Control and Prevention. Publication #2009-100. Retrieved November 18, 2009, from http://www.cdc.gov/niosh/docs/2009-100/pdfs/2009-100.pdf

National Institute for Occupational Safety and Health. (2009, February 11). *Nine Career Fire Fighters Die in Rapid Fire Progression at Commercial Furniture Showroom—South Carolina*. Morgantown, WV: Department of Health and Human Services, Centers for Disease Control and Prevention. Publication #F2007-18. Retrieved November 18, 2009, from http://www.cdc.gov/niosh/fire/reports/face200718.html

National Institute of Standards and Technology. (2003, November). *Trends in Firefighter Fatalities Due to Structural Collapse, 1979–2002*. Gaithersburg, MD: U.S. Department of Commerce. NISTIR #7069. Retrieved November 18, 2009, from http://fire.nist.gov/bfrlpubs/fire03/PDF/f03024.pdf

National Institute of Standards and Technology. (2004, August). *The Economic Consequences of Firefighter Injuries and Their Prevention: Final Report*. Gaithersburg, MD: U.S. Department of Commerce. NIST GCR#05-874. Retrieved November 18, 2009, from http://www.fire.nist.gov/bfrlpubs/NIST_GCR_05_874.pdf

U.S. Fire Administration. (2002, April). *Firefighter Fatality Retrospective Study.* Emmitsburg, MD: National Fire Data Center. Report #FA-220. Retrieved November 18, 2009, from http://www.usfa.dhs.gov/downloads/pdf/publications/fa-220.pdf

U.S. Fire Administration. (2009a, September). *Firefighter Fatalities in the United States in 2008.* Retrieved November 19, 2009, from http://www.usfa.dhs.gov/downloads/pdf/publications/ff_fat08.pdf

U.S. Fire Administration. (2009b, October 1). *Firefighter Casualties, 2000–2009.* Retrieved October 17, 2009, from http://www.usfa.dhs.gov/fireservice/fatalities/statistics/casualties.shtm

U.S. Fire Administration. (2009c, October 14). *Historical Overview.* Retrieved October 17, 2009, from http://www.usfa.dhs.gov/fireservice/fatalities/statistics/history.shtm

1

Fire and Emergency Services Culture

John B. Tippett, Jr.

Courtesy of Travis Ford, Nashville Fire Department

KEY TERMS

accountability, *p. 17*

culture, *p. 8*

defensive mode, *p. 22*

leadership, *p. 17*

management, *p. 17*

normalization of deviance, *p. 15*

norms, *p. 9*

offensive mode, *p. 23*

responsibility, *p. 17*

standard operating procedures/guidelines (SOPs/SOGs), *p. 9*

tradition, *p. 8*

OBJECTIVES

After reading this chapter, the student should be able to:

- Define *culture* as it applies to the safety and survival behavior of the fire and emergency services.
- Discuss the attitudes and behaviors that contribute to an unsafe culture within the fire and emergency services.
- Describe the elements that impact cultural change within the fire and emergency services.
- Understand the importance of changing culture when cultural elements inhibit improved safety.
- Discuss how to implement the National Fallen Firefighters Foundation 16 Firefighter Life Safety Initiatives to support changing the culture for a safer fire and emergency services.

PEARSON

myfirekit™

For practice tests and additional resources, visit **www.bradybooks.com** and follow the **MyBradyKit** link to register for book-specific resources.

Register for **MyFireKit** by following directions on the **MyFireKit** student access card provided with this text. If there is no card, go to **www.bradybooks.com** and follow the **MyBradyKit** link to Buy Access from there.

Culture and the Fire and Emergency Services

culture ■ The all-encompassing set of traditions, procedures, norms, attitudes, behaviors, and personnel demographics shared by an emergency services department and in reference to the community served.

The fire and emergency services, whether paid, volunteer, or a combination of both, is one arm of public service that generally enjoys a high level of respect within the community it serves. The result of that respect is a high sense of self-esteem among its members. The respect also has given rise to the development of the fire and emergency services **culture**. Just what is the culture of the fire and emergency services? As used in this chapter, it is the all-encompassing set of traditions, procedures, norms, attitudes, behaviors, and personnel demographics shared by an emergency services department and in reference to the community served. Each of those parts of the fire and emergency services culture will be addressed in this chapter.

There are some generalizations that we can use as a springboard in defining a common organizational culture in the fire and emergency services as well as the role an individual firefighter plays in achieving change within that culture; however, defining a national fire and emergency services department culture is nearly impossible.

TRADITIONS

tradition ■ An act, action, policy, or procedure that is handed down from generation to generation and is considered key to an organization's cultural identity.

Traditions abound in the fire and emergency services. A **tradition** is an act, action, policy, or procedure that is handed down from generation to generation and is considered key to an organization's cultural identity. A tradition is generally performed or adhered to with unquestioning loyalty. There are three types of fire and emergency services traditions, and they develop differently:

■ *Broad fire and emergency services traditions.* These are the traditions that define the fire and emergency services as a whole. They include:
   ■ Serving the public before self
   ■ Maintaining a high state of operational readiness
   ■ Having pride in the organization
   ■ Looking out for each other
■ *Policy-based traditions.* Some traditions are established by policy. For example, shift-change schedules, crew meeting nights, valor awards ceremonies, and promotion ceremonies are traditions established by policy. (See Figure 1.1.)

FIGURE 1.1 Awards ceremonies are an example of a policy-based tradition. *Courtesy of Gary Layda, Metropolitan Government of Nashville*

- *Traditions of unknown origin.* Finally, there are those traditions, the origins of which have been lost to the ages, that are adhered to with a steadfast commitment. Examples include rookies doing the dishes, the pinning of the insignia of rank by family members on a newly appointed officer, the annual crew outing, and pushing a new piece of apparatus into the station. (See Figure 1.2.)

What traditions do for the fire and emergency services can be likened in some ways to the importance of wearing the appropriate personal protective ensemble while on an emergency. Traditions are the strings that tie a department's history together, connecting the members of the past with the members of the present. Traditions are preserved because of what they symbolize. The drawback to traditions is the risk of preserving a past practice without assessing what effect it has in the present; for example, keeping old ways just because "we've always done it that way." When a tradition adversely impacts worker safety or morale, or pushes people to risk injury just to prove themselves (e.g., recruit hazing, mounting interior attacks without SCBA, treating bleeding patients without using universal precautions, or wearing dirty, worn-out protective ensemble), the tradition needs to be ended.

PROCEDURES

Although fire and emergency services departments share similar general operating approaches, each department has its own specific procedures. **Standard operating procedures (SOPs)**, or **standard operating guidelines (SOGs)**, and preferred tool inventories are among the different procedures that define a department's particular culture. SOPs, or SOGs, are methods defining the functional limitations in which an organization performs emergency and nonemergency operations.

NORMS

All fire departments operate with **norms**. As used here, norms are the actions and beliefs that are part of routine operating principles for a fire and emergency services department. Although the norms are often based on tradition (root of the well-worn phrase "250 years

standard operating procedures/guidelines (SOPs/SOGs) ▪ A method defining the functional limitation in which an organization performs emergency and nonemergency operations.

norms ▪ Actions and beliefs that are part of routine operating principles for a fire and emergency services department.

of tradition unimpeded by progress"), others develop out of a conscious effort on the department's part to study its experiences and learn from them. Examples of norms include:

- A particular work shift
- The monthly drill night
- Riding assignments
- Progression of seniority
- Start of the workday
- Apparatus and equipment check-off procedure
- Colors of apparatus and markings
- Department shoulder patches
- Crew patches
- Types of uniforms, and so on (See Figure 1.3.)

(a)

(b)

(c)

(d)

FIGURE 1.3 The culture of an individual crew is defined by the patch that each member proudly displays on his or her shoulder, emergency response vehicle, or helmet. *(a, b, c) Courtesy of Martin Grube, (d) Courtesy of Danny R. Yates, Deputy Chief Fire Suppression, Nashville Fire Department*

Each of the examples defines the culture of an individual department, shift, battalion/district, or crew, not an all-encompassing national culture. Even within the department, each shift, battalion/district, and crew have their own norms.

The norms can take on an ominous tone when risk is not considered during an assessment of conditions. Norms such as "*always* make an interior attack," "open *all* roofs," *not* wearing a seat belt, searching *every* vacant structure "in case there is a homeless person inside," and *not* wearing exam gloves on an EMS call, are just a few examples that ended up becoming the catalyst or root cause of a line-of-duty death (LODD) or injury. The lack of a risk management approach to the conditions firefighters face has proven to be a formula for tragedies that unfortunately repeat themselves with alarming regularity. The repetition of actions that end with the same result is a by-product of a cultural norm that believes "it won't happen to me."

ATTITUDES

In assessing firefighter culture, hazardous attitudes play a vital role that defines who emergency responders are and why they do what they do. The hazardous attitudes that firefighters and company officers fall prey to include:

- *Invulnerability.* "It won't happen to me."
- *Anti-authority.* "I've been on the job [fill in blank for the appropriate time in grade]; you can't tell me what to do."
- *Impulsivity.* "Act first; think later."
- *Macho.* "I can last longer than you. Throw more pallets on the fire to see who the real firefighters are."
- *Resignation.* "Nothing ever changes around here. Why bother?"
- *Air show syndrome.* "The public expects a good show," or, "The TV crews just got here; we need to really push things."
- *Pressing.* "Let's go lights and sirens. I'm getting off in a half hour."

Those attitudes become a part of the culture that individuals, crews, battalion/districts, and shifts adopt. Within each attitude, every organizational element (i.e., crew, station, battalion/district, shift, etc.) can find causes of behaviors that lead or lure firefighters into situations from which they may return unscathed countless times. Each time nothing adverse happens, the behavior is reinforced as okay to repeat the next time. Eventually, luck runs out and a firefighter is killed or injured performing an act he or she has engaged in countless times before. Fortunately, there is a shift away from the pervasive attitude in the fire and emergency services up to the mid-1970s that getting killed or injured is just "part of the job." This philosophy of getting the job done by "doing whatever it takes, no matter the consequences" has been a part of the fire and emergency services attitude for years. (See Figure 1.4.)

In some respects, the 1970s became the turning point for the U.S. fire service, because that is when the National Commission on Fire Prevention and Control (NCFPC) first realized the significant number of firefighter line-of-duty deaths and injuries. Years later, based upon the awareness created by the Commission's *America Burning* report (NCFPC, 1973), the National Fire Protection Association created NFPA 1500, *Standard on Fire Department Occupational Safety and Health Program*.

In more recent years, there is an emerging philosophy that risk taking in the fire and emergency services needs to be balanced with a benefit. This is a sign that a cultural shift toward a safer fire and emergency services is emerging. If the risk is great, such as a firefighter putting his or her life on the line, the reward must be great as well. An example of this emerging mind-set is the philosophy of putting firefighters at maximum risk only when there is a "savable" life at stake. (See Chapter 3, Risk Management.)

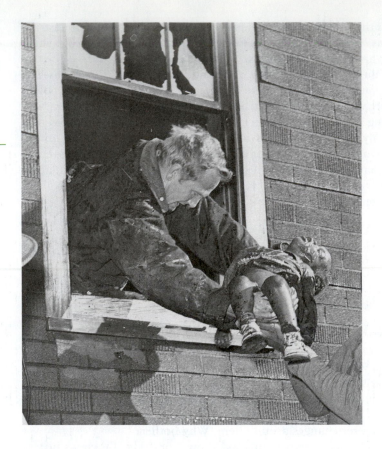

FIGURE 1.4 Doing whatever it takes to get the job done regardless of the consequences has been a part of the U.S. fire service attitude for years. *Courtesy of George Russell/Nashville Fire Department*

BEHAVIORS

Behaviors are the actions firefighters demonstrate in response to stimuli. Behaviors are in many ways influenced by attitudes, training, and the mentoring that a firefighter receives from coworkers and company officers. Those behaviors that are most beneficial for firefighters are intention based, not just automatic responses to specific instances. In many emergency situations, the actions that firefighters take to correct what they see at the very second of engagement, place them behind the curve of incident escalation—and place them at risk. (See Figure 1.5.)

For example, imagine that a fire department receives a call about a sick person. The information provided by dispatch says only that the patient is not feeling well. The crew assumes the patient has no real medical emergency and forgoes taking the oxygen cylinder and medic bag into the house. When the firefighters enter the front door, a family member advises that the spouse "is coughing up a lot of blood." The firefighters then scramble back to the unit for protective gear and proper equipment that would have been standard had they followed proper departmental SOPs/SOGs. (See Figure 1.6.)

The behavior exhibited by the crew is borne out of a number of factors. Complacency is one crucial behavior that routinely lures firefighters into bad situations. Although complacency can be the result of fatigue, it is often a by-product of poor decision making and a lack of conscious risk assessment. Fire and emergency services leaders are on guard for the behavioral effects of complacency and strive to ensure that their crews maintain a behavioral pattern of *heightened situational awareness,* or being on guard for the unexpected. When firefighters develop and maintain heightened situational awareness, there are fewer surprises at the incident scene because they anticipate and predict the outcome of the incident rather than react to changes.

FIGURE 1.5 The company officer's job is to mentor new members and to train them to work as a team. *Courtesy FPP/IFSTA*

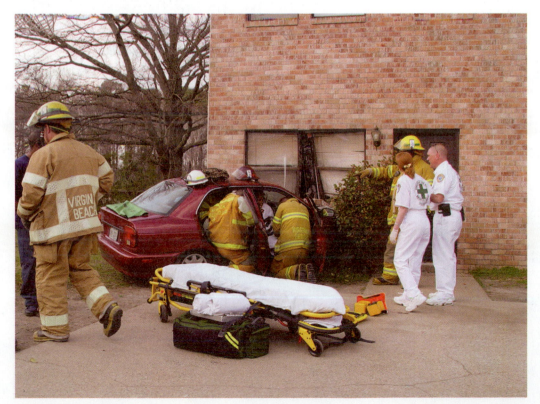

FIGURE 1.6 You are solely responsible for wearing the proper gear and taking all the proper equipment with you on every emergency run. *Courtesy of Martin Grube*

PERSONNEL DIVERSITY

Fire departments are becoming more diverse, including individuals with different cultural backgrounds. Regardless of the differences in individual values and preferences, once an individual becomes a part of the fire and emergency services, the need to operate safely, efficiently, and effectively as a crew should be the number-one priority. Although the increasingly diverse fire and emergency services of today may look, think, and act differently than those in the past, safety is still considered everybody's responsibility.

One way to get an understanding of the dilemma in defining fire and emergency services culture is to look at a common fire department term used to describe the relationship of its members. Among firefighters, the term *brotherhood* is used as an expression of fraternity. The fire and emergency services is like being a part of something that can sometimes be considered larger than life. It takes different individuals and throws them into situations that sometimes create the "do whatever it takes to get the job done" mentality. Those individuals then become closer and closer after each incident experienced and form bonds that cannot be broken. (See Figure 1.7.)

COMMUNITY SERVED

Fire protection and emergency medical care are local concerns. Therefore, one critical element that determines a fire department's culture is the community the department serves. Some departments enjoy a loyal, admiring population. Those departments are well funded, are oftentimes well trained, and usually enjoy high morale. Other departments are afterthoughts or not thought of at all until their services are needed. Those departments can struggle desperately with funding (by tax levies, fund-raisers, or contributions), have trouble recruiting and retaining members (including employees), and sometimes adopt an attitude of "us versus them."

FIGURE 1.7 Firefighters come from a vast array of cultures; regardless, safety is still everyone's responsibility. *Courtesy of Travis Ford, Nashville Fire Department*

Changing the Fire and Emergency Services Culture

Implementing cultural change in the fire and emergency services is driven by three factors: a pivotal event or tragedy, a technological advancement (e.g., the automatic transmission, engine brakes, and integrated personal alert safety system devices), and legislation (e.g., federal regulations regarding two in/two out, use of traffic safety vests on highways, etc.).

However, some consider visionary leadership as another way of changing attitudes and promoting safety within the fire and emergency services culture. Visionary leadership is about setting the tone for change by both actions and repeated communication of what must be done in hopes of inspiring others throughout the department. Visionary leaders have a clear idea of where the department needs to go and how things should be. Max Depree, in *Leadership Is an Art*, states that "The first responsibility of a leader is to define reality" (1989, p. 11). Visionary leaders have the vision to see things differently than others and the ability to communicate it throughout the department.

PIVOTAL EVENTS AND CULTURAL CHANGE

Change driven by tragedy or other pivotal event is the most readily accepted by those closest to the firefighter killed or severely injured. The line between cause and effect is never clearer or more laser sharp than when a firefighter dies or is injured or is otherwise harmed in the line of duty. The reaction is primeval:

- "One minute he was there, and the next he fell into the burning basement."
- "She was just telling me we needed to get an additional unit to block the road when she was struck and killed by a passing motorist."
- "The crew was only responding to a routine medical emergency when they all contracted an infectious disease."

At moments such as these, the need for change is undisputed. The change implemented is often named for the deceased individual.

The further removed from the tragedy one is, the less of an impact it has on some members of the fire and emergency services. The chorus, "that it couldn't happen here," is heard in many circles. The truth is that the members of the affected department or crew believed "that it couldn't happen here," but it did. In the end, "it" could and does happen everywhere. (See Figure 1.8.)

Countless examples in fire and emergency services history demonstrate that individuals, crews, and whole departments are changed by a seminal event. One needs to spend only a few minutes on any of the fire and emergency services media outlets, including video clips, to observe such change taking place. However, examples abound of repeated practices that serve as precursors to replication of well-known tragedies. In many cases individuals, crews, and fire departments experience the same events repeatedly; but if there is no adverse outcome (the threshold being death or severe injury), they continue responding (a norm) as they did in the past, which is called **normalization of deviance**.

normalization of deviance ■ Repeating a practice that lowers safety or other standards of performance until the practice becomes a norm.

TECHNOLOGICAL ADVANCEMENT AND CULTURAL CHANGE

Technological advances provide a second catalyst to cultural change. For example, as technology makes a personal protective ensemble (PPE) lighter and easier to don, firefighters' resistance to its use generally decreases.

One of the baffling factors in the industry is the relationship between the technological improvements in the fire and emergency services and the lack of significant improvement in the firefighter death-and-injury rate. One hypothesis about this phenomenon is the way in which the fire and emergency services reacts to technological improvement. For example, the PPE industry has made galactic leaps in protecting firefighters entering

hostile fire environments. The improvements in the protective clothing and breathing apparatus were designed to keep firefighters out of harm's way if fire conditions changed rapidly and unexpectedly. Firefighters came to realize that in the new PPE they were able to enter fire situations that previously would have been untenable. As a result, firefighters were still getting killed in fire situations despite huge advances in the gear designed to protect them. Here the culture of the fire and emergency services (risk taking versus risk assessing) took the technology designed for improved safety and used it to increase firefighters' exposure to risk. It needs to shift its view to value that PPE is designed to protect firefighters in an emergency, not permit longer exposure to higher risk incidents. More emphasis must be given the importance of assessing risk constantly as the determining factor for exposure. (See Figure 1.9.)

LEGISLATION AND CULTURAL CHANGE

The final driver of cultural change is legislation. Typically, legislative change is the most dynamic and far-reaching because of the force of law. Legislated change is driven by tragedy, labor unrest, and political activism. Several of the more significant legislated changes the fire and emergency services has experienced include the following: the regulation of hazardous materials mitigation (OSHA, n.d., *CFR 1910.120*), how many firefighters are required to be on the scene before interior operations can begin (OSHA, n.d., *29 CFR 1910.134*), the standard for the construction and operation of ambulances (General Services Administration, 2007), a national incident management system (FEMA, 2008), and the required use of safety vests (Federal Highway Administration, 2006).

Each of those legislated changes is mandatory, which creates a strong incentive to comply because failure to comply often includes a stiff financial penalty (fines, restricted access to grants). The heavy-handed penalty moves organizations and individuals to rapid changes of practice, thereby infusing cultural change by the force of law. Where there is

FIGURE 1.9 New, improved personal protective ensembles allow firefighters to increase their exposure to risk. *Courtesy of Travis Ford, Nashville Fire Department*

opposition to complying or making changes in the wake of new legislation, the handing down of just a few stiff penalties expedites change.

TIMETABLE FOR CULTURAL CHANGE

Ultimately, the timetable for cultural change is in the minds and hands of the individuals impacted by the shift in culture. Some individuals accept change readily; others object with all their might. The psyche of an individual is the most difficult obstacle to change because humans resist change when they are comfortable with their performance. Therefore, cultural change first involves impacting the psyche of an individual, crew, and then department as a whole.

CONTRIBUTING ELEMENTS

There are four elements that impact and contribute to cultural change: **accountability**, **responsibility**, **leadership**, and **management**. Those elements are equally important and, in fact, often occur simultaneously. They may be described as follows:

- *Accountability.* The individual's act of taking ownership for attitudes and behaviors related to safety for fire and emergency responders. As used here, accountability differs from the term *personnel accountability,* which refers to the location of fire and emergency responders.
- *Responsibility.* The shared burden and opportunity to achieve the cultural changes needed for the safety of fire and emergency services responders.
- *Leadership.* Leaders impact cultural change by being unsatisfied with the current conditions and motivating individuals *toward safe attitudes and behaviors.*
- *Management.* In contrast with leadership, management consists of activities that achieve results from establishing and enforcing the fire and emergency services organization's safety policies and procedures.

accountability ■ An individual's act of taking ownership for attitudes and behaviors related to safety for fire and emergency responders. As used here, accountability differs from the term *personnel accountability,* which refers to the location of fire and emergency responders.

responsibility ■ The shared burden and opportunity to achieve the cultural changes needed for the safety of fire and emergency responders.

leadership ■ Activities that motivate individuals toward safe attitudes and behaviors.

management ■ Activities that achieve results from establishing and enforcing the fire and emergency services organization's safety policies and procedures.

TABLE 1.1	Examples of Culture Elements and Subcategories	
ELEMENT	**SUBCATEGORY**	**EXAMPLE**
Accountability	Personal accountability	When safety failures occur, assess what your role was before assigning blame.
	Crew accountability	Remind members they control their destiny. Safety is a priority.
	Departmental accountability	Monitor departmental performance. When failures occur, assess the department's role.
Responsibility	Personal responsibility	Recognize and acknowledge the role you play in the department's overall safety performance.
	Crew responsibility	Recognize and acknowledge the role the crew plays in the department's overall safety performance.
	Departmental responsibility	Communicate to crews and individuals the department's vision of safe performance and the role each plays in contributing to the vision.
Leadership	Personal leadership	Personally follow safety rules and set them as a condition for working for you.
	Crew leadership	Set the example by following and enforcing safe practices. Establish your own such practices in the absence of departmental support.
	Departmental leadership	Participate in committees in which safety enhancements will be noticed (apparatus, health and safety, PPE, etc.).
Management	Personal management	Incorporate safety principles and practices into daily management decisions.
	Crew management	Make safety a defining criterion for crew cohesion.
	Departmental management	Demonstrate commitment by developing standards and goals that contain safe practices as measurement.

Within those four elements are three subcategories that evaluate and modify, if necessary, to foster change. They are:

- Personal culture
- Crew culture
- Departmental culture

Table 1.1 provides safety examples of each element and subcategory.

Ultimately, cohesion must be achieved within each element in order for cultural change to be effective. Once the elements are in agreement, cultural change is evident.

Accountability

PERSONAL CULTURE ACCOUNTABILITY

Holding oneself to a high standard for all actions is irrefutable if a change in departmental culture is expected to succeed at the personal level. Introspective self-assessment related to how the individual is performing for the betterment of the crew and the department is necessary. Ensuring that personal accountability is measured falls somewhat on the shoulders of the crew supervisor. But individual firefighters also can assess their personal

contribution and any other personal practices (e.g., heightened vigilance at the incident scene, maintaining the "60 foot" view of situations, recalling previous training and experiences, etc.) to keep the crew safe by measuring their own performance. Essentially, personal culture accountability is about meeting individual responsibilities relative to one's role on the crew.

To monitor their personal accountability, firefighters and officers can ask themselves questions such as:

- Is my SCBA in good working order?
- Have I checked my turnout gear pockets for hood and gloves?
- Are the tires on the emergency response vehicle I am driving inflated to the proper pressure with the proper thread depth?
- As a company officer, have I met the expectations of those I supervise and of my supervisors?
- Do I train frequently?
- Is the medic bag full of needed supplies? (See Figure 1.10.)

Chief officers must develop a sense of personal accountability in the face of the daunting responsibility of being responsible for others. Supervisory positions can be vise-like in their demands, but extremely rewarding to those who hold themselves to a high standard. Whatever an individual's position in the department, personal accountability must be in line with a culture that promotes ensuring every member of the department goes home at the end of duty.

CREW CULTURE ACCOUNTABILITY

One of the greatest sources of information on how people work together is the military. No organization on Earth takes individuals from all walks of life and forges all the individual personalities into effective, cohesive work crews. As one source notes: "Evidence of cohesion in an army must be sought where it occurs—at the small-unit level among the intimate, face-to-face groups that emerge in peacetime as well as in war" (Air University, n.d.).

The same logic can and should be applied to the fire and emergency services. The basic element in any fire department is the crew. The success or failure of a department, including its safety, hinges on the actions of the crew. When the crew arrives at the scene of an emergency and functions well, firefighters return safely, incidents are mitigated, and the department fares well in the public eye.

What happens when the crew does not function well, or when a situation is resolved in poor fashion, or when a member is injured? Those are the points at which the crew's cultural accountability is tested. Crews that are cohesive, perform a self-evaluation, and fix the problem through training, education, and hard work demonstrate that they are in tune with external as well as internal expectations. Ensuring the safety of the crew and concentrating on operating safely is a conscious decision introduced to the crew by its leader and supported by all the members. Every individual on the crew is held to the same level of accountability for his or her decisions.

DEPARTMENTAL CULTURE ACCOUNTABILITY

Any successful department functions with a sense of trust among its members and a mutual understanding of roles and responsibilities. Changing the culture of a department is difficult due to traditions, norms, and history. Entire departments are going to have to instill trust up and down the line for the fire and emergency services to adjust its culture to a more safety-conscious approach. The department must insist that members at the crew and personal levels are accountable for actions that do not support a safety-focused performance model.

Achieving the culture shift that puts safety first depends on top management and leadership in the organization being accountable for making the goals, expectations, performance, and discipline part of every day's operation.

Responsibility

PERSONAL CULTURE RESPONSIBILITY

The individual is the root of any cultural change. Within each person lives a level of personal responsibility. Some people just take ownership of their actions, the jobs they perform, the things they say, and to a certain degree, the things that happen to them. People with this strong sense of personal responsibility are generally self-sufficient and steer clear of blaming others for what happens to them. Individuals with strong personal responsibility are excellent catalysts for cultural change. Because they have a high degree of personal responsibility, as a measurement of success, safety is a high priority for them. Steve Thorne (2008) notes: "Taking personal responsibility for the safety of each other is one of the most powerful and most effective attributes in an active safety culture." Generating and sustaining a culture shift toward reducing injuries and fatalities will have the greatest chance for success when individuals with a strong sense of personal responsibility are enjoined or appointed to become change agents.

What of the individuals who lack a strong sense of personal responsibility? Can they be included or convinced to "get on the train" to a safer fire and emergency services culture? The answer to that question lies in the commitment to sustain change on the part of other elements of the department, starting first with the crew's responsibilities.

CREW CULTURE RESPONSIBILITY

The functioning work group in the fire department is a crew, which can be from as few as two members on an EMS transport crew up to a five- or six-member engine, truck, or rescue crew. The cohesiveness of the crew depends on the individual members developing

some common ground. A like-minded view and approach to their work is fundamental to success and safety.

It is not by accident that someone can visit a fire station and within minutes determine a crew's "culture." The way crew members dress, the condition of the apparatus, how their gear is hung on the emergency response vehicle, the way they interact, all play a role in revealing the crew's approach to fostering a safe operating culture. Work crews that remain vigilant and focused tend to determine their own destiny. The crew members perform constant risk assessment, individually, and then provide input to the crew leader, allowing the leader to make better decisions. The end result at the crew cultural responsibility level is Dumas' *Three Musketeers* pledge of "All for one and one for all." (See Figure 1.11.)

DEPARTMENTAL CULTURE RESPONSIBILITY

Cultural responsibility at the departmental level is probably the most difficult to infuse given today's society. Members of the "greatest generation," those who rolled up their sleeves and joined the war effort in World War II to rid the world of totalitarianism, are well into their golden years. The war effort helped to create a culture of having the responsibility to follow certain rules. Therefore, in order to succeed, individuals had to understand and adopt the same responsibilities. The succeeding generations have not had the same "pull togetherness" that existed then.

The fire and emergency services is no exception because it attracts its members from the general population. However, that is not to say departments have not been successful in setting the tone and expectations for operations. For example, the Phoenix Fire Department's (PFD) "Be Nice" philosophy demonstrates a department's successful infusion of cultural responsibility from the top down (now known as customer service excellence; 2009). It is a simple philosophy that requires all PFD members to be considerate at all times. The slogan is a major part of the department's mission—not just posted on the walls of every work site, but infused in every member at every level.

As a result of the Phoenix Fire Department's commitment to this simple philosophy, department personnel accomplish many things at a variety of levels: on the incident scene, during training, and within labor relations at the fire station. The success of the culture of "Be Nice" lies in the commitment from senior staff to uphold the ideal. The message sent to the troops is "We do as we say." Carrying that philosophy to the safety culture, an expectation that the industry's culture needs to change sets the tone for future advances. The departmental responsibility piece of shifting the department to a safety-focused culture is rooted in a constant repetition of the message that safety is an expectation, not a goal.

Leadership

PERSONAL CULTURE LEADERSHIP

Members in positional leadership roles are the instigators and enforcers of change. At the personal level, a leader must believe that the shift to a safer fire and emergency services culture is a necessity. Leaders who passively approach the "safety first, safety always" policy betray themselves and the individuals they lead. Nothing alienates firefighters more than a leader who does not look out for them or demonstrates hollow commitment to the elements of the safety-focused culture: constant risk assessment, risk analysis, and assurance of personal and personnel safety. Even though the department has an expectation of safety, each and every day leaders must demonstrate by words and actions that safety is an expectation, not just a goal. Examples can be found throughout the fire and emergency services of individuals who have broken from the norm to implement change within their sphere of influence. They are determined to introduce new ideas and change traditions that are steeped in doing something wrong long enough so everyone begins to think it is right. (See Figure 1.12.)

Two examples are the individual leadership approaches to an offensive mode in an abandoned building fire and the requirement that seat belts are to be used any time an individual is inside an emergency vehicle.

Abandoned Building Example

Officers in departments across the country have begun to look at abandoned building fires as total losses from the start. Instead of approaching the structures as offensive fires, they switch to **defensive mode** and approach with the reverse mentality: no department SOPs/SOGs, no general order, just a personal decision to look out for their members by making sure the risk is worth the gain. At the departmental chief officer level, leaders are

defensive mode ■
Firefighting actions that place crews outside of a burning structure and outside of an established building collapse zone. Defensive operations are called for when the risk to firefighters' lives outweighs the benefit of putting firefighters inside or near a burning structure.

FIGURE 1.12 Firefighter safety must be a departmental priority. *Courtesy FPP/IFSTA*

FIGURE 1.13 Officers should never be afraid to make the right decision when it involves sending everyone home at the end of the shift. *Courtesy of Daniel A. Nelms, Emergency Service Training*

doing the same. When members in those leadership positions observe crews engaging in **offensive mode** operations for marginal or misguided reasons, evacuations are called for and a shift to defensive mode should commence. At early junctures, there is resistance. However, those leaders demonstrate a personal commitment to changing culture within their departments without experiencing the overwhelming personal losses other departments have suffered. When the years of "doing the wrong thing until you believe it is right" catch up, the end result is almost always accompanied by a chorus of "I told you so" or "I knew this was going to happen." (See Figure 1.13.)

offensive mode ■
Firefighting actions that place crews inside a burning structure and inside an established building collapse zone, thereby placing the safety of fire and emergency responders at risk.

Seat-Belt Example

A very similar approach can be seen regarding seat-belt use. For many leaders, it is a personal decision to make sure all members are seated and belted while any emergency response vehicle is moving. They are visionary enough not to have to personally see a member of their crew lying broken on the street. Therefore, they ensure all members are safely seated and belted before any emergency vehicle moves. The leader demonstrates a personal commitment to changing a cultural norm without having to experience a tragedy. This level of proactive leadership fosters change more quickly because of the face-to-face value a leader demonstrates with the individual members of the crew.

CREW CULTURE LEADERSHIP

A crew takes its cue from the leader, who might or might not be an officer. When a leader demonstrates genuine concern for the crew's members, cultural change is more readily accepted. The trust that the crew develops in the leader becomes infectious. The dynamics of human behavior are such that when people form a crew and become cohesive, they as a unit take direction from the leader. Members of the crew adopt the leader's values and often begin using the same tools and criteria to make decisions as the leader does. Encouraged by the leader's praise and respect, the crew accepts cultural change and often enthusiastically supports it.

One case in point is the practice of storage and use of personal protective ensemble. There was a time in the fire department when members in PPE would walk through the fire station living areas, sit at the kitchen table after a response, and store their turnout pants next to their bunks at night, not realizing that PPE is impregnated with carcinogenic contaminates at its worst, and a magnet for dirt and other debris that is carried into the living areas of the station. Forward-thinking leaders at the crew level began requiring their members to keep PPE out of the living areas. Crew members began convincing members of other crews in the same fire station to do the same. Eventually, entire organizations issued orders to refrain from entering the living areas in PPE. This resulted in cleaner living quarters in general, less cleaning required, and an overall improvement in the health of all members. That was change instigated by a leader and fostered by a crew.

DEPARTMENTAL CULTURE LEADERSHIP

In order for cultural change to take place, leadership has to have a mind-set that supports open communication and an open-minded approach to change. This approach sets a tone for a willingness to try new things on the one hand and obtain buy-in from the membership on the other. A departmental leadership culture that is steeped in "no" does not progress.

A frequently cited example is the reaction of the chief when firefighters return from conferences or visits to other departments. The returning firefighters often will bring back new ideas or concepts that are in place and working very successfully for other departments (e.g., large-diameter hose, single-stage pumps, new extrication techniques, patient protocols, etc.). When the ideas are brought to the chief, he or she immediately closes the door by saying, "We don't need that stuff here. What we have has been working fine." Stifling experimentation leads to a disgruntled workforce, antiquated equipment, and a failure of leadership fulfilling its primary mission, which is to move the department into the future while ensuring the members' safety.

Management

PERSONAL CULTURE MANAGEMENT

"Lead people, manage time," is an old axiom that is frequently twisted in today's fire and emergency services. Members in management roles get wrapped up in the minutiae of the job and forget they are dealing with people. At the personal level of changing culture within management, remembering this axiom will ensure success when implementing change in culture. Changing culture is not about adjusting a machine; it is about adjusting a person's values and perceptions.

CREW CULTURE MANAGEMENT

Within the crew, culture management becomes a function of the leader's approach and the crew's acceptance of needed change. Managing the change can be difficult at first if the change goes against long-standing practices. Using self-contained breathing apparatus (SCBA) for interior firefighting is a prime example. Previous generations of firefighters established firefighting prowess by the amount of smoke a firefighter could "eat." In the 1960s a shift in the mind-set began to appear. Firefighting crews went from having one member in SCBA "in reserve," in case the nozzle team could not get the fire out, to having all members start wearing SCBA.

The 1970s and 1980s saw more crews not only donning but also actually using SCBA in smoke-filled situations. The sight of firefighters using SCBA became more prevalent in the 1990s. From 2000 on, firefighters observed not wearing SCBA in smoke-filled situations are considered behind the times. This change, often mandated by the chief's office, is implemented and truly enacted at the crew level by company officers and individual members, who start to hold individuals accountable. However, managing change is essential for a permanent

FIGURE 1.14 Moving toward a safer fire and emergency services relies on individuals following policies and procedures. *Courtesy of Kevin L. Neville, Fire Investigator, Nashville Fire*

shift to a safer culture. Moving the crew's culture to a more safety-focused approach or fighting the complacency that develops when crew cohesiveness breaks down requires constant vigilance on the formal as well as the informal leader's part. (See Figure 1.14.)

One element to address in the management of cultural change is time. An assessment of the criticality of the change is necessary. Is the culture shift one that is required immediately because of the high risk involved (e.g., wearing SCBA throughout all structure fire operations, even through overhaul, and wearing exam gloves on all emergency medical calls), or is it a change that can be implemented and reinforced over a specified period of time (e.g., a mop leaning against an infrequently used door)? The leaders' judgment determines whether the cultural shift is measured with a stopwatch or a calendar.

DEPARTMENTAL CULTURE MANAGEMENT

Ultimately, departments are affected by culture. The most prevalent symbols of the culture in the fire and emergency services arena range from the visual (patches, uniforms, and apparatus) to the audible (language, terminology, and public perception) to the expectations and perceptions (mission statement, personality, and reputation). The department's role in culture management is a function of how its members perceive change that occurs (or does not occur). Departmental culture management is most successful when every element of the leadership team convincingly professes to subscribe to a given change and believes in the change. In order for the fire and emergency services industry to adopt a true cultural change to a safer fire and emergency services, more departments are going to have to adopt the tenets of change and expect (not wish) their organization to follow suit. An example of departmental culture management was the seat-belt pledge program initiated by Dr. Burton Clark (2005), adopted by fire departments across the country. Many departments register 100% sign-up and boast 100% compliance. Those departments demonstrate a departmental culture management that shares the same vision from the chief's desk to the youngest firefighter riding in a jump seat.

Life Safety Initiatives and Cultural Change

In 2004, the National Fallen Firefighters Foundation (NFFF) convened professional fire and emergency services leaders in Tampa, Florida, to draft a set of firefighter safety initiatives. The resulting 16 Firefighter Life Safety Initiatives (NFFF, 2005–2009) follow:

1. Define and advocate the need for a cultural change within the fire service relating to safety, incorporating leadership, management, supervision, accountability and personal responsibility.
2. Enhance the personal and organizational accountability for health and safety throughout the fire service.
3. Focus greater attention on the integration of risk management with incident management at all levels, including strategic, tactical, and planning responsibilities.
4. All firefighters must be empowered to stop unsafe practices.
5. Develop and implement national standards for training, qualifications, and certification (including regular recertification) that are equally applicable to all firefighters, based on the duties they are expected to perform.
6. Develop and implement national medical and physical fitness standards that are equally applicable to all firefighters, based on the duties they are expected to perform.
7. Create a national research agenda and data collection system that relate to the initiatives.
8. Utilize available technology wherever it can produce higher levels of health and safety.
9. Thoroughly investigate all firefighter fatalities, injuries, and near misses.
10. Grant programs should support the implementation of safe practices and/or mandate safe practices as an eligibility requirement.
11. National standards for emergency response policies and procedures should be developed and championed.
12. National protocols for response to violent incidents should be developed and championed.
13. Firefighters and their families must have access to counseling and psychological support.
14. Public education must receive more resources and be championed as a critical fire and life safety program.
15. Advocacy must be strengthened for the enforcement of codes and the installation of home fire sprinklers.
16. Safety must be a primary consideration in the design of apparatus and equipment.

Those initiatives resulted in a unification of principles rarely seen in the fire and emergency services. Since that time, the 16 Firefighter Life Safety Initiatives have evolved and gained in popularity—and controversy. The fact that the fire and emergency services is talking about the initiatives more frequently indicates that the seeds of a cultural shift have truly been planted.

In March 2007, fire and emergency services leaders convened again to address the problem of stagnated firefighter fatality and injury numbers. The "Novato Summit" reviewed the initial 16 Firefighter Life Safety Initiatives and the results of six mini-summits that had been held since 2004. That summit drew over 200 fire and emergency services representatives and resulted in over 100 recommendations to attack the fire and emergency services fatality and injury problem (NFFF, 2007). A monumental feature of the Novato Summit was its focus on the culture of the fire and emergency services and the need for the culture to change if fatalities and injuries are truly to be reduced.

The summit's recommendations called for actively promoting a shift to a more safety-focused culture by:

- Infusing safety in all department levels by defining expectations
- Assigning people with the appropriate safety attitudes and skills to positions of leadership and training

- Integrating the concepts of risk management into all department activities
- Rewarding and providing incentives for safe behaviors while no longer rewarding unsafe, inappropriate behaviors
- Redefining the cultural definition of "hero" to take safety practices and attitudes into account

Maybe fire departments should suggest that individuals think about safety before they begin thinking about promotion.

This task will not be easy. The challenge is to reduce the number of line-of-duty deaths by 50% before 2014. The cultural change called for will require time, limitless energy, constant prodding, and persistent accountability, personal responsibility, leadership, and management. Many of the recommendations regarding cultural change go against the grain of the fire and emergency services that is as dense as a 250-year-old oak. But it is the cultural change that will move the fire and emergency services to a safer performance, more so than improvements in tools and equipment, emergency response vehicles, or PPE. "Any revolutionary movement requires committed and unwavering leadership to bring about this type of major change" (NFFF, 2007, p. 3). However, it is important that every individual make a conscious effort to individually change. (See Figure 1.15.)

The Novato Summit perpetuated a new movement that renounces dying as "part of the job" because we can no longer pretend we do not know better; renounces expecting injuries because we know better; and renounces crashing emergency response vehicles every so often because we know better. Taking a proactive approach by using the 16 Firefighter Life Safety Initiatives as a template to introduce the elements of a safer fire and emergency services does justice to why the initiatives were created. They serve as a means of affecting culture in a positive sense or sustain the efforts of others already embarking on making sure "everyone goes home."

FIGURE 1.15 Make a conscious effort to be safe and do the right thing regardless of the incident that you respond to.
Courtesy of Martin Grube

Summary

The definitive, universal definition of what constitutes a line-of-duty death or injury may be years away. What is needed immediately is attention to factors the fire and emergency services can control: the culture that places firefighters in perilous situations with no quantitative reward.

The shift to a safer fire and emergency services will require a significant and sustained effort to overcome "250 years of tradition unimpeded by progress." That tradition has created an environment that tolerates more than 100 firefighter line-of-duty deaths and even more injuries each year from repeated events. However, the call for change is becoming louder and louder. As a result of more individuals subscribing to change, there is hope that firefighter fatalities and injuries will be reduced. The cultural change necessary to become more effective risk managers not only would raise the bar of the fire and emergency services in the eyes of the rest of the industrial world but also would make good the promise of getting everyone home at the end of his or her shift.

This chapter has defined key terms related to cultural change and provided examples from the fire and emergency services. The chapter also addressed the role of four elements—accountability, responsibility, leadership, and management—that impact cultural change. Finally, the chapter reported on the National Fallen Firefighters Foundation 16 Firefighter Life Safety Initiatives and the need for everyone to go home safely. Creating a safe fire and emergency services culture will not be easy, but it is doable, if it addresses each of the four elements that impact cultural change.

Review Questions

1. Define *culture* as it applies to the safety and survival behavior of the fire and emergency services, and discuss what actions are needed to promote a safety-and-survival culture.
2. Discuss the hazardous attitudes that contribute to an unsafe culture within the fire and emergency services.
3. List the four elements that impact cultural change in the fire and emergency services.
4. Discuss how personal, crew, and departmental accountability affect cultural change.
5. Discuss how to implement the National Fallen Firefighters Foundation 16 Firefighter Life Safety Initiatives to support changing the culture for a safer fire and emergency services.

References

Abrashoff, D. M. (2002). *It's Your Ship: Management Techniques from the Best Damn Ship in the Navy.* New York: Warner Books.

Air University, Maxwell-Gunter AFB, Maxwell, AL. (n.d.). *Characteristics of a Cohesive Army.* Retrieved April 15, 2010, from http://www.au.af.mil/au/awc/awcgate/ndu/cohesion/ch02.pdf

Brunacini, A. (2003). *Timeless Tactical Truths.* Peoria, AZ: Uptown Graphics and Design.

Cialdini, R. (1993). *Influence, the Psychology of Persuasion.* New York: William Morrow.

Clark, B. A. (2005, June 21). *Leadership: We Killed Firefighter Brian Hunton.* From *Firehouse.com.* Updated June 14, 2007. Retrieved December 13, 2009, from http://www.firehouse.com/node/46085

Coleman, R. J. (1978). *Management of Fire Service Operations.* North Scituate, MA: Duxbury Press.

DePree, M. (1989). *Leadership Is an Art.* New York: Dell.

Federal Emergency Management Agency. (2008, December). *National Incident Management System (NIMS).* Washington, DC: Author. Retrieved April 3, 2009, from http://www.fema.gov/pdf/emergency/nims/NIMS_core.pdf

Federal Highway Administration. (2006, November 24). *23 CFR Part 634, Worker Visibility.* Washington, DC: Author. Retrieved April 3, 2009, from http://www.workzonesafety.org/files/documents/laws_regulations/federal/nov_24_06.pdf

General Services Administration. (2007, August 1). *Federal Specification for the Star of Life Ambulance, KKK-A-1822*. Washington, DC: Author. Retrieved April 3, 2009, from http://www.deltaveh.com/f.pdf

Giuliani, R. (2002). *Leadership*. New York: Miramax Books.

Johnson, G. (1977). *F.D.N.Y.: The Fire Buff's Handbook of the New York Fire Department, 1900–1975*. Belmont, MA: Western Islands.

Kaltman, A. (1998). *Cigars, Whiskey and Winning: Leadership Lessons from General Ulysses S. Grant*. Upper Saddle River, NJ: Prentice Hall.

Kern, A. T. (2001). *Controlling Pilot Error Culture, Environment, and CRM (Crew Resource Management)*. New York: McGraw-Hill Professional.

Maclean, N. (1993). *Young Men and Fire*. New York: University of Chicago Press.

Managing Fire Services. (1988). Washington, DC: Published for the ICMA Training Institute by the International City Management Association.

Mullane, M. (2007). *Riding Rockets: The Outrageous Tales of a Space Shuttle Astronaut*. New York: Scribner.

National Commission on Fire Prevention and Control. (1973, May 4). *America Burning: The Report of the National Commission on Fire Prevention and Control*. FA-264. Washington, DC: U.S. Government Printing Office, #1973-O-495-792. Retrieved April 15, 2009, from http://www.usfa.dhs.gov/downloads/pdf/publications/fa-264.pdf

National Fallen Firefighters Foundation. (2005–2009). *Everyone Goes Home: 16 Firefighter Life Safety Initiatives*. Emmitsburg, MD: Author.

Retrieved November 26, 2009, from http://www.everyonegoeshome.com/initiatives.html

National Fallen Firefighters Foundation. (2007, March 3–4). *2007 National Firefighter Life Safety Summit, Novato, California*. Retrieved April 15, 2010, from http://www.everyonegoeshome.com/summit/novato.pdf

Occupational Safety and Health Administration. (n.d.). *CFR 1910.120, Hazardous Waste Operations and Emergency Response (HAZWOPER) Standard*. Washington, DC: United States Department of Labor. Retrieved April 3, 2009, from http://www.osha.gov/pls/oshaweb/owadisp.show_document?p_table=STANDARDS&p_id=9765

Occupational Safety and Health Administration. (n.d.). *29 CFR 1910.134, Respiratory Protection Standard (Two In/Two Out)*. Washington, DC: United States Department of Labor. Retrieved April 3, 2009, from http://www.osha.gov/pls/oshaweb/owadisp.show_document?p_table=STANDARDS&p_id=12716

Okray, R., and Lubnau, T. (2004). *Crew Resource Management for the Fire Service*. Tulsa, OK: PennWell.

Page, J. O. (2002). *Simple Advice*. Carlsbad, CA: Jems Communications.

Phoenix Fire Department. (2009, March). *Phoenix Regional Standard Operating Procedures: Customer Service Management*. Retrieved April 3, 2009, from http://phoenix.gov/FIRE/20100.pdf

Smith, D. (2002). *Report from Ground Zero*. New York: Viking.

Thorne, S. (2008). *Attributes of an Active Safety Culture and Why They Apply to the Fire Service*. Retrieved April 15, 2010, from http://www.riskinstitute.org/peri/images/file/S908-D4-Thorne.pdf

2

Personal and Organizational Accountability

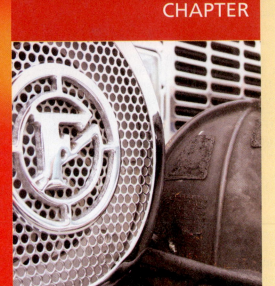

Nicol P. Juratovac

Courtesy of Eric Melcher, Photographer, Volunteer State Community College

KEY TERMS

crew integrity, *p. 46*

discipline, *p. 45*

liability, *p. 47*

negligence, *p. 47*

situational awareness, *p. 44*

OBJECTIVES

After reading this chapter, the student should be able to:

- Demonstrate a general knowledge about the roles and responsibilities that individuals and organizations have to ensure safety.
- Identify traditions that have compromised fire and emergency services safety and the role that the individual plays in eliminating the unsafe and promoting the safe traditions.
- Recognize dangerous situations that have resulted in fire and emergency services line-of-duty deaths (LODDs) and injuries, and be able to draw from (as well as build on) his or her education, training, and experience in order to be safe while operating on the emergency scene.
- Identify changes that need to occur in order to reduce fire and emergency services LODDs and injuries.

PEARSON
myfirekit™

For practice tests and additional resources, visit **www.bradybooks.com** and follow the **MyBradyKit** link to register for book-specific resources.

Register for **MyFireKit** by following directions on the **MyFireKit** student access card provided with this text. If there is no card, go to **www.bradybooks.com** and follow the **MyBradyKit** link to Buy Access from there.

Safety Is Everyone's Responsibility

The department must accept responsibility for the safety of every member. All members must realize the internal consequences of departmental actions, as well their own personal responsibility. This individual and departmental approach to firefighter safety requires structure and discipline.

Tradition in the U.S. fire service has been an integral part of its culture and history. When viewing the fire service from a global perspective, it is evident that many noble traditions have been passed down for hundreds of years. Traditions such as graduating from the academy, promotional ceremonies, and line-of-duty-death (LODD) funerals continue to be part of what the fire and emergency services are about: honor, duty, courage, and brotherhood. Many see the department for which they work as a second family. (See Figure 2.1.)

However, certain traditions within the fire and emergency services have not changed and need to. The traditions center on accountability. There is a saying in the fire and emergency services that "change occurs one funeral at a time." The black eye of the fire and emergency services for many years has been the alarming rate of firefighter line-of-duty deaths and injuries among those who have taken on this noble profession. Indeed, emergency response is inherently dangerous. However, many LODDs and injuries are in fact preventable. One must ask why the numbers of deaths and injuries have not decreased despite years of advancement in the fire and emergency services that would indicate otherwise, including advancement in education, training, equipment, and information technology. (See Figure 2.2.)

FIGURE 2.1 Tradition in the U.S. fire service has been built around honor, duty, courage, and brotherhood. *Courtesy of Michael Conder*

The resistance to change in the fire and emergency services might help explain why so many LODDs and injuries still occur. Although progress has been made with changes to equipment and standards, changes that relate to personal and organizational accountability have met with resistance. Traditionally, firefighters have learned to fight fire better and respond to other emergencies through trial and error. That approach, however, is not without its casualties, which result when different firefighters make the same mistakes. In other words, despite knowing exactly how a firefighter dies in the line of duty, firefighters are still committing the same fatal errors. If firefighters would actually implement what they have learned by training and experience, death or injury could be prevented.

What must occur to reduce fire and emergency services LODDs and injuries is the development of a complete learning process in which the end user—the emergency responder—actually incorporates what has been learned, albeit vicariously from another's experience, and applies the newly acquired knowledge, skills, and abilities to the emergency incident at hand. More importantly, the emergency responder must accept and appreciate the personal responsibility involved in completing the learning process.

Equally vital is the departmental accountability for firefighter safety. The department itself must correct errors made on the emergency scene or training ground and ensure that safety discipline is practiced at all times by all members. Only when harmful traditions are dwarfed by new and better techniques of operating on the emergency scene will firefighter safety be increased so that every firefighter goes home after each shift, every time.

Safety in the fire and emergency services is everyone's responsibility, from the commissioner or director and fire chief to the individual firefighter. (See Figure 2.3.)

COMMISSIONER/DIRECTOR

Many major departments in the fire and emergency services refer to the head of the department as either the commissioner or director. The commissioner or director is responsible for the overall management of the department, ensuring that the chief performs

FIGURE 2.3 Safety on the emergency scene is everyone's responsibility. *Courtesy of Martin Grube*

the business of the department, effectively implementing its mission of protecting lives, property within the community, and the environment, and in so doing, protecting the health, safety, and welfare of every member. The commissioner or director is accountable for providing leadership in the development of policies, procedures, and programs that champion excellence and continual improvement in the department. The policies, procedures, and programs include, but are not limited to, providing strategic planning and direction for the delivery of financial, administrative, and information technology services that support the safe, effective, and sustainable operation of the department.

CHIEF

In smaller departments, the chief assumes the responsibilities of the commissioner/director. The chief provides for the health and safety of all department members by enforcing the use of personal protective ensembles (PPEs) and standard operating procedures and guidelines (SOPs/SOGs).

It is the chief's responsibility to ensure that department members are safe. Systems must be in place to ensure that firefighters follow safety practices, policies, and procedures. If firefighters do not follow them, the chief then must hold the firefighters and their immediate supervisors accountable. Without enforcement systems in place, there can be no accountability. And without accountability, firefighters will not be held to perform at a certain standard, often resulting in death or serious injury to firefighters.

DEPUTY/ASSISTANT CHIEF

The deputy or assistant chief oversees vital operational roles during emergency incidents as well as in daily departmental life. Deputy or assistant chiefs work closely with battalion/district chiefs at emergency incidents utilizing the incident management system (IMS) to ensure safety and accountability of all firefighters. They ensure that training to incorporate departmental SOPs/SOGs—including those related to safety—is conducted. Failure to perform within the standards should result in discipline and/or other punitive action by the department.

FIGURE 2.4 Battalion/district chiefs play a vital role during emergency scene operations. *Courtesy of Wayne Haley*

BATTALION/DISTRICT CHIEF

The battalion/district chief is responsible for leading and interacting with company officers during emergency scene operations. This responsibility includes probationary firefighters during any drill assigned by the company officer. As the voice and selling point of the department administration's safety policies, procedures, and programs, the battalion/district chief's buy-in and ongoing support are critical for the success of the fire and emergency services safety program. If the battalion/district chief is "bad-mouthing" instead of enforcing a safety policy, procedure, or program for members, then there is no incentive for members to comply with such standards. (See Figure 2.4.)

SAFETY OFFICER

Through the chief, the safety officer has the authority and responsibility to identify and recommend safety practices. The safety officer has the authority to stop any unsafe act and ensure that firefighters on the emergency scene have properly donned their appropriate department-issued PPE, including firefighting helmet, hoods, turnout/bunker gear with liners attached, safety boots, gloves, and so forth. The proper level of PPE is required during vehicle extrication, emergency medical calls, and in other hazardous situations. Another responsibility is to make sure that the PPE is periodically cleaned and inspected according to the standards adopted by the department.

COMPANY OFFICER

The job of the company officer is to keep the crew prepared and safe. This can be accomplished by activities such as conducting daily drills, area orientation, and walking through newly constructed buildings in the crew's first-alarm response area. The company officer is considered the most influential position in the department.

Company officers should be the most qualified and not just the most senior on the job, because the lives of personnel are on the line. The focus of officers should exemplify the U.S. Fire Administration's (USFA) learning triad of training, education, and experience. All three areas make the complete fire and emergency responder. On every shift, a crew should be learning something new in the fire and emergency services. (See Figure 2.5.)

According to NFPA 1021, *Standard for Fire Officer Professional Qualifications*, officers of all levels must possess fundamental skills that are critical at the emergency scene, such as the ability to perform a scene size-up and draw appropriate conclusions. The observations of company officers during size-up will draw from their thorough understanding of building construction, fire behavior, abilities of their crew, timing of the arrival of additional resources, mechanism of injury of patients, hazardous materials,

FIGURE 2.5 Crew training is an important part of being safe, effective, and efficient on the emergency scene. *Courtesy of Travis Ford, Nashville Fire Department*

wind direction, topography, demographics of the company's first-alarm response area, and other observations relevant to the type of emergency incident.

Effective company officers appreciate getting back to the basics with training in order to ensure that their crew is trained and prepared for all types of emergencies. Successful company officers must relay to their crew what their expectations are both on and off the emergency scene. To do so, company officers must ask themselves questions such as:

- Does my crew know what tools to bring and what their roles are in case we encounter a fire?
- Will my driver/operator know how to secure an uninterrupted supply of water?
- Who will be going to the roof for vertical ventilation tasks? Who is doing forcible entry?
- Is my crew prepared to initiate rapid intervention crew (RIC) operations?
- Are my crew members prepared to initiate mayday or emergency traffic procedures in the event they go down or become injured, lost, missing, or trapped? Do they know what to do to escape an immediately dangerous to life and health (IDLH) environment?
- Is my crew prepared to triage patients at a multi-casualty incident?
- Have my crew members been adequately trained?

Those questions, of course, deal not only with firefighter effectiveness but also with fire and emergency responder safety.

Such practices lead to ensuring that the company officer's crew members are proficient in their skills. This is achieved by the performance of daily drills, including drills that emphasize the "back to basics" duties, the "bread and butter" of emergency scene operations, such as deploying hose lines, throwing ladders, properly donning a self-contained breathing apparatus (SCBA), and the ABCs of patient care.

Once members become proficient in the basic tasks and in working together with their assigned crew, officers may then advance to teaching other more specialized skills, such as trench rescue shoring, bridging of ladders, and rope ties that are associated with high-angle rescue, and lockout-tagout for confined space rescue. The use of scenario-based drills is an excellent way for companies to practice working in a coordinated manner within a battalion/district. In all cases, training should be aimed at improving safety, effectiveness, and efficiency. (See Figure 2.6.)

Officers are ultimately responsible for ensuring the safety of their crew, so that each trip out to the incident is a round-trip back home to the station and, ultimately, to their families. This, in a practical sense, may be achieved by officer activities such as communicating any and all safety messages, conducting roll call in the morning, and confirming riding assignments.

The company officer's safety responsibility translates into holding the crew accountable for following the department's SOPs/SOGs. The officer holds members accountable for safety by ensuring that they practice discipline during the emergency incident. Freelancing and other independent actions that can jeopardize the mission and other firefighters may not be tolerated.

There are three types of company officers: the inefficient, the frightened, and the capable. Whereas the inefficient company officer has had very little training as a company officer and, as a result, often gives the crew inadequate instructions and explanations, the frightened company officer tries to be effective, but is clearly ill at ease with the crew. In contrast, the capable company officer appreciates his or her responsibility of being a good company officer in the fire department and knows and completely accepts his or her role as teacher, leader, and counselor.

Both the inefficient and frightened officers have neglected to read and take advantage of the wealth of educational and training materials that are readily available to every firefighter. Equally important, such officers have disregarded the experience that could be

FIGURE 2.7 Company officers are responsible for leading by example and making sure that all training drills are conducted safely. *Courtesy of Martin Grube*

passed on by more senior and experienced mentors in the fire and emergency services. The capable officer instead has the confidence to lead his or her crew; this confidence is generated through hard work, diligence, and a constant desire to improve and master the skills necessary for the job. (See Figure 2.7.)

Drilling is a critical component of the company officer's responsibility. Yet even more important is the correction of errors while drills are occurring. Because it is practically impossible to take the time to correct errors committed on the emergency scene, the responsible company officer must ensure that drills are conducted applying current policies and procedures. Errors committed during drills must be corrected so that fewer errors may be committed on the emergency scene.

Mandatory training, then, becomes the paramount vehicle to develop competent company officers. Training is considered to be among the most important functions in a department because it is so critical to safety as well as operational effectiveness. A primary reason for training is the need to maintain skills at peak levels. Only with a comprehensive training program is a department able to establish and maintain a staff that is prepared and fully skilled. Continuing training then, such as that provided by in-service training programs, is necessary for fire officers, because unused skills fade. Lapses can be avoided only by continuous training and repeated drilling. Good departments devote part of every day to drills and training work.

Company officers must be the ones exemplifying the same safety policies adopted by their crew. To ignore, disobey, or be critical of an organizational policy is to negate everything that the policy intends to do with regard to preventing deaths and serious injuries. The company officer must convincingly relay buy-in to the crew; otherwise, the safety policy will never be effective. Company officers must then lead by example and perform as trained, putting into practice what they expect from their crew. As such, company officers must be consistent with their own behaviors and attitudes in regard to safe practices

on any emergency incident. Company officers cannot be only 95% credible with their crew—they must be 100% credible to be effective. (See Figure 2.8.)

DRIVER/OPERATOR

The fact that motor-vehicle collisions account for almost 25% of all firefighter line-of-duty deaths accentuates the importance of the driver/operator position. Not everyone is meant to be a driver/operator. Good driver/operators know their initial response area. Because they must possess a calm demeanor to be able to act in pressure situations, the driver/operator is a promotional position in many departments. NFPA 1002, *Standard for Fire Apparatus Driver/Operator Professional Qualifications,* specifies the technical skills required for driver/operators. Some fire departments refer to the driver/operator position as the rank of engineer. Safe driver/operators are professionals and act accordingly.

Members must want not only to become a driver/operator but also to be the best and safest driver/operator. At the very least, the driver/operator must be able to arrive safely on the scene, position the apparatus properly, and either operate the aerial device or place the apparatus in pump operations. Effective driver/operators must be able to secure their own water supply when needed and not just wait for the second due engine that might not arrive in a timely manner or in fact never arrive at all. (See Figure 2.9.)

Driver/operators are constantly mindful of firefighter safety, both when responding to and returning from the emergency scene and during emergency scene operations. They must keep their eye on their crew, who are at the end of hose lines that must be regulated to maintain the proper pressure or operating on aerial ladders above or next to power lines. In addition, driver/operators are the eyes and ears of what is happening outside of the structure fire or emergency scene that could affect operations and safety inside (e.g., a garage door closing on fire attack crew inside or an individual approaching the scene of

FIGURE 2.9 Regardless of any obstacles, driver/operators must be able to secure their own water supply when needed. *Courtesy of Captain Terry L. Secrest*

a medical call with a weapon). Good driver/operators know who is working in the first-alarm response area.

Effective driver/operators take care of the apparatus. A good rule of thumb for an effective driver/operator to follow is from a popular sticker placed on many pieces of equipment: "If you do not take care of it, it won't take care of you." Good driver/operators are cognizant of reporting any and all deficiencies or problems to the officer as well as to the incoming driver/operator during shift change. (See Figure 2.10.)

FIGURE 2.10 One way driver/operators take care of their emergency response vehicle is by checking the tire tread depth. *Courtesy of Martin Grube*

Certain LODDs of Boston and Houston firefighters occurred in 2009 when ladder truck brakes failed (KTRK TV, 2009; NIOSH, 2010). Such occurrences illustrate the importance of emergency response vehicle inspection. Driver/operators must inspect the equipment and emergency response vehicles each shift. As the driver/operators of the truck, they deploy the aerial. As the tillers, they start the saws. Rainy days are perfect for auto extrication drills, and driver/operators must ensure that each extrication tool is ready for use. After all, the emergency incident cannot be the training ground where the driver/operators discover—when it is too late—that they are unfamiliar with the equipment, that a piece of equipment has failed for lack of inspection and maintenance, or that the emergency response vehicle will not start. The driver/operators know what to expect and what is expected.

In some departments, especially in larger ones, driver/operators do not actually maintain their vehicles. NFPA 1071, *Standard for Emergency Vehicle Technician Professional Qualifications*, addresses requirements for those who conduct actual maintenance of the apparatus.

FIREFIGHTER

The fire and emergency response profession provides a high degree of job satisfaction. The key then is to last in a career that could easily be cut short by the sheer danger of the job. An even better situation is to make it to the "30/30 club," which consists of 30 years of a healthy career and at least 30 years of enjoying a healthy retirement.

The onus rests on the firefighter. Although the firefighter may never be completely "ready," he or she certainly can be "prepared" through training, education, and experience. Individual firefighters must embrace self-preparedness and be ready to perform every day they come to work.

To be prepared and safe on the fireground or in other emergency situations, consider weather and road conditions, PPE, situational awareness, firefighter training, and accountability.

Weather and Road Conditions

Preparation begins with observing the weather. Is it rainy, snowing, or windy? What are the conditions of the road on which one is driving?

Personal Protective Ensemble

Upon arrival at the fire station, safe firefighters switch out their gear with that of the firefighter previously on duty. This is a golden opportunity to inspect one's PPE to ensure that it is in good condition and meets safety standards. Remember, it is called *personal* protective ensemble for a reason; the firefighter is responsible for inspecting and maintaining in a prepared state his or her own equipment.

For example, in what condition are your turnout coat and bunker pants? Do they still have soot and burnt tar on them? To wear old, tattered turnout gear may appear macho and give one a "veteran" look, but firefighters who wear such tattered PPE are essentially walking toxic time bombs that expose themselves to cancer-causing products of combustion.

When was the last time your hood was swapped out or properly cleaned? What condition are your gloves in? Are they easily accessible so that when you don your PPE you do not forget to wear them? And when they are accessible, are you forgetting them anyway because you are in such a rush to fight the fire? What happens when you fight a fire without your hood or gloves, and you receive steam burns? This might be a career-ending injury that could have easily been prevented.

Safe firefighters become proficient in donning their PPE. They develop a routine at work to respond to emergencies in a prepared state. They arrange their ensemble so that they develop a system to be able to put it on in less than 60 seconds. Remember, the emergency scene is a very unpredictable environment so everyone must bring as much predictability as possible to it.

Situational Awareness

Emergency responders need to ask questions such as the following: "Who is on duty and on my crew for the shift? Who is my officer for the shift? Are there any safety-related announcements that I should be concerned with?" Safety-conscious firefighters will ask themselves:

- Are there plenty of exam gloves to protect me and my patient?
- Did I check at shift change to make sure there are enough medical supplies for treating patients?
- Do I have proper eye and face protection shields for incidents involving unforeseen events?
- Do I have a mask to protect against airborne diseases?

Training

Prepared firefighters ask, "What incidents has the station responded to since my last shift? For what incidents do I feel I need additional training?" For example, automobile technology is changing so quickly that if crew members are not taking an auto extrication training class at least once a year, they are probably falling behind. Effective firefighters maintain their skills and competencies through training and education. Experience alone will not make someone into a professional firefighter who can keep him or herself safe, the crew safe, and ultimately, the community safe.

Accountability of Firefighters

Because firefighters must have the courage to be safe and disciplined, departments must have the courage to have SOPs/SOGs in place that address and prevent independent action. Departments must be able to lead in this area and hold all members accountable, regardless of service time or rank. Without such systems in place, firefighters will be prone to act independently, jeopardizing their own safety and the safety of those around them. Essentially, departments must be willing to hold their members accountable for unsafe acts.

ACCOUNTABILITY TO THE CREW

Chief John Salka of the Fire Department of New York has adopted the slogan, "If you see something, say something," to empower all firefighters to be the eyes and ears at an emergency incident. Gone are the days when probationary or younger firefighters were expected simply to follow their officers or senior crew members and not say a word when they saw something unsafe. The emergency scene is a dynamic, sometimes hostile, and volatile environment. Although good officers try to bring some amount of predictability to what is deemed at times an unpredictable environment, all members have a responsibility to inform each other of any and all dangers they may witness at the emergency incident.

Too many deaths and serious injuries have resulted from incidents at which a firefighter may have seen something dangerous and refrained from reporting it under the (wrong) assumption that the incident commander (who is positioned in front of the building) knew or should have known of the hazard. Empowering the firefighter to speak up on the emergency scene will help prevent avoidable deaths and injuries. (See Figure 2.11.)

ROLE OF NFPA 1500 AND THE U.S. FIRE ADMINISTRATION

The roles of the U.S. Fire Administration (USFA) and the National Fire Protection Association (NFPA) in reducing the number of firefighter line-of-duty deaths and injuries are similar in many ways. The USFA is an entity of the Department of Homeland Security's Federal Emergency Management Agency with a mission to foster a solid foundation in prevention, preparedness, and response by providing national leadership to local fire and emergency services. The NFPA is an international nonprofit organization whose purpose is to reduce the worldwide burden of fire and other hazards on the quality of life by providing and advocating consensus codes and standards, research, training, and

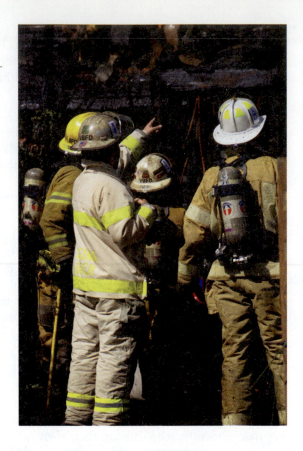

FIGURE 2.11 It is your responsibility to speak up if you see something that will make the emergency scene safer. *Courtesy of Martin Grube*

education. The primary goal of both organizations is to advocate safety for the citizen, as well as the firefighter or emergency responder.

Role of NFPA 1500

The role of NFPA 1500, *Standard on Fire Department Occupational Safety and Health Program,* is to ensure the implementation of practices by fire departments for improving firefighter health and safety. Compliance with all the provisions of NFPA 1500 is important to ensure consistently high levels of safety and health for firefighters. A thorough understanding and complete implementation of NFPA 1500 are credited with reducing firefighter fatalities and limiting emergency scene injuries.

NFPA 1500 addressed for the first time such issues as health and safety programs, training requirements, operational requirements of vehicles, and protective clothing requirements. Many fire departments in the past felt that the guidelines were too rigid and would affect a firefighter's ability to work successfully at the emergency scene.

Many years after its first publication in 1987, NFPA 1500 still is having an effect, as more and more firefighters are aware of the safety component of the profession and how such standards will help their department. In particular, as departments are seeing lower and lower staffing levels, the impact of firefighter death or injury becomes even more significant.

Younger firefighters, who have grown up with NFPA 1500, are having an impact on the new and positive culture that awareness brings. Many firefighters can now say, "I like the job and I like what I do, but I don't want to die doing it. Instead, I want to live and I want to be around for my family." Those firefighters are now asking, "Why would we *not* wear our PPE?"

Role of the U.S. Fire Administration

One of the missions of the U.S. Fire Administration (USFA) is to reduce LODDs. Faced with an average of 100 firefighter LODDs and approximately 80,000 injuries each year,

FIGURE 2.12 Sharing information with other firefighters about close calls helps everybody identify areas of concern. *Courtesy of Eric Melcher, Photographer, Volunteer State Community College*

the USFA sponsors a myriad of training programs and projects in order to reach its objective: to reduce the line-of-duty deaths by 50% by 2014. This is to say nothing of the many more "near misses" and "close calls" that we do not hear about. According to the National Fallen Firefighters Foundation (see www.everyonegoeshome.com), "for every serious incident that we hear about, there are 300 near misses that we never hear about."

ROLE OF NEAR-MISS AND CLOSE CALL REPORTING

Firefighter Near Miss (see firefighternearmiss.com) is a voluntary reporting system that allows firefighters to report real-life incidents so that others may learn from those experiences by the collection of confidential information. The goal is to identify the types of close calls that repeatedly are occurring in the fire and emergency services to identify the causes for concern. This nonpunitive process allows firefighters to share information so others can recognize similar situations. Weekly reports provide a way for improvement to be made without personally being involved in the incident. Situational awareness and decision making have been two top areas of concern for those submitting reports. (See Figure 2.12.)

Firefighter Close Calls (see firefighterclosecalls.com) allows individuals to access information about line-of-duty deaths and injuries. The site provides SOPs/SOGs, weekly fire drills, and so on. However, its main focus is issues that are often ignored, quickly forgotten, or just not talked about.

Another prominent site is Honor Their Sacrifices (see honortheirsacrifices.com) at which individuals can search line-of-duty deaths by building type, factors, incident type, location, reporting agency, and year. The site promotes the lessons learned from past tragedies in order to prevent their reoccurrence.

Personal Responsibility and Risk Management

Departments must develop and implement programs that instruct all firefighters about the importance of personal responsibility and risk management for their own safety and survival.

ROLE OF PERSONAL RESPONSIBILITY IN SAFETY

The only way to begin to break the ugly cycle of firefighter LODDs is to focus on personal responsibility. Each member of the fire and emergency services must make it a personal mission to prevent unsafe acts on the emergency scene and ensure that his or her own performance and behavior are within the parameters of safe practices. Crew members must evaluate and reevaluate their attitudes and behavior on the emergency scene and thoughtfully examine why such attitudes may exist. For example:

■ Are personnel acting or reacting to the peer pressures of their crew, their officer, their department?

■ Do they have some inherent belief that firefighters are to risk their lives at all times without regard to the risk of dangerous behavior?

Most LODDs are caused by human error, poor decision making, and lack of **situational awareness**. The constant awareness of one's surroundings in a dynamic, immediately dangerous to life and health (IDLH) environment includes the perception of the environment critical to decision makers in other complex, dynamic fields such as aviation, air traffic control, power plant operations, and military command and control. When firefighters are making decisions that can lead to serious injury or death, they must have the courage to be safe and self-disciplined. (See Figure 2.13.)

Although being aggressive is a positive characteristic in the fire and emergency services, firefighters are individually responsible for bringing hazards to their officer's attention. For example, firefighters must be able to say "stop" or that this emergency incident is a "no go" due to situations such as:

■ A lightweight truss structure that is threatened by heavy fire involvement

■ A structure's imminent probability of lighting up (flashover) or smoke explosion (backdraft)

Saying "stop" or "no go" is a way for firefighters to act in a personally responsible manner.

FIGURE 2.13 Firefighters must maintain situational awareness when making decisions. *Courtesy of Buddy Byers, District Chief, Nashville Fire Department*

FIGURE 2.14 Firefighters must master the task of an operation before being called upon to perform it on the emergency scene.
Courtesy of Guy E. Willette

One of the ways to prevent LODDs and injuries is if firefighters perform as they have been trained to perform. Although the department is responsible for providing training, firefighters are personally responsible for participating fully in all recruit, in-service, or off-site training. For example, firefighters need to arrive prepared for training. Because departments spend a considerable amount of public money to train firefighters and provide them with appropriate PPE and other equipment, firefighters must do their part to constantly practice and perfect their skills and care for their equipment. Fire departments enforce safety SOPs/SOGs by providing firefighters an alternative that is less than desirable: failure to perform as trained and according to SOPs/SOGs (which ensure safety and accountability for all) will result in **discipline**, a system of rules of conduct or method of practice that may include termination.

Another facet of personal responsibility for preventing firefighter LODDs and injuries is for firefighters to know their strengths and weaknesses. At a minimum, firefighters should obtain various certifications, which require them to train. Training never ends for firefighters; they must constantly test themselves and try to improve those areas in which they are weak. In fact, firefighters should strive for mastery of their basic assigned tasks and then expand to more advanced tasks. All in all, firefighters should not let an emergency incident become an avenue for discovery and on-the-job training. (See Figure 2.14.)

discipline ■ The maintenance of a level of control and adherence to one's role and responsibility on the emergency scene; staying within one's incident management system's position; a system of rules of conduct or method of practice.

ROLE OF PERSONAL RISK MANAGEMENT

Only through honest self-assessment can one begin to unveil some of the whys of risky behavior on the emergency scene. Sometimes a safety-conscious department is able to recognize the truth about firefighter LODDs and serious injuries.

It is time that firefighters look seriously at their attitudes of apathy, sense of invincibility, ignorance, and "it will never happen to me" as they work in one of the world's most dangerous occupations. Those types of rogue approaches are bound to get the firefighter or fellow firefighters killed or seriously injured. The rogue attitude or behavior does not produce success on the emergency scene and has been proven to have serious consequences.

Firefighters must work together and follow SOPs/SOGs, which almost always involves working in teams with a minimum of two firefighters, looking after one another, and not performing beyond a reasonable "risk versus reward" analysis. Everyone must work to maintain **crew integrity**, which is the practice of staying together as a crew; working with at least one other emergency responder; or staying within voice, visual, or physical contact of one another.

Although the most honorable thing a human being can do is to give one's life for the life of another, this sacrifice must still be performed with a reasonable idea that the life is savable. Risking a life (high risk) to save property (low reward) is rarely, if ever, acceptable. If, for example, it is clear that there are no victims in a vacant fast-food restaurant fire whose structural features include a lightweight truss roof and floors that happen to be heavily threatened by fire, the emergency incident does not warrant an interior aggressive fire attack—and the associated risk to firefighting crews. Where the rescue profile is negligible, so too must be the risk that firefighters are willing to take.

Indeed, property conservation ranks second in priority after life safety. However, firefighters risk a lot to save a lot, but little to save little, and nothing to save nothing. A vacant property is one structure firefighters would risk very little to preserve. (See Figure 2.15.)

Firefighters must understand and appreciate the risks associated with the profession and be encouraged in behaviors that hold high the life safety of themselves, their fellow firefighters, and the public. The only way to curb behaviors related to freelancing, independent action, cutting corners, and complacency is if departments make it unequivocally

crew integrity ■ The practice of staying together as a crew; working with at least one other emergency responder; or staying within voice, visual, or physical contact of one another.

FIGURE 2.15 Discussing the incident can be done at the scene. *Courtesy of Martin Grube*

clear that those types of behaviors are no longer accepted or tolerated. Firefighters must know that anyone committing behaviors that are clearly outside the scope of their desired and expected performance may be out of a job. The departmental responsibility to affect this type of punitive action must be clear and consistent all across the board and across all ranks. Consistency must show no boundaries or barriers when it comes to behaviors that have been proven to cost the lives of firefighters.

In turn, personal responsibility must be such that each firefighter accepts his or her role in a department that is free of rogue behaviors on the emergency scene. That department should be rich in safe practices that have a near 100% chance of ensuring that each firefighter makes it back to the station after responding to an incident and ultimately makes it back to his or her family.

Legal Consequences of Unsafe Behavior

Departments must teach all personnel the potential legal consequences of their unsafe behavior. In doing so, organizations must establish systems whereby members may be held accountable for their unsafe actions. Fire departments and members may avoid being held legally liable by following department SOPs/SOGs related to safety.

In particular, fire departments must ensure that every supervisor is aware of his or her personal and departmental accountability for the health and safety of all subordinates. To be aware is to have knowledge about something or expected knowledge about something based on one's education, training, experience, and circumstances. Failure to act when possessing such knowledge or expected knowledge exposes the firefighter and the department to legal liability.

Negligence is the failure to act when one has a duty to act. A firefighter may be deemed negligent if he or she acted in a manner in which a duty was breached or failed to act in a manner that was required of him or her. When a duty is breached, it may result in injuries or damages as a proximate cause of the breach.

negligence ■ The failure to act when one has a duty to act. When that duty is breached, it may result in injuries or damages as a proximate cause of the breach.

On the emergency scene, to avoid being negligent simply means to perform as trained and according to SOPs/SOGs. The emergency scene is not the training ground where one hopes that the incident will be safely resolved. The firefighter must be able to perform his or her tasks with competency and, ideally, with mastery. For example, a firefighter assigned to the engine must be able to deploy a hose line in a safe and adequate manner. A firefighter assigned to the ladder company must be able to ventilate, allowing the smoke and heat to be removed from the building to make conditions viable for firefighters and victims inside.

Litigation against fire departments is on the rise; most plaintiffs allege negligent firefighting or search-and-rescue operations. Fire departments may be unwilling to expose themselves to legal liability by conducting after-action reviews about what went well and what went wrong in an effort to improve operations. Still, it can be suggested that the overall benefits of conducting a lessons learned review outweigh any and all potential **liability** to a fire department. Liability is exposure to responsibility or accountability at a personal or professional level; it is the state of being legally responsible.

liability ■ Exposure to responsibility or accountability at a personal or professional level; the state of being legally responsible.

To have known about a safety policy and to violate that policy anyway may be deemed intentional negligence or gross negligence. Regardless of the legal findings, the individual firefighter or officer will have demonstrated some willingness to avoid practicing a safety policy that he or she knew or should have known to have negative consequences to the individual or crew. Such reckless actions disregard the intent of the policy or procedure, which is to ensure that firefighters are safe. They demonstrate a rogue type of behavior on the part of the offender toward safety in general, and this personality trait could expose the individual and department to more serious legal consequences.

After-Action Review as a Form of Accountability

The safety accountability of a firefighter and the entire department extends beyond the life of a single emergency incident. In fact, the manner in which a firefighter or department handles after-action reviews is a measure of the individual's and department's accountability. As this term is used in this chapter, *after-action review* includes the (1) incident review, (2) after-action review, and (3) investigation of serious fatalities and serious injuries.

Vincent Dunn, a 42-year veteran of the Fire Department of New York, states in *Command and Control of Fires and Emergencies* (1999, p. 11), "You learn something new at each fire, and so does everyone else if you share your experience. This way, everybody gets more experience from the fire." Dunn often speaks of a fire he responded to as a new lieutenant where 12 firefighters were killed at a building collapse. Dunn is certain that he would have been one of the 12 who perished if it were not for the battalion chief assigning him to check an exposure building.

INCIDENT REVIEW

After each emergency incident, officers should at the very least summarize any positive and negative issues they observed at the incident in a learning-friendly environment. Reinforcing proper performance on the emergency scene or correcting issues that need immediate attention can be effectively done during the incident review. Many in the fire and emergency services consider this "tailboard debriefing" to be very effective, as one individual holds it immediately subsequent to any emergency incident and includes all those who participated in the incident's mitigation.

AFTER-ACTION REVIEW

After-action review (AAR) usually occurs after significant events, basically providing the "big picture" look at what went right and what can be done differently to improve emergency response operations on similar types of incidents. For an after-action review to be effective, it should be conducted with all involved parties as soon as practically possible after the incident. Even if an incident requires an extensive investigation (e.g., as in the case of firefighter injury), every effort must be made to analyze the emergency incident as soon as possible. Delaying the analysis may give some members time to develop defensive behaviors or mount attacks against other members, thus compromising any constructiveness intended. (See Figure 2.16.)

It is natural for many departments to shy away from any lessons learned from after-action review due to fear that they will look bad or that there will be dissent and trouble if they talk about how badly firefighters underperformed (not to mention fear of lawsuits). However, progressive departments learn from their mistakes and the mistakes of others; whereas average departments do not learn from anything or anyone. Every incident is an opportunity to learn, and holding a "lessons learned" after-action review is the key to success.

If a department allows too much time to go by, memories fade and stories change. Regular and frequent after-action reviews can reduce freelancing because independently acting firefighters soon learn that what they do at an incident will become part of the record. After-action reviews keep incident commanders (ICs) aware of what is happening on the emergency incident. After all, the purpose of the AAR is not to blame or look the other way, but to improve operations, which ultimately reduces firefighter LODDs and injuries.

But most importantly, any changes derived from the lessons learned through the AAR must receive total support from the highest levels in the department in order to make all of the necessary improvements. Avoiding the blame game is difficult, which is why an

FIGURE 2.16 After-action review is an important part of improving emergency scene operations. *Courtesy of Daniel A. Nelms, Emergency Service Training*

after-action review needs to be approached as an open and honest forum with overall improvement as its goal. For example, it needs to ensure a safe environment, analysis of task and actions, and the like. If the environment is truly considered safe, firefighters will be more likely to take responsibility for mistakes and accept criticism.

The after-action review is relevant to every member of the department, because the analysis can have far-reaching effects on each member and future operations. Since there has been a decline in the number of actual structure fires, this transfer of information becomes invaluable. With the reduction in actual working fires, many of today's new officers and firefighters have less opportunity to develop their skills through actual experience. The ability to relay information through the after-action review is one way to fill this gap.

An after-action review is useful only to the extent that fire departments understand and practice it. This ventures into the areas of firefighter accountability (i.e., whether officers and chiefs are willing to take responsibility by holding personnel accountable), personal responsibility (i.e., to one's self and one's buddy), and of course, the culture of the specific department (i.e., how the fire department and members actually view safety). After all, unless recommendations from an after-action review are followed, they serve no purpose. They serve no purpose if recommendations end up sitting on a shelf somewhere without having been shared with the firefighters.

INVESTIGATION OF SERIOUS INCIDENTS

Departments must investigate fatalities and serious injuries and take corrective actions in order to thrive and sustain their safety messages and vision for their members. Such investigations are far more rigorous and formal than the more routine AAR. (See Chapter 9, Fatality and Injury Investigations, for more detailed discussion.)

The lessons learned approach may not always be kind. It does not have to include blame, but it must be clear about identifying the desired behaviors at any given incident that resulted in a death or serious injury to a firefighter. It must be emphasized that the behavior exemplified in an incident that resulted in dire consequences must not continue, even if it means that an interior aggressive attack must cease in favor of a defensive strategy. This is the only way that the behavior of unreasonable risk taking will begin to undergo a paradigm shift, one that is more focused on safety rather than ego or adrenaline rush.

The Role of Cultural Change in Accountability

Chapter 1, Fire and Emergency Services Culture, addresses fire and emergency services culture in some detail. This section focuses more specifically on the role of cultural change in accountability for safety.

A CULTURE OF UNNECESSARY RISK

The fire and emergency services must cease rewarding a culture of unnecessary risk taking and focus on rewarding those who incorporate safety practices in their daily activities. Everyone appreciates and understands the inherent dangers of this profession. There is no need to reward "heroes" by placing them on the cover of its periodicals when it is known that the individual took unreasonable and dangerous risks that but for time and sheer luck resulted in saving a human life. This does not reward safe behavior, but actually perpetuates behavior that will result in LODDs and serious injuries to firefighters.

TOOLS FOR CULTURAL CHANGE

Fire departments can promote safety and accountability by focusing on empowerment, ownership, and motivation. (See Figure 2.17.)

To permit is to promote. If a fire department permits unsafe behavior, it is in effect promoting unsafe behavior. One must have the courage to be safe in a department whose culture may not treat safety as a priority. Safety must be a shared vision, and firefighters are constantly being reminded of their role in achieving that vision. The vision can be reflected in chief officers and company officers along with firefighters leading by example, posting signs in fire stations that read "safety is an attitude" and "safety is no accident."

FIGURE 2.17 Fire departments should be focusing on cultural change by promoting safety and accountability through empowerment, ownership, and motivation. *Courtesy of Martin Grube*

Some departments are so committed to accountability that any member found to be violating Occupational Safety and Health Administration (OSHA) standards is responsible for paying any OSHA fine associated with the violation.

Empowerment and Ownership

How does a young firefighter make the transition from one who is trained to follow officers and veteran firefighters without question to one who says something when he or she sees an unsafe act? And how does a firefighter learn to speak up when that unsafe act involves anybody on the emergency scene? The transition may be accomplished through empowerment. The department must support members of all ranks to have the courage to be safe and to be confident of support whenever they stand up for safety. The organizational culture within the department must not make a safety advocate a "bad guy." Instead, the fire department must instill confidence in its goal that unsafe behaviors must stop—and that it is everyone's responsibility to stop them. This is part of the ownership principle. The fire department must allow the firefighters to own their own safety and do something about it.

Motivation

What is the motivation for safety? The carrot at the end of the stick or the "what's in it for me?" factor is clearly one that benefits all firefighters. When one member demonstrates safety practices, the individual is safer, the entire crew is safer, and the emergency scene is safer. This correlates with the incident getting mitigated, and everyone is that much closer to going home at the end of the shift. And that ultimately translates into going home to the most important family of all, one's own.

A recruit arriving at the fire academy on the first day of his or her career brings a host of skills. In fact, most of the recruit's personal skills, habits, aptitudes, and attitudes have already been instilled. However, the majority of the fire-related skills a firefighter needs still must be developed. In hiring a recruit, it is safe to assume that the fire department enters into an implied partnership in that the department will guide and train the individual, and the individual, in turn, will work to become an asset to the department and community.

CHAPTER REVIEW

Summary

Fire departments must put systems in place to have each department member accountable for his or her actions on the emergency scene. Without systems such as SOPs/SOGs, firefighters will not know to which practices to adhere. SOPs/SOGs do not exist to get firefighters into trouble, but instead keep firefighters safe.

Health and safety begin with personal responsibility. Firefighters are constantly applying the risks versus gain benefit analysis on the emergency scene, which requires the need to practice good habits and always keep safety in mind. Firefighters must have a structured and disciplined approach to everything they do on the job so that they are able to return to their families at the end of each shift.

Fire departments must ensure that every supervisor is aware of his or her personal and organizational accountability for the health and safety of all subordinates. It is virtually impossible to expect firefighters to accept the responsibility of safety unless everybody in the department is held personally and organizationally accountable. The personal and organizational environment should encourage and acknowledge everybody's efforts when it comes to health and safety.

Fire departments must teach all personnel the potential legal consequences of their behavior. In doing so, departments must establish systems that hold members accountable for their actions and inactions related to safety. Fire department members can minimize being held legally liable by following departmental SOPs/SOGs.

Departments must investigate fatalities and serious injuries and evaluate the information collected to take corrective actions. The after-action review process plays an important part in collecting the information needed. This review enables departments to identify which areas related to safety need corrective action. Fire departments must promote accountability by focusing on empowerment, ownership, and motivation. The real tragedy in personal and organizational accountability is not utilizing the one thing that all of us have, and that is the ability to speak up and stop an unsafe act and take corrective action when the need arises.

Review Questions

1. List the three areas of the U.S. Fire Administration's learning triad.
2. Explain the difference between personal and departmental accountability.
3. Compare the roles of the U.S. Fire Administration and the National Fire Protection Association in reducing the number of firefighter and emergency responder line-of-duty deaths and injuries.
4. Discuss each level of responsibility for safety within a department.
5. Describe the safety purpose of having standard operating procedures/guidelines.

References

Ashworth, K. (2001). *Caught Between the Dog and the Fireplug; How to Survive Public Service.* Washington, DC: Georgetown University Press.

Avillo, A. (2002). *Fireground Strategies.* New York: PennWell.

Bachtler, J. R., and Brennan, T. F. (1995). *The Fire Chief's Handbook* (5th ed.). Saddle Brook, NJ: PennWell.

Carter, H., and Rausch, E. (2008). *Management in the Fire Service* (4th ed.). Sudbury, MA: Jones and Bartlett.

Chetkovich, C. (1997). *Real Heat: Gender and Race in the Urban Fire Service.* New Brunswick, NJ, and London: Rutgers University Press.

Conger, J. A. (1998a, May–June). "The Necessary Art of Persuasion." *Harvard Business Review.*

Conger, J. A. (1998b). *Winning 'em Over: A New Model for Management in the Age of Persuasion*. New York: Simon & Schuster.

Delmar Cengage Learning. (2004). *Firefighter's Handbook: Essentials of Firefighting and Emergency Response* (2nd ed.). Florence, KY: Author.

Dunn, V. (1999). *Command and Control of Fires and Emergencies*. Saddle Brook, NJ: PennWell.

Gordon Graham Research Consultants. (2000). *Fire Service Operations: Why Things Go Right—Why Things Go Wrong*. Long Beach, CA: Gordon Graham Productions.

International Association of Fire Chiefs. (2003). *International Association of Fire Chiefs Officer Development Handbook*. Oklahoma City, OK: Fire Protection Publications.

International Fire Service Training Association. (1989). *Fire Department Company Officer* (2nd ed.). Stillwater, OK: Fire Protection Publications.

International Fire Service Training Association. (2004). *Chief Officer*. Stillwater, OK: Board of Regents, Oklahoma State University.

International Fire Service Training Association. (2007). *Fire and Emergency Services Company Officer*. Stillwater, OK: Fire Protection Publications.

KTRK TV, Houston, TX. (2009, April 1). *Monday's Fire Truck Wreck Being Investigated*. Retrieved January 13, 2011, from http://abclocal.go.com/ktrk/story?section=news/local&id=6735569

Montagna, F. C. (1999). *Responding to "Routine" Emergencies*. Saddle Brook, NJ: PennWell.

Moore, B. (1985). *Risk Management Today: A How-to Guide for Local Government*. New York: Practical Management Series by International City Management Association.

National Fire Academy. (2000). "Unit 2. Developing Self as a Leader." *Executive Leadership Student Manual*. Emittsburg, MD: Author.

National Institute for Occupational Safety and Health (NIOSH). (2010, February 26). *A Career Lieutenant Dies and Three Fire Fighters Are Injured in Ladder Truck Crash—Massachusetts*. Fatality Assessment and Control Evaluation Investigation Report #F2009-05. Morgantown, WV: Author. Retrieved January 13, 2011, from http://www.cdc.gov/niosh/fire/reports/face200905.html

Office of the State Fire Marshal. (2008, May.). *State Fire Training Procedures Manual*. Retrieved May 2, 2009, from http://osfm.fire.ca.gov/training/pdf/sftproceduresmanual.pdf

Page, J. O. (1973). *Effective Company Command*. Alhambra, CA: Borden Publishing Company.

3
Risk Management

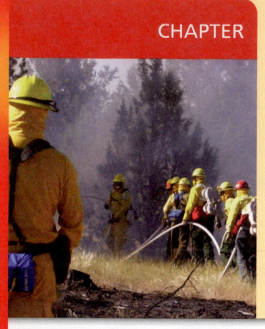

Photo by Miranda Simone

Marc A. Revere

OBJECTIVES

After reading this chapter, the student should be able to:

- Identify the three principles of risk management.
- Discuss the risk management process.
- Explain the need for a continual evaluation process.
- Explain the keys for implementing a successful risk management program.

Risk and Risk Management Defined

The goals of this chapter are to illustrate how the concepts of risk and risk management apply to the fire and emergency services, to identify attributes of an effective risk management process and programs, and to provide a methodology of assessing and mitigating risk with the ultimate goal of helping the organization improve firefighter safety and reduce LODD and injuries. In addition, the intent is to encourage and assist firefighters to follow the three principles of risk management:

- Integrate risk management into all aspects of the fire and emergency services.
- Make risk decisions at the appropriate level in the chain of command.
- Accept no unnecessary risk.

The first step is to define risk and risk management.

RISK

The fire and emergency services are faced with numerous risks daily. Some of these risks include everything from the actual workplace, to responding to and returning from emergencies, and emergency scene operations. Each risk carries consequences that need to be managed, because failure to manage risks properly can create problems throughout the department. Even though those consequences vary for each risk, some involve death and injury of firefighters when they are not properly planned for and managed. No single method or solution exists to effectively manage risks. Determining how to manage each risk is a decision made by each individual fire department and every member of that department.

The term *risk* has come to be used interchangeably in widely disparate settings. As a result, its meaning may have been blurred in the minds of many. Generally, the term *risk* is understood as the possibility of a loss or injury. However, NFPA 1500, *Standard for Fire Department Occupational Safety and Health Program*, defines risk as "a measure of the probability and severity of adverse effects." The adverse effects result from an exposure to some type of hazard. In this chapter, **risk** is defined as the potential negative impact of the exercise of vulnerability, considering both the probability and the impact of occurrences. As used here, risk can be individual (such as to an individual firefighter) or departmental (such as to a crew). The word *exercise* means to accidentally trigger or intentionally exploit the vulnerability. And the word **vulnerability** is a weakness that can be exercised. Finally, a **threat** is the potential for a threat source to exercise (accidentally trigger or intentionally exploit) a specific vulnerability or basically the act necessary to create the loss.

Managing fire department and community risk is common in the fire and emergency services, because various types of nonemergency and emergency operations are conducted on a daily basis. Table 3.1 provides emergency response examples of risk, vulnerability, and threat. Table 3.2 identifies examples of common threats.

FREQUENCY/SEVERITY MODEL

Risk combines two concepts: the frequency of an event and the severity of that same event. Frequency is an estimate of how often an event has occurred or is likely to occur. Severity, the second half of the evaluation phase, estimates the potential losses to the individual or department posed by the identified risk. Any event can have a high or low frequency and a high or low severity.

Frequency:
- High frequency means something will probably occur in most circumstances.
- Low frequency means something might occur at some time.

risk ■ The potential negative impact of the exercise of vulnerability, considering both the probability and the impact of occurrences, and applies to both organizational and individual risk.

vulnerability ■ A weakness that can be accidentally triggered or intentionally exploited.

threat ■ The potential for a threat source to exercise (accidentally trigger or intentionally exploit) a specific vulnerability or basically the act necessary to create the loss.

TABLE 3.1	Emergency Response Examples
TERM	**EMERGENCY RESPONSE EXAMPLES**
Risk	Conflagration, flooding, earthquake
Vulnerability	Levees, lack of vegetation from burned hillside, truss construction
Threat	Open match, spills, explosion, collapse

TABLE 3.2	Common Threats
TYPE OF COMMUNITY THREAT	**EXAMPLES**
Natural threat	Floods, earthquakes, tornadoes, landslides, avalanches, electrical storms, and similar events
Human threat	Unattended campfire, motor-vehicle collision, bombing, campus shooting
Environmental threat	Long-term power failure, pollution, chemicals, liquid leakage

Severity:

- High severity means a major impact on the individual or department.
- Low severity means a minor impact on the individual or department.

Figure 3.1 illustrates the relationship between frequency and severity. The **frequency/severity model** is a technique for conceptualizing the frequency and severity of a given event or hazard and thereby establishing priorities for risk management actions. Events with expected high frequency and severity are priorities for risk management, whereas events with expected low frequency and severity are not. Using the frequency/severity model can help an individual or department establish proper safety priorities, such as making a decision on an interior attack to an abandoned building or going into a neighborhood where shots have been fired without police protection. It requires firefighters to become more aware of responding to incidents in which the frequency is low but the severity is high. Of course, certain risks are so severe that they should be avoided regardless of frequency. (See Figure 3.2.) However, risk management in high-frequency and low-severity incidents such as responding to medical emergencies and properly lifting the patient also should be considered.

The challenge to fire departments is to identify the exposure to risk and then take the proper action to reduce the potential for that risk before the risk occurs or to reduce the severity or impact of the risk once it has taken place. Some of the reasons

FIGURE 3.1 Risk management matrix.

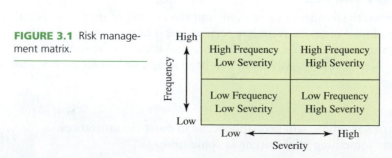

firefighters get killed or injured may fall into one of the risk categories in Table 3.3. It could be because firefighters did not follow proper policy and procedures, did not utilize safety equipment that was provided, had poor training or no training at all, or had an inappropriate attitude toward safety. Table 3.3 gives the frequency and severity of the associated risks.

RISK MANAGEMENT

To improve the performance of fire departments and reduce the risks that communities, departments, and individual members face virtually every day, incorporate a risk management approach into the delivery of fire and emergency services. Consider the terms *management* and *risk management*: the term *management* suggests a direct approach that uses tried-and-true techniques to control systems, events, and people. In contrast, Kipp and Loflin (1996, p. 7) define the *risk management process* as a "system for treating pure risk; identification and analyses of exposures, selection of appropriate risk management techniques to handle exposures, implementation of chosen techniques, and monitoring of the results." As used in this chapter, **risk management** is the process of identifying risk, assessing risk, and taking steps to reduce risk to an acceptable level. Risk management

risk management ■ The process of identifying risk, assessing risk, and taking steps to reduce risk to an acceptable level.

TABLE 3.3	Sample Risks and Associated Frequency and Severity		
RISK		**FREQUENCY**	**SEVERITY**
Becoming trapped in building with truss construction		Low	High
Becoming a victim yourself during shooting incident		Low	High
Emergency vehicle crash while responding		Low	High
Sprains and strains from emergency operations		High	Low
Back strains from improper lifting of patients onto stretchers		High	Low

TABLE 3.4	Level of Departmental Management Involvement Needed for Risk Management		
	LOW FREQUENCY	MODERATE FREQUENCY	HIGH FREQUENCY
High Severity	Upper management involvement required	Upper management involvement required	Upper management involvement required
Moderate Severity	Middle management may accept and monitor risk	Middle management effort needed	Middle management involvement required
Low Severity	First-level management will accept risks	First-level management will accept and monitor risks	First-level management involvement required

attempts to identify and then manage threats that could severely impact individuals and the department. Generally, risk management involves reviewing departmental operations, identifying potential threats to the department and the likelihood of their occurrence, and then taking appropriate actions to address the most likely threats. Risk management is a structured approach to manage the uncertainty related to a threat. It involves a sequence of human activities: **risk assessment**, the focused assessment of potential risks to the organization; strategic considerations to manage the risk; and **risk mitigation**, a systematic methodology used by senior management to reduce mission risk.

The objective of risk management is to reduce the potential for loss associated with risks to a level deemed acceptable by the department. It may refer to numerous types of threats caused by the environment, technology, human organizations, and politics.

Risk management implements a proactive rather than a reactive approach to solving problems or limiting risks. The term implies a systematic effort to identify, evaluate, and control risk(s) to reduce both the probability that something might go wrong and the adverse effects (magnitude) if something should go wrong. Risk management is one way that a fire department can reduce firefighter line-of-duty deaths and injuries by reviewing departmental policies and procedures, job tasks, and equipment. In developing proper methods to provide guidance on risk management, the different levels of readiness and experience in a department, as well as available resources, need to be considered. Examples of departmental management involvement in risk management are presented in Table 3.4.

risk assessment ■ The focused assessment of potential risks to the organization.

risk mitigation ■ A systematic methodology used by senior management to reduce mission risk.

The Purpose, Components, and Tools of Risk Management

The purpose of risk management is the conservation and preservation of resources and assets by identifying, evaluating, and controlling risks. All activities conducted within a department should have controls that help minimize adverse financial effects of inadvertent losses to the department, including property loss, net income loss, liability loss, and key personnel loss. Risk management incorporates a full range of control measures that may be used to limit, reduce, or eliminate an undesirable outcome. However, because there is no single method or solution for effectively managing risk, risk management becomes a continual process that involves organizing and verifying that risks are being controlled.

Departmental leaders must understand what a risk management process does *not* do to fully understand its role in the fire and emergency services. Risk management:

■ Does NOT replace sound tactical decision making by management and subordinate leaders
■ Does NOT inhibit flexibility, initiative, or accountability

FIGURE 3.3 The risks and consequences involving truss construction collapse must be considered early on in the incident. *Courtesy of Travis Ford, Nashville Fire Department*

- Does NOT remove all risks or support a zero defect mentality
- Does NOT sanction illegal or unethical behavior and activity
- Does NOT remove the need for training, tactics, techniques, and SOPs/SOGs

U.S. ARMY FIVE-STEP RISK ASSESSMENT PROCESS

The U.S. Army developed a five-step risk assessment process during World War II, when General William Tunner assumed command of airlift operations in the India-Burma-China Theater of Operation. The process has influenced contemporary risk management and is very similar to that outlined in NFPA 1250, *Recommended Practice for Emergency Service Organization Risk Management.*

The army process includes the following:

- Identification of hazards
- Assessment of those hazards
- Development of controls to aid with decision making
- Implementation of controls
- Supervision and evaluation of safe execution

Truss construction continues to be a major issue in the number of firefighters killed and injured in the line of duty. Using the risk assessment process can help in addressing the risks and consequences involved during emergency scene operations involving truss construction. (See Figure 3.3.) See Table 3.5 for an application of the five-step U.S. Army risk assessment process to fire departments associated with truss construction incident responsibilities.

The incident commander (IC) during any truss construction structure fire, regardless of rank, will make tactical decisions and strategic plans based upon known and reasonably expected risks. However, safe operations require that the incident commander and every individual on the emergency scene continually evaluate changing conditions. Evaluation of the emergency scene should be a continuous process to determine whether

TABLE 3.5	Application of Risk Assessment to Truss Construction
U.S. ARMY STEP	**FIRE AND EMERGENCY SERVICES TRUSS CONSTRUCTION EXAMPLE**
Identification of hazards	Identification of truss construction can primarily be recognized during the construction phase of the first due response area or through the preplanning phase, when older buildings that might have truss construction can be identified.
Assessment of those hazards	Assessing the hazards should be done by asking the following two questions: 1. What is the potential of frequency of occurrence? Modern construction methods primarily use truss construction, so the potential of frequency may depend on the response area. 2. What is the potential severity and exposure of its occurrence? Truss construction is severely affected once fire has reached the truss assembly area, and firefighters should be aware of the potential for total collapse or failure with little or no warning.
Development of controls to aid with decision making	Control options may include one or all of the following: 1. Training all emergency responders about the dangers of truss construction. 2. Changing organizational policies and procedures to mandate new code provisions for installing sprinkler systems. 3. Developing a policy of not allowing an interior fire attack upon arrival.
Implementation of controls	Departments need to ensure the controls are integrated into SOPs/SOGs and other forms of written guidance. 1. Do the firefighters have proper education on truss construction and the fire's impact on a truss-framed structure? 2. Do the firefighters know how to identify truss construction during preplanning operations? 3. Before performing a task on the emergency scene, do the firefighters know which policies, procedures, and guidelines related to operations involving truss construction are in place to help make proper decisions?
Supervision and evaluation of safe execution	The last step in the risk assessment process is evaluating to see whether the chosen corrective action did what it was expected to do. 1. Determine whether the action taken on the emergency scene was successful and effective. 2. Establish a way to monitor performance and progress on the emergency scene. 3. All actions should be evaluated routinely in truss construction fires to ensure that they are achieving the desired outcomes and are not creating other unanticipated problems.

the control measures are effective. Whenever firefighter safety is involved, actions should be on the side of caution. This means when in doubt, stay out or get out when truss construction is involved. (See Figure 3.4.)

COMPONENTS AND TOOLS

The risk management process helps to identify appropriate components and tools for reducing or eliminating risk. For example, officers have always been called upon to make decisions that weigh the risks of a particular course of action against the potential benefits.

Risk management involves the following components:

- Routine evaluation of risk in all situations
- Well-defined strategic options
- SOPs/SOGs
- Effective training
- Personal protective ensemble
- Effective incident management communications
- Safety policies, procedures, and guidelines

FIRES ON THE RAMPAGE

STAR is an acronym for stop, think, act, and review, a risk management approach for individuals. The STAR approach helps an individual focus his or her attention on the task at hand to consciously and deliberately review the intended action and expected responses before performing the task. The self-checking method should be used for any task that has a potential to impact the overall scene outcome.

Here is an outline showing how to use STAR. Notice that "Stop" is the most important step of any self-checking technique: the simple act of pausing increases the likelihood of performing the task correctly and eliminates current or potential distractions.

Stop:

■ Briefly pause before performing a task to enhance the mental focus on the specific task, such as arriving on scene or performing a size-up.

Think:

■ Understand specifically what is to be done before acting; for example, know which floor is to be searched.
■ Question the situation by trying to identify the information necessary to perform the task correctly; for example, what is the construction of the roof to be ventilated?
■ Do not proceed in the face of uncertainty. Instead, seek help.

Act:

■ Gather the necessary tools identified during the previous phase.
■ Be sure all protection schemes are in place (protective gear, RIT team, etc.).
■ Perform the intended action; for example, for physical action, ensure continuous hand contact with a ladder, hose line, or chainsaw.

Review:

■ Verify that the actual response is the expected response.
■ If an unexpected response is obtained, take action as previously anticipated/determined. (Call for assistance if victims are found.)
■ Ensure all actions are on the side of caution.

STAR ■ Acronym for stop, think, act, and review, a risk management approach for individuals.

- Rapid intervention
- Adequate resources
- Rest and rehabilitation
- Regular reevaluation of conditions
- Pessimistic evaluation of changing conditions
- Experience based on previous incidents and critiques

Process of Risk Management

Risk management is a process of identifying, evaluating, prioritizing, and controlling risk that can impact the assets of a department as well as its members and activities. The risk management plan incorporates a full range of control measures that may be used to limit, reduce, or eliminate the probability that an undesirable outcome may occur. The plan includes control measures that can be used to limit or eliminate anticipated hazards even if the event does occur. The risk management process is not only essential but also a very challenging five-step process that includes the following steps:

- Risk identification
- Risk evaluation
- Risk prioritization
- Risk control measures
- Risk monitoring (See Figure 3.5.)

RISK IDENTIFICATION

Identify the worst-case scenario for what might go wrong regarding each individual risk. This can be done by reviewing both emergency and nonemergency operational risks that may be encountered in the following areas: fire suppression, emergency medical service, technical rescue, vehicle operations, fire prevention, public fire education, communications, and general operations. Develop a list, and review past data to identify areas of concern.

FIGURE 3.5 The risk management plan may be used to limit, reduce, or eliminate the probability of an undesirable event.
Courtesy of Wayne Haley

RISK EVALUATION

After identifying each risk, evaluate risks from a frequency and severity standpoint for classification. The risk evaluation step is particularly important if the department has experienced a death or injury, because if the department has not developed a plan to manage the risk, the chance of experiencing the same result is likely should the same event be encountered again.

RISK PRIORITIZATION

Understanding the prioritizing of risks is not a simple task. However, any risk that has been evaluated as high frequency and high severity should be considered first and those evaluated as low frequency and low severity later. Risk prioritization may depend upon several factors including available resources for addressing each individual risk.

RISK CONTROL MEASURES

Risk control measures involve finding the proper solution for each identified risk. Although risk occurring during emergency scene operations sometimes cannot be controlled compared to nonemergency risk, the severity of the risk can be taken into consideration when selecting methods used to control risks, such as avoiding, control, and accepting. Former FEMA Director and former U.S. Fire Administrator R. David Paulison provides a message for the fire and emergency services: a simple, yet significant message that takes only 13 seconds to deliver but, if applied nationally, could save lives on the emergency scene. Paulison states:

> What we're trying to do is change the culture of the fire service. It's no longer acceptable to put your life on the line for a piece of property. Yes, we're going to save lives and we're going to put our lives on the line if we have to save somebody else. But stop and think what you're doing before you go into a burning building. (Mora, 2009)

RISK MONITORING

A provision for evaluating the effectiveness of the controls implemented should be periodically reviewed and required modifications made to the plan.

The risk management process is a toolbox for managing risk just as the incident command system is a toolbox for managing emergency incidents. You use the tools that you need. There are many components of risk management including financial, security, liability, and safety factors. The fire and emergency services focuses primarily on the safety component.

Strategies and Techniques of Risk Management

Choosing the right strategy and technique can allow a fire department to maintain effective and efficient daily operations. In order to ensure that all of the risk management strategies and techniques that are chosen provide safety, they should be monitored for performance and evaluated for progress. Therefore, one of the important aspects of this process is to ensure there is a methodical approach to the development and implementation of the chosen strategies and techniques of risk management.

GENERAL STRATEGIES

Risk management is the process of measuring or assessing risk and then developing strategies and techniques to manage that risk. The general strategies of risk management include the following:

- Avoiding the risk
- Controlling the negative effect of the risk
- Accepting some or all of the consequences of a particular risk

Table 3.6 links risk management strategies, approaches for accomplishing the strategies, and hypothetical fire and emergency services examples.

TABLE 3.6 — Risk Management Strategies, Approaches, and Examples

STRATEGY	HOW TO ACCOMPLISH	HYPOTHETICAL FIRE AND EMERGENCY SERVICES EXAMPLE(S)
Avoid	■ Eliminate the risk altogether.	■ Cancel response.
Control	■ Mitigate (reduce) the risk.	■ Enforce seat-belt usage. ■ Implement firefighter training programs. ■ Enforce firefighter fitness and health policies. ■ Enforce SOPs/SOGs on use of personal protective ensembles.
Accept	■ Balance high risk and reward with low risk and reward.	■ Risk much to accomplish much. (Begin rescue operations when residents are known to be inside a building.) ■ Risk little to accomplish little. (Perform exterior fire attack on an abandoned vacant building.)

Some traditional risk management processes focus on risk stemming from physical or legal reasons (e.g., natural disaster or fires, accident, ergonomics, death, and lawsuits). Whatever the type of risk, risk management is simply the process of identifying and prioritizing risks, then taking specific steps to control or accept them. How a firefighter or department controls and accepts risk is central to firefighter safety and survival.

INCIDENT COMMAND SYSTEM AS A RISK MANAGEMENT STRATEGY

A basic emergency scene strategy is the use of the incident command system (ICS) or incident management system (IMS). As stated in NFPA 1561, *Standard on Emergency Services Incident Management System,* "The IMS shall provide structure and coordination to the management of emergency incident operations to provide for the safety and health of emergency services organization responders and other persons involved in those activities. With respect to risk management the IMS shall integrate risk management into the regular functions of incident command" (2008).

The current edition of NFPA 1561 recommends that fire departments develop and implement a qualification process for individuals who function in the IMS. The qualification system would be incident specific to department response capabilities and responsibilities. The qualification system for ICS is broken down into five types, from local and discipline specific to a national organized team.

Strategic and tactical decisions occur very quickly based on the information available and changing conditions. The incident commander should not wait until an emergency occurs to use the risk management process. The preplanning phase should be the beginning of the risk management process in order to help determine the amount of acceptable risk that firefighters will be allowed to take. (See Figure 3.6.)

It is recommended that the department establish and implement an incident management system with written SOPs/SOGs for all firefighters. The system should be in accordance with nationally accepted practices (such as the National Incident Management System, NIMS, as set forth in NFPA 1561, *Standard on Emergency Services Incident Management System*) and provide for the following:

■ A well-coordinated approach to the emergency
■ Accountability of all firefighters
■ Overall safety of all firefighters at the scene of the emergency

Departments are responsible for training firefighters in this system and providing periodic refresher courses to review policies and procedures. Firefighters must always be fully aware of SOPs/SOGs and their roles and responsibilities. ICS is conditional on

how soon one can focus on the risk and take the proper corrective action that fully addresses that risk.

NFPA Standards Related to Risk Management and Safety

The National Fire Protection Association publishes several standards and recommended practices that relate to risk management and firefighter safety. There are many other relevant NFPA standards to help organizations establish safety policies, procedures, and guidelines. Of particular interest are NFPA 1250, *Recommended Practice in Emergency Service Organization Risk Management,* and NFPA 1500, *Standard on Fire Department Occupational Safety and Health Program.*

NFPA 1250, *RECOMMENDED PRACTICE IN EMERGENCY SERVICE ORGANIZATION RISK MANAGEMENT*

As outlined in its scope, the recommended practice "establishes minimum criteria to develop, implement, or evaluate a fire department risk management program for effective risk identification, control, and financing." Elements of the standard include the following, found in its Chapters 4 through 9:

- Risk Management as a Function of Management
- Identifying and Analyzing Risk Exposure
- Formulating Risk Management Alternatives
- Selecting Risk Management Alternatives
- Implementing Risk Management Solutions
- Monitoring the Risk Management Program

NFPA 1500, *STANDARD ON FIRE DEPARTMENT OCCUPATIONAL SAFETY AND HEALTH PROGRAM*

In 1987, the National Fire Protection Association adopted NFPA 1500, *Standard on Fire Department Occupational Safety and Health Program.* It was the first consensus standard to directly address many issues related to the avoidance of fatalities, injuries, and occupational illnesses firefighters experience in the performance of their duties. NFPA 1500 specifies the minimum requirements for a fire and emergency services occupational safety and health program, and the safety procedures for members involved in rescue, fire suppression, and related activities (NFPA, 2007). Considered radical by some at the time, some of the principles that form the foundation of the standard have since been accepted and adopted throughout the U.S. fire service.

The risk management plan as outlined in NFPA 1500 requires, at a minimum, that a department's risk management plan address the risks associated with the following functions:

- Administrative
- Facilities
- Training
- Vehicle operation
- Personal protective ensembles
- Operations at emergency incidents
- Operations at nonemergency incidents
- Other related activities

According to NFPA 1500, in emergency operations, the incident commander should utilize the following risk management principles:

- Activities that present a significant risk to the safety of members shall be limited to situations where there is a potential to save endangered lives.
- Activities that are routinely employed to protect property shall be recognized as inherent risks to the safety of members, and actions shall be taken to reduce or avoid these risks.
- No risk to the safety of members shall be acceptable when there is no possibility to save lives or property.
- In situations where the risk to fire department members is excessive, activities shall be limited to defensive operations. (NFPA, 2007, §8.3.2) (See Figure 3.7.)

NFPA 1500 includes a sample risk management plan, presented as Figure 3.8.

FIGURE 3.7 Risk to firefighters should be limited to situations in which the potential to save lives is a real possibility. *Courtesy of Barry Byers*

Purpose

The fire department has developed and implemented a risk management plan. The goals and objective of the plan are as follows:

1. To limit the exposure of the fire department to situations and occurrences that could have harmful or undesirable consequences on the department or its members
2. To provide the safest possible work environment for members of the fire department, while recognizing the risks inherent to the fire department's mission

Scope

The risk management plan is intended to comply with the requirements of NFPA 1500, *Standard on Fire Department Occupational Safety and Health Program.*

Methodology

The risk management plan uses a variety of strategies and approaches to address different objectives. The specific objectives are identified from the following sources of information:

1. Records and reports on the frequency and severity of accidents and injuries in the fire district
2. Reports received from the fire district's insurance carriers
3. Specific occurrences that identify the need for risk management
4. National trends and reports that are applicable to the fire department
5. Knowledge of the inherent risks that are encountered by fire districts and specific situations that are identified in the fire department
6. Any additional areas identified by fire district staff and personnel

Responsibilities

The fire chief has the responsibility for the implementation and operation of the district's risk management plan. The district's health and safety officer has the responsibility to develop, manage, and annually revise the risk management plan. The health and safety officer also has the responsibility to modify the risk management plan when warranted by changing exposures, occurrences, and activities.

All members of the fire department have the responsibility for ensuring their own health and safety based upon the requirements of the risk management plan and the department's safety and health program.

Plan Organization

The risk management plan includes the following:

1. Identification of the risk members of the fire department could actually or potentially encounter, both emergency and nonemergency
   a. Emergency risk includes those presented at emergency incident, both fire and no fire (e.g., hazardous material), Emergency Medical Service incidents, and emergency response.
   b. Nonemergency risk includes those encountered while performing function such as training, physical fitness, nonemergency (e.g., vehicle maintenance and station maintenance, daily office functions),
2. Evaluations of the identified risks based upon the frequency and severity factors
3. Development and implementation of an action plan for controlling each of the risks in order of priority
4. Provision for monitoring the effect of the controls implemented
5. A periodic review of the plan with modification made as needed

The plan requires a monitoring process that may be done by the health and safety committee or the health and safety officer.

Risk Management Plan Monitoring

1. The (anytown) fire department risk management program will be monitored annually, in January, by the health and safety officer.
2. Recommendation and revision will be made based on the following criteria:
   a. Annual accident and injury data for the preceding year
   b. Significant incidents that have occurred during the past year
   c. Information and suggestions from department staff and personnel

Every three years, the risk management program will be evaluated by an independent source. Recommendation will be sent to the fire chief, the health and safety officer, and the occupations safety and health committee.

FIGURE 3.8 Sample risk management plan. Reproduced with permission from NFPA 1500-2007, *Fire Department Occupational Safety and Health Program,* Copyright © 2006, National Fire Protection Association. This reprinted material is not the complete and official position of the NFPA on the referenced subject, which is represented only by the standard in its entirety.

States with OSHA Plans

Since the Occupational Safety and Health Act of 1970 was signed into law, 23 states and two territories have developed agreements with the Department of Labor to enforce federal standards through state agencies. The following have state OSHA plans that cover public employees: Alaska, Arizona, California, Connecticut, Hawaii, Indiana, Iowa, Kentucky, Maryland, Michigan, Minnesota, Nevada, New Mexico, New York, North

Carolina, Oregon, Puerto Rico, South Carolina, Tennessee, Utah, Vermont, Virginia, Virgin Islands, Washington, and Wyoming.

Even though the Department of Labor does not have jurisdiction over state and local government agencies, designated state agencies in the states with OSHA plans are required to enforce federal regulations on public agencies.

OSHA regulations establish a minimum standard. Individual states with OSHA plans may adopt equivalent or more stringent regulations. Those states determine whether they will enforce regulations for volunteer fire departments and other emergency response organizations, or if they will apply them to paid workers only. Increasingly, the trend is to enforce the same regulations on volunteer, combination, and fully paid departments. OSHA itself enforces federal regulations in the remaining states; states that do not have agreements to enforce the federal regulations generally adopt their own regulations and determine how to apply them. For example, in some states, the regulation of fire departments is assigned—not to an occupational safety and health agency—but to the state fire marshal or a state agency responsible for fire protection regulations. Policies and enforcement programs differ significantly from state to state, so it is important that each fire and emergency services organization identify the minimum applicable OSHA regulation or standard.

IAFC Rules of Engagement as Risk Management Tools

Beginning in 2008, the Safety, Health, and Survival Section of the International Association of Fire Chiefs (IAFC) began developing rules of engagement for structural firefighting, with participation by representatives from other fire service organizations. The IAFC team identified the need for two separate sets of rules of engagement: one for responders and one for incident commanders. In effect, the rules function as national standard operating procedures and translate the general principles of risk management into action items for use at the emergency scene. (See Figure 3.9.)

IAFC RULES OF ENGAGEMENT FOR FIREFIGHTER SURVIVAL

IAFC rules of engagement for firefighter survival are summarized in Table 3.7. The rules of engagement for firefighter survival are meant to provide each individual firefighter

FIGURE 3.9 Rules of engagement for firefighter safety and survival apply to every emergency scene.
Courtesy of Wayne Haley

TABLE 3.7	**Rules of Engagement for Firefighter Survival**
Size Up Your Tactical Area of Operation	*Objective:* To cause the company officer and firefighters to pause for a moment and look over their area of operation and evaluate their *individual* risk exposure and determine a safe approach to completing their assigned tactical objectives.
Determine the Occupant Survival Profile	*Objective:* To cause the company officer and firefighter to consider fire conditions in relation to possible occupant survival of a *rescue event* as part of their initial and ongoing *individual risk assessment* and action plan development.
DO NOT Risk Your Life for Lives or Property That Cannot Be Saved	*Objective:* To prevent firefighters from engaging in high-risk search-and-rescue and firefighting operations that may harm them when fire conditions prevent occupant survival and significant or total destruction of the building is inevitable.
Extend LIMITED Risk to Protect SAVABLE Property	*Objective:* To cause firefighters to limit risk exposure to a reasonable, cautious, and conservative level when trying to save a building.
Extend VIGILANT and MEASURED Risk to Protect and Rescue SAVABLE Lives	*Objective:* To cause firefighters to manage search-and-rescue and supporting firefighting operations in a calculated, controlled, and safe manner, *while remaining alert to changing conditions* during high-risk primary search-and-rescue operations where lives can be saved.
Go In Together, Stay Together, Come Out Together	*Objective:* To ensure that firefighters always enter a burning building as a team of two or more members and *no firefighter is allowed to be alone at any time* while entering, operating in, or exiting a building.
Maintain Continuous Awareness of Your Air Supply, Situation, Location, and Fire Conditions	*Objective:* To cause all firefighters and company officers to maintain constant situational awareness of their SCBA air supply, where they are in the building, and all that is happening in their area of operations and elsewhere on the fireground that may affect their risk and safety.
Constantly Monitor Fireground Communications for Critical Radio Reports	*Objective:* To cause all firefighters and company officers to maintain constant awareness of *all* fireground radio communications on their assigned channel for progress reports, critical messages, or other information that may affect their risk and safety.
You Are Required to Report Unsafe Practices or Conditions That Can Harm You. Stop, Evaluate, and Decide	*Objective:* To prevent company officers and firefighters from engaging in unsafe practices or exposure to unsafe conditions that can harm them and *allowing any member to raise an alert about a safety concern without penalty* and <u>mandating</u> the supervisor to address the question to ensure safe operations.
You Are Required to Abandon Your Position and Retreat Before Deteriorating Conditions Can Harm You	*Objective:* To cause firefighters and company officers to be aware of fire conditions and cause an early exit to a safe area when they are exposed to deteriorating conditions, unacceptable risk, and a life-threatening situation.
Declare a May Day as Soon as You THINK You Are in Danger	*Objective:* To ensure the firefighter is comfortable with, and there is no delay in, declaring a May Day when a firefighter is faced with a life-threatening situation and the May Day is declared as soon as they THINK they are in trouble.

Source: IAFC Safety, Health and Survival Section. Used with permission.

with a set of "model procedures" that can help reduce the number of firefighter deaths and injuries on the fireground. (For more complete discussion, see www.iafc.org.)

THE INCIDENT COMMANDER'S RULES OF ENGAGEMENT FOR FIREFIGHTER SAFETY

IAFC rules of engagement for the incident commander are summarized in Table 3.8. During risk assessment on the fireground, incident commanders must minimize unsafe practices and stop unsafe acts by making proper decisions. (For more complete discussion, see www.iafc.org.)

TABLE 3.8	The Incident Commander's Rules of Engagement for Firefighter Safety
Rapidly Conduct, or Obtain, a 360-Degree Size-up of the Incident	*Objective:* To cause the incident commander to obtain an early 360-degree survey and risk assessment of the fireground in order to determine the safest approach to tactical operations as part of the risk assessment and action plan development *and before firefighters are placed at sub-stantial risk.*
Determine the Occupant Survival Profile	*Objective:* To cause the incident commander to consider fire conditions in relation to possible occupant survival of a *rescue event* before committing firefighters to high-risk search-and-rescue operations as part of the initial and ongoing *risk assessment* and action plan development.
Conduct an Initial Risk Assessment and Implement a SAFE ACTION PLAN	*Objective:* To cause the incident commander to develop a safe action plan by conducting a size up, assess the occupant survival profile, and complete a risk assessment *before* firefighters are placed in high-risk positions on the fireground.
If You Do Not Have the Resources to Safely Support and Protect Firefighters—Seriously Consider a Defensive Strategy	*Objective:* To prevent the commitment of firefighters to high-risk tactical objectives that cannot be accomplished safely due to inadequate resources on the scene.
DO NOT Risk Firefighter Lives for Lives or Property That Cannot Be Saved—Seriously Consider a Defensive Strategy	*Objective:* To prevent the commitment of firefighters to high-risk search-and-rescue and firefighting operations that may harm them when fire conditions prevent occupant survival and significant or total destruction of the building is inevitable.
Extend LIMITED Risk to Protect SAVABLE Property	*Objective:* To cause the incident commander to limit risk exposure to a reasonable, cautious, and conservative level when trying to save a building that is believed, following a thorough size up, to be savable.
Extend VIGILANT and MEASURED Risk to Protect and Rescue SAVABLE Lives	*Objective:* To cause the incident commander to manage search-and-rescue and supporting firefighting operations in a highly calculated, controlled, and cautious manner, *while remaining alert to changing conditions,* during high-risk search-and-rescue operations where lives can be saved.
Act Upon Reported Unsafe Practices and Conditions That Can Harm Firefighters. Stop, Evaluate, and Decide	*Objective:* To prevent firefighters and supervisors from engaging in unsafe practices or exposure to unsafe conditions that will harm them and *allowing any member to raise an alert about a safety concern without penalty* and <u>mandating</u> the incident commander and command organization officers to promptly address the question to ensure safe operations.
Maintain Frequent Two-Way Communications and Keep Interior Crews Informed of Changing Conditions	*Objective:* To ensure that the incident commander is obtaining frequent progress reports and all interior crews are kept informed of changing fire conditions observed from the exterior by the incident commander, or other command officers, that may affect crew safety.
Obtain Frequent Progress Reports and Revise the Action Plan	*Objective:* To cause the incident commander, as well as all command organization officers, to obtain frequent progress reports, to continually assess fire conditions and any risk to firefighters, and to regularly adjust and revise the action plan to maintain safe operations.
Ensure Accurate Accountability of Every Firefighter's Location and Status	*Objective:* To cause the incident commander, and command organization officers, to maintain a constant and accurate accountability of the location and status of all firefighters within a small geographic area of accuracy within the hazard zone and be aware of who is presently in or out of the building.
If After Completing the Primary Search, Little or No Progress Toward Fire Control Has Been Achieved, Seriously Consider a Defensive Strategy	*Objective:* To cause a benchmark decision point, following completion of the primary search, requiring the incident commander to consciously determine if it is safe to continue offensive interior operations where progress in controlling the fire is not being achieved and there are no lives to be saved.
Always Have a Rapid Intervention Team in Place At All Working Fires	*Objective:* To cause the incident commander to have a rapid intervention team in place ready to rescue firefighters at all working fires.
Always Have Firefighter Rehab Services in Place at All Working Fires	*Objective:* To ensure all firefighters who endured strenuous physical activity at a working fire are rehabilitated and medically evaluated for continued duty and before being released from the scene.

Source: IAFC Safety, Health and Survival Section. Used with permission.

After-Action Review as a Risk Management Tool

The after-action review (commonly called the AAR) is a discussion of an event that compares the outcome of an event against organizational policies, procedures, and guidelines for a task being performed. Individuals and organizations cannot wait to act until a near miss or line-of-duty death or injury occurs. The AAR process (covered in Chapter 2) requires a constant search for ways to improve training/education, ability to properly perform tasks safely and efficiently, and continually update policies, procedures, and guidelines.

Firefighters must know the standard against which they are performing, or there is no regard for success or failure for making proper decisions involving risk management. Critical to any AAR process is an understanding that the information is to be used for future improvements within the department for similar situations. Everyone involved on the emergency scene should participate in the review. Firefighters yield insight, observations, and questions that help the department identify and correct safety policies, procedures, and guidelines. One of the most important values of an AAR comes from the capacity to replay (in light of the known outcome) the different thought processes that went into making strategic and tactical decisions. Individuals who were involved in the decision-making process should discuss what they saw and how they interpreted the situation, and then explain the strategic and tactical decisions they made based on that information. (See Figure 3.10.)

The review serves as an option for departments to incorporate lessons learned into future planning. According to *Risk Management Practices in the Fire Services*, FEMA/USFA (1996) states:

> Participation in incident critiques should be an important part of the learning process for everyone who responds to emergency incidents. Fire officers should carefully review every incident in which they have been involved and use each new experience to expand their personal risk evaluation skills. Looking back after having seen the outcome at each significant incident allows the participants to focus on the accuracy of their observations and their analysis of the situation.

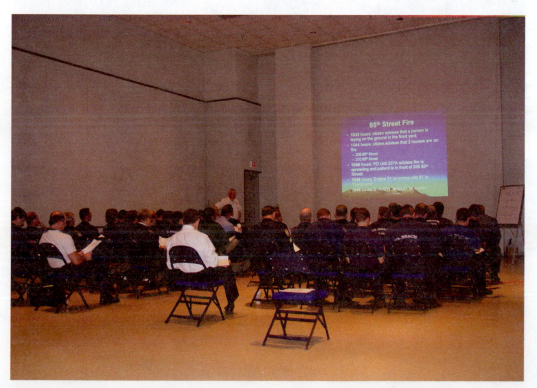

FIGURE 3.10 The after-action review is an important process for making future improvements in departmental policies and procedures. *Courtesy of Martin Grube*

Although the fundamental goal of any risk management process is the enhancement of operational capabilities while minimizing loss, the review of contributing factors regarding successes and shortfalls must be evaluated and incorporated into future undertakings. Departments should seek to answer the following four general questions regarding a mission and the potential for improvement:

- What was the incident?
- What went well?
- What happened that could have been improved? (Must be evaluated against established departmental standards)
- How can the lessons learned be incorporated into future SOPs/SOGs and training policies?

One of the most important values of an after-action review comes from the capacity to replay in light of the known outcome the thought process that went into making strategic and tactical decisions.

Risk Management During Wildland Operations

LCES ▪ Acronym for lookouts, communications, escape routes, and safety zones; used in wildland firefighting.

In the wildland fire environment, four basic safety hazards confront the emergency responder: lightning, fire-weakened timber, rolling rocks, and entrapment by running fires. Each firefighter must know the established **LCES**, which stands for lookouts, communications, escape routes, and safety zones, as described in NFPA 1500, *Standard on Fire Department Occupational Safety and Health Program*. LCES should be established before any emergency scene operations begin: select lookouts, set up communications, establish safety escape routes, and select safety zones. (See Figure 3.11.)

FIGURE 3.11 LCES should be established before any wildland operation begins. *Photo by Miranda Simone*

Assigning individuals to be lookouts should be done first. Lookouts are responsible for maintaining the overall safety of everybody on scene by monitoring fire behavior changes, crew and equipment locations, and weather changes. Effective communications is essential for providing a safe working environment. Communications must be properly established with everyone on the incident to allow for any significant updates about changing conditions and the actions needed to be taken.

Escape routes are another area of concern that should be addressed early on. Identifying the appropriate location to which to move all incident personnel and equipment should be considered in case conditions change. Those safety zones should be areas large enough for personnel and equipment on the incident scene to stay safe if conditions change. They should be based on the flame length of the oncoming flame front and at least four times the potential size of that front (Welles, 2010).

LCES functions sequentially; it is a self-triggering mechanism. Lookouts assess and reassess the fire environment and communicate the threat to safety not visible by ground forces. Firefighters then use escapes routes to safety zones. All firefighters should be alert to changes on the emergency scene and have the authority to initiate communication. The two basic guidelines of LCES are:

- Before safety is threatened, each firefighter must know how the LCES system will be used.
- LCES must be reevaluated continuously as conditions change.

The LCES system approach is an outgrowth of an analysis of wildland fatalities and near misses for over 20 years of wildland emergency incidents (U.S. Forest Service, n.d.). (See Figure 3.12.) LCES simply focuses on the essential elements of the 10 Standard Firefighting Orders and the 18 Watchout Situations. (See Figure 3.13.)

FIGURE 3.12 LCES was developed after the analysis of wildland firefighter fatalities and near misses. *Photo by Miranda Simone*

Use of the 10 Standard Firefighting Orders should be automatic at all times in any wildland fire operation and should never be broken or bent. The 18 Watchout Situations are equally important. (See Figure 3.14.)

FIGURE 3.13
10 Standard Firefighting Orders.

1. Keep informed on fire weather conditions and forecasts.
2. Know what your fire is doing at all times.
3. Base all actions on current and expected behavior of the fire.
4. Identify escape routes and safety zones and make them known.
5. Post lookouts when there is possible danger.
6. Be alert. Keep calm. Think clearly. Act decisively.
7. Maintain prompt communications with your forces, your supervisor, and adjoining forces.
8. Give clear instructions and insure they are understood.
9. Maintain control of your forces at all times.
10. Fight fire aggressively, having provided for safety first.

FIGURE 3.14
18 Watchout Situations.

1. Fire not scouted and sized up
2. In country not seen in daylight
3. Safety zones and escape routes not identified
4. Unfamiliar with weather and local factors influencing fire behavior
5. Uninformed on strategy, tactics, and hazards
6. Instructions and assignments not clear
7. No communication link with crew members/supervisors
8. Constructing line without safe anchor point
9. Building fire line downhill with fire below
10. Attempting frontal assault on fire
11. Unburned fuel between you and the fire
12. Cannot see main fire; not in contact with anyone who can
13. On a hillside where rolling material can ignite fuel below
14. Weather is getting hotter and drier
15. Wind increases and/or changes direction
16. Getting frequent spot fires across line
17. Terrain and fuels make escape to safety zones difficult
18. Taking a nap near the fire line

Summary

Most fire departments have heard of risk management, but few have actually developed a risk management program. This is due in part to the commitment required to maintain the program. Risk management encompasses three processes: risk assessment, risk mitigation, and evaluation and assessment. The risk management process includes five steps: identification, evaluation, prioritization, control measures, and monitoring. The continual evaluation process is the key for maintaining a successful risk management program. Evaluation should occur at least twice a year by a team that represents all the major functions and is well versed in the process.

When specifically focusing on the fire and emergency services and inherent risks, both NFPA and FEMA/USFA view risk management at three separate and distinct levels: the community as a whole, the emergency response organization, and the emergency response operations of individuals.

A successful risk management program will rely on the following: (1) the commitment of senior management; (2) the full support and participation of the department leaders, who must have the expertise to apply the risk assessment methodology to specific mission risks and provide cost-effective safeguards that meet the needs of the organization; (3) the awareness and cooperation of members of the user community, who must follow SOPs/SOGs and comply with the implemented controls to safeguard their department's mission; and (4) an ongoing evaluation and assessment of the related mission risks.

There always will be areas in the fire and emergency services that need to be improved upon, and certain practices need to be continually reevaluated to increase firefighter safety. If each fire department were to develop a risk management plan, firefighter safety would be positively impacted.

Review Questions

1. Define risk management.
2. List the three principles of risk management.
3. Identify the general strategies of risk management.
4. Describe what STAR stands for and who should use it.
5. Describe the relationship of the IAFC rules of engagement to risk management.

References

International Association of Fire Chiefs. (2010, July). *Rules of Engagement Project: Increasing Firefighter Survival.* Retrieved August 1, 2010, from http://www.iafcsafety. org/downloads/Rules_of_Engagement.pdf

Kipp, J. D., and Loflin, M. E. (1996). *Emergency Incident Risk Management: A Safety and Health Perspective.* New York: Van Nostrand Reinhold.

Mora, W. R. (2009, February). "Not for a Piece of Property." *Everyone Goes Home Newsletter.* Retrieved January 6, 2011, from http://www.everyonegoeshome. com/newsletter/2009/february/property.html

National Fire Protection Association. (2007). *NFPA 1500: Standard on Fire Department Occupational Safety and Health Program.* Retrieved April 10, 2010, from http://www.nfpa.org/aboutthecodes/AboutTheCodes. asp?DocNum=1500

National Fire Protection Association. (2008). *NFPA 1561: Standard on Emergency Services Incident*

Management System. Retrieved April 10, 2010, from http://www.nfpa.org/aboutthecodes/AboutTheCodes. asp?DocNum=1561

National Fire Protection Association. (2010). *NFPA 1250: Recommended Practice in Emergency Service Organization Risk Management.* Retrieved April 10, 2010, from http://www.nfpa.org/aboutthecodes/ AboutTheCodes.asp?DocNum=1250

U.S. Fire Administration. (1996, December). *Risk Management Practices in the Fire Service.* Retrieved May 1, 2010, from http://www.usfa.dhs.gov/ downloads/pdf/publications/fa-166.pdf

U.S. Forest Service. (n.d.). *Risk Management: Standard Firefighting Orders and 18 Watchout Situations.* Retrieved June 20, 2010, from http://www.fs.fed.us/ fire/safety/10_18/10_18

Welles, D. (2010, February). "Applying LCES to All-Hazard Incidents." *Fire Engineering,* pp. 107–109.

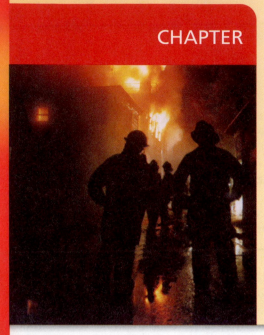

John F. Sullivan

Courtesy of Roger B. Conant

OBJECTIVES

After reading this chapter, the student should be able to:

- Describe the circumstances that constitute an unsafe practice.
- Explain the concept of empowerment as a way to stop unsafe practices.
- Compare and contrast the concepts of "challenge and confirm" and "speak up."
- Identify the three-step process of implementing crew resource management (CRM).
- Describe the factors that determine an individual's ability to develop and utilize situational awareness.
- Illustrate how situational awareness and recognition primed decision making (RPD) can enhance individual and group safety.

Feeling Empowered to Stop Unsafe Practices

Over the past two decades, great strides have been made in the way in which the fire and emergency services addresses the issue of health and safety. For example, since 1966 and every tenth year since, national leaders in the fire and emergency services have met to discuss issues significant to the profession in the United States. However, it was only at the 1986 Wingspread Conference that "death and injuries of firefighters continue at an unacceptably high level" came into focus (Wingspread Conference Report, 1986, p. 10). One year later the National Fire Protection Association (NFPA) issued NFPA 1500, *Standard on Fire Department Occupational Safety and Health Program,* a blueprint for fire and emergency services health and safety, which is now in its fifth (2007) edition. The far-reaching standard provides the solid organizational structure necessary to make a lasting and dramatic institutional change in the way the fire and emergency services approach departmental and individual safety issues.

In addition, in the spirit of the landmark Wingspread Conference series, in 2004 the National Fallen Firefighters Foundation (NFFF) convened a significant gathering of subject-matter experts in Tampa, Florida, specifically to address the root causes of line-of-duty deaths (LODDs) and serious injuries. (See Figure 4.1.) The 2004 conference produced a revolutionary consensus document—the 16 Firefighter Life Safety Initiatives (NFFF, 2004, 2005–2009), which outlines a series of urgent steps to immediately reduce the alarming number of firefighter deaths and serious injuries. In 2007 the NFFF met again to discuss the the 16 Firefighter Life Safety Initiatives.

Perhaps one of the most controversial initiatives establishes that "All firefighters must be empowered to stop unsafe practices." The major debate within this initiative lies in the interpretation of the phrase "empowered to stop." At first glance, it might seem dangerous, something just short of anarchy. There is some fear that the notion of widespread empowerment might compromise the incident commander's plan by allowing too much independent action, to the detriment of the intrinsic safety of the emergency scene. The overriding goal is a safe and coordinated response to an emergent situation, which implies

FIGURE 4.1 Worcester Cold Storage and Warehouse Fire is where six firefighter line-of-duty deaths occurred on December 3, 1999. *Courtesy of Roger B. Conant*

that firefighters must all be working off of the same plan. However, even when a plan is in place, everyone has the responsibility to consider safety first, well before the plan is acted upon.

Historically, it has not been the practice of firefighters, especially non-officers, to intervene in the decision-making process at the scene of an emergency, even regarding matters of safety. Instead it was assumed that the officers "knew what they were doing" or were somehow aware of *all* of the real or potential dangers that any situation might afford. It was assumed that "they" would deal with it, and that any input from the lower ranks was both unnecessary and unwanted. Thus, when it comes to stopping unsafe acts beyond an individual's own limited scope of practice on the emergency scene, a laissez-faire culture of noninterventionism developed.

In realistic application, however, empowerment is intended to be nothing more than the fostering of open communication pathways. The old axiom "Two heads are better than one" establishes the idea that a collective approach to information gathering and decision making is a more practical, efficient use of resources and experiences than relying on a single point of reference. The chief or company officer cannot possibly process every bit of available information at the emergency scene, especially the information beyond one's awareness or simply out of sight. Allowing for timely input from the on-scene crew will help chiefs or company officers make better decisions, a concept more fully developed later in this chapter in the practice of crew resource management (CRM).

An unsafe act can occur within or outside of an individual's scope of practice. It may be an action of one's own doing or one engaged in by a fellow responder who may not see a dangerous condition unfolding. Firefighters trained in the practice of situational awareness are more acutely alert to potential dangers and better able to recognize and stop or alter such unsafe practices.

Situational awareness is not an ability limited to the chief or company officer, but a skill set that can be developed through training and role-play exercises at all levels of the organization. Certain critical skills associated with situational awareness must be targeted—starting in recruit training and reinforced throughout a responder's career—that better prepare responders to assume the role of command when they promote up.

All of the 16 Firefighter Life Safety Initiatives are predicated upon the concept of cultural change. Specifically, they depend on changing the way firefighters have traditionally viewed the concepts of safety and accountability. The fire and emergency services is beginning to grasp the extent to which *personnel* accountability depends upon *personal* accountability. Every responder's action or inaction on the emergency scene is connected to every other responder's **locus of safety**, which means being in tune with one's environment beyond one's immediate scope of responsibility. Through empowerment every responder becomes invested with the authority to speak up when witnessing a situation that he or she believes may be unsafe.

Nothing is more important than firefighter safety. Why? Well, no one joins the emergency services to become a line-of-duty death or injury statistic. If responders are not thinking safety first, then no difference will be made by better emergency response vehicles and equipment, better policies and procedures, better personnel protective ensemble, and better command and control of the incident scene.

What Are Unsafe Practices?

Unsafe practices are any actions that directly or indirectly compromise the physical well-being of a firefighter. They arise from decisions that are made without situational awareness or without having all the information necessary to make a decision. Unsafe practices usually are not committed with the intent to cause harm. Most often, they are spontaneous

locus of safety ■ Being in tune with one's environment beyond one's immediate scope of responsibility.

unsafe practices ■ Any actions that directly or indirectly compromise the physical well-being of a firefighter. They arise from decisions that are made without situational awareness or without having all the information necessary to make a decision.

acts, driven by the circumstances of the moment, that cause a responder to stray outside standard operating procedures (SOPs)/standard operating guidelines (SOGs). Usually, the decision is based on a seemingly credible task necessary to accomplish a specific goal. The decision is often made first, without thinking through all possible consequences, including the potential negative results. Unfortunately, all too often those quick decisions do come with negative results.

Some individuals, crews, and indeed entire departments have displayed a total disregard for established safety practices in both policy and practice. It is a heartbreaking truth that after so many tragic examples and events have taken place in the professional and volunteer history of the fire and emergency services, many still choose to cling to the perilous idea that "it won't happen to us" or "it won't happen to me." It can, and it will if you or your organization ignores commonly recognized safety practices because of apathy or for the sake of nostalgia.

Yes, it may be true that fires have been going out in your town since Bessie the mare first pulled a hand pump, and no one has been seriously killed or injured in your city in decades, if ever. It also may be true that no one in your department has died from an infectious disease exposure from patient care. However, there are a million and one "buts" that are just waiting to find a hole in your organizational armor and send one or more firefighters to the morgue. The buts do not care about the past. The fire that finally kills is not reading the history of your department. Any emergency scene is unforgiving. It does not care!

The difference between unsafe acts committed today and those committed 20 or more years ago is knowledge. In the past, most training and experience was locally grown. Firefighters were bred and trained by the departments or communities they served in the same practices and theories that had been handed down for generations. Those practices were adequate for the limited scope of practice engaged in by departments of that era, namely structural firefighting. The products of combustion and building materials were made from natural products, such as wood, cotton, and wool. And, despite the lack of modern fire protection and signaling equipment, building construction principles were well established and well known by responders.

Today's emergencies are far more complex and varied than those of our forefathers, because they include standard structural firefighting, hazardous materials response, emergency medical care, technical rescues, and wildland fires that race throughout entire communities. Even though the total number of building fires is significantly lower than in previous decades, the nature of the enemy has changed. For example, most interior furnishings are derived from plastics or other hydrocarbon products, producing a hotter, faster moving fire than wooden furniture would. Together with the dramatically increased use of lightweight building construction components, interior structural firefighting has become a more dangerous prospect than ever before.

Many more intricacies to the LODD and serious injury problems exist than just building construction components and plastics. However, it is sufficient to acknowledge here that modern-day firefighters must be far more educated about the emergency scene than their predecessors were, because any department or firefighter that continues to gamble with safety, including personal safety, in this day and age of information and education is not only ignorant but also quite possibly negligent.

PUSHING THE LIMITS

Often firefighters find themselves caught in a situation in which their objective is just at the edge of their capabilities. Especially today, with limited staffing and station closures, the modern firefighter finds him or herself trying to accomplish the same objectives within a prescribed time frame with fewer resources. This can lead to shortcuts being employed that may be more dangerous than if there were full staffing. It is within a firefighter's

nature to solve problems, even when the resources or equipment is not available to do so, as in the following scenario: (See Figure 4.2.)

Firefighter Smith is working from a ground ladder 20 plus feet up, overhauling a soffit on the "C" side of a residential structure, where fire has traveled from the exterior walls to the attic space. The objective is to pull as much of the soffit facing from the building as possible to gain access to this hidden area with the hose stream. Smith is locked into the ladder and is using a six-foot hook pole to pull the soffit. The soffit boards have been partially burned through and do not pull from the building easily. In fact, one section near the "B/C" corner of the roof, which has some visible fire pulsing from underneath the boards, is just beyond the firefighter's immediate reach.

Being locked into the ladder is holding him back from the few inches that he needs, so he pulls his leg out from the rungs of the ladder. His supervisor, Lieutenant Jones, is butting the ladder and directing Smith's actions from the ground. As he looks up and sees Smith unlocking from the ladder, he believes that Smith is preparing to descend and reposition the ladder for the last piece of soffit. Lieutenant Jones braces the ladder and momentarily looks away from the overhead to see how well the footing might be where he intends to set the butt of the ladder next.

Lieutenant Jones has already assessed the current position for safety issues and taken note of the residential power line that is attached to the building just around the corner on the "B" side, fairly close to the corner where they are working. Suddenly, Lieutenant Jones hears someone shout and feels the ladder beginning to fall, the tip sliding across the roof's edge toward the corner of the building. Above, Smith is seen stretched from the ladder, hook pole in hand, pulling at the corner board.

In this scenario, Smith is an excellent firefighter with good intentions, but makes a poor decision in order to accomplish his task. He should have taken the time to descend the ladder and reposition it so that he would not have to exceed his safe reach. The Lieutenant's only slip was to get ahead of himself. His focus was momentarily on the next task and away from the current one.

There are several possible endings to this simple example, which may or may not lead to a firefighter being killed or seriously injured. Think about how many times you have been in Smith's or Jones' boots!

MAKING THE RIGHT DECISION

Every day firefighters, company officers, and chiefs are called upon to make quick decisions with very limited information. The preceding example is an all too common scenario on the fireground. "I just looked away for a second," Lieutenant Jones might say. "I thought he was getting ready to come down the ladder." Such unsafe practices can occur in a split second. Your ability to stop an unsafe practice before it becomes physically harmful may involve an even shorter time frame. The skill of anticipating possible outcomes prior to an event actually occurring is developed with experience and supported in training and practice. It is the very foundation of situational awareness.

Obviously, Firefighter Smith did not fully consider his actions. He did, however, make a poor decision: first by compromising his safety by "unlocking" from the ladder and then by reaching beyond the safe limits of the position. Such behavior might be considered an intentional and deliberate act by Smith. Lieutenant Jones failed to carry out the responsibility of being safe by looking away. Every assigned task on the emergency scene requires the responder's full attention. How does that scenario end? Here it is:

> Fortunately, for both Firefighter Smith and Lieutenant Jones, Firefighter Ramirez, who was operating a nearby hose line, recognized what was about to occur and yelled as Smith started to reach out: "Smith! Don't!" Smith felt that familiar rush of adrenaline and fear that comes when your life is suddenly put in danger. Lieutenant Jones was startled into a heightened sense of awareness of the unfolding event, and though the ladder had already begun to slide, Lieutenant Jones was able to brace it; and Firefighter Smith was able to reposition himself centered on the rungs.

In the scenario, Firefighter Ramirez made the right decision to speak up immediately when he recognized an unsafe practice unfolding and helped to avert a potentially tragic ending to this story. His quick action was critical to the resulting positive outcome. Ramirez's peak situational awareness came as a result of his experience and being in tune with his environment beyond his immediate scope of responsibility. His locus of safety was expanded to include an operation adjacent to his own, thus increasing the entire safety factor of the emergency scene.

Empowerment: Speak Up!

The NFFF is a zealous advocate for firefighter safety. Its mission is clearly founded on the premise of keeping firefighters safe in their work and reducing the possibility of line-of-duty deaths and injuries through education and research. The concept of "speak up" is really simple, yet in practice, it has proven to be difficult to initiate. Such things as organizational culture, generational differences, and tradition, and a lack of confidence in one's own knowledge need to be overcome. Although those aspects are not insurmountable, as complex issues they must be addressed through institutionalization of the concept of **empowerment**. To empower is to enable, or to equip or supply with an ability. Speaking up is directly related to the practice of empowerment. Every firefighter has the right and responsibility to voice concern when he or she perceives an unsafe act is about to take place, without fear of negative consequences for doing so.

empowerment ■ To enable, or to equip or supply with an ability.

The concept of speaking up is not new. On one level, it is simply the act of raising one's voice loud enough to be recognized and heard. Sometimes it can be difficult to communicate through the chaos of the emergency scene. One literally does have to speak up to be heard. A less literal interpretation of the concept is being given permission to voice an opinion or observation. When granted permission, one is empowered to engage in not only the physical act of speaking aloud but also the underlying authority to do so.

Authority is an important factor of empowerment because without it, an employee is at risk for discipline or retribution. However, a specific mandate can minimize that risk. The concept of empowerment is applicable to all levels of a department and is part of the cultural paradigm shift that is occurring across the spectrum of fire and emergency services. Note: Realizing that empowerment relates mainly to unsafe acts is not an invitation to defy rank structure and chain of command in all situations.

How does a young firefighter make the transition from following officers without question to speaking up when he or she sees an unsafe act, even when that act involves a senior officer? That transition may be accomplished through empowerment. The International Association of Fire Fighters (IAFF) supports the idea with the following message: "Fire departments and unions have an obligation to adopt and enforce standard operating procedures that enhance the safety of firefighters. Each firefighter must take personal responsibility for [his or her own] safety. We must also watch out for each other and not be afraid/intimidated to stop an unsafe action" (2005, 2010, p. 2). The department must support members of all ranks to have the courage to be safe, and in turn members should be confident of support when they stand up for safety. The organizational culture within the department must not make the safety advocate a "bad guy." Instead, it must instill the idea that unsafe behaviors must stop and that it is everyone's responsibility to stop them. This outlook is part of the ownership principle; that is, firefighters must be allowed to own their own safety.

CULTURAL SHIFT

Like any other profession, fire departments continuously develop and improve. The old ways are constantly being challenged with new ideas and innovative techniques. As a new generation of employees takes a more active role in building the future of a department, inevitably changes will occur. Over the past several decades, fire and emergency services has undergone a profound change, and empowerment of subordinate employees to speak up is an example of that cultural shift.

A tremendous influx of innovative ideas and techniques has occurred as a result of the more open exchange of information. The fire and emergency services is filled with talented individuals who have made an enormous impact on firefighter safety over the past several decades, all because the culture allowed them to voice their ideas, and progressive departments gave them the latitude to explore their vision and make it a reality. Innovations in personnel safety and survival techniques are being taught across all spectra. The foundation of the future lies in the concept of empowerment.

GENERATIONS

As those retiring from fire departments increase in number, and younger and at times less experienced personnel take their place, the function most important to a department is management of human resources. The timing of the latest shift in personnel is not accidental, but instead contiguous to the coming of age of a new generation of employees. Sometimes called *Generation Xers* (people born between 1965 and 1980), the new employees are a major driving force behind the cultural shift occurring in society and consequently within fire departments.

Just as the first wave of Baby Boomers (people born between 1946 and 1964) is reaching retirement age, the Generation Xers are rising into positions of influence and authority. The major influences of the group are far more technically advanced than those of their predecessors. Generation Xers have been brought up in an educational system that values exploration of ideas and an open learning environment, in stark contrast to their predecessors who were taught never to question authority.

The reality is that even before a firefighter starts the probation period with the department, his or her accrued experiences are different from others' experiences. Empowerment can therefore cause tension in the ranks. The Generation X employee is innately inquisitive and embraces empowerment as a normal part of personal interaction and not as a cultural shift at all. The Baby Boomer boss cannot get accustomed to being asked so many questions or having a subordinate seemingly question his or her judgment. However, the discord is slowly being overcome through education and training. The Baby Boomers have come to recognize the merits of fresh ideas, and the Generation Xers are maturing to the necessity of linear authority.

TRADITION: A DOUBLE-EDGED SWORD

Fire departments are fiercely proud of their long and storied heritage. Many of the icons prominently displayed on uniforms, vehicles, and buildings are representations of that long-standing legacy. This type of institutional pride can strengthen an organization and its members by connecting present members with the distinctive legacy of the past. Many of those traditions are timeless and have as much relevance today as they did over one hundred years ago, such as the tolling of the bell signifying a line-of-duty death. (See Figure 4.3.)

FIGURE 4.3 Tolling of the bell at a firefighter's funeral signifies a line-of-duty death. *Courtesy of Terry Hood*

Much of what is considered tradition is not really traditional, but simply past practice. "We have always done it *that* way" is a common refrain when a long-standing practice is challenged. Even if a past practice has a direct connection to a historically relevant custom, more often than not current circumstances render the practice obsolete and sometimes dangerous. The widespread use of the "booster hose" as an interior attack line is a prime example. "We've put out a lot of fires with that hose" is often the excuse for preserving the practice, despite the well-known fact that the nature and configuration of combustibles is dramatically different today than in decades past. However, it is important to remember that in changing tradition, today's firefighters must arm themselves with the necessary training, education, and experience in order to be prepared to be safe.

CHALLENGE AND CONFIRM

An excellent method to bolster personnel empowerment and strengthen the generational convergence is a process called "challenge and confirm." It is an easily identifiable process for implementing the speak-up policy. Challenge and confirm allows any member of a team or crew to voice a "challenge" to a supervisor regarding a potentially life-threatening action or situation that the supervisor must then consider. The supervisor must willingly entertain the cause of the challenge, without feeling threatened or undermined, consider any additional information, and "confirm" the safety and appropriateness of the action or situation. The crew leader, after quick consideration of the facts, can either confirm the original plan of action or alter it to meet the challenges of the added information.

The challenge-and-confirm method can be utilized at any level within the department. Chief officers are subject to challenge and confirm as well. (See Figure 4.4.) Individual crew members sometimes have a better view, or more timely and accurate information, which an incident commander (IC) may not initially have when offering tactical assignments. A division/group leader or company officer has the right and in fact the duty to challenge and confirm with the IC, if conditions are questionable for implementing a

FIGURE 4.4 Decision making is not always an individual process. When an individual is given an assignment in which conditions are questionable, he or she should challenge and confirm.
Courtesy of Scott LaPrade

tactical assignment. His or her input is vital to the safety of all emergency scene personnel, because the IC will make tactical assignments based on an accurate accounting from all sources.

Crew Resource Management

The initial concept of crew resource management (CRM) (originally called *cockpit resource management*) was developed from a 1979 aviation industry summit sponsored by NASA (Helmreich, Merritt, and Wilhelm, 1999). The summit was convened in the wake of a series of tragic airline disasters. The initial crash investigation reports (similar to the present after-action reports) identified several human factors that had critical consequences in the tragic outcome. The summit's mission was to correlate the human factors and develop a plan of action to reduce or eliminate those causes. A unique, comprehensive training program, CRM was the final product developed from that summit.

In 2003, the International Association of Fire Chiefs (IAFC) issued its own version of CRM. The publication (IAFC, 2003) draws unmistakable similarities between the factors leading to the tragedies in the airline industry and the tragic loss of firefighters in the line of duty every year. The CRM model and its chronicled success in reducing human errors in the decision-making process have been adopted and duplicated in other fields such as the military, the maritime industry, and the medical profession. The same model is now being championed and implemented through training programs developed to benefit fire departments.

WHAT IS CRM?

Crew resource management (CRM) is simply a system of utilizing the collective experience, knowledge, and attentiveness of all members of an emergency crew in the decision-making process to promote safety. The officer maintains his or her authority to make the final decisions, but allows for input from all members in order to gain a broader perspective on a given situation. The communal accumulation of information is then processed to make a more well-rounded decision. However, the CRM is not about a need for better communications. Instead, its focus is on the need for better situational awareness, or being aware of what should be happening versus what is happening.

A core concept of CRM is the institutionalization of empowerment. That is, it establishes both the right and the responsibility of every responder not only to maintain situational awareness but also to take action when circumstances dictate. Those actions must not be viewed as subversive and cannot be subject to negative consequences, or else the system is bound to fail. No firefighter should feel or be led to feel that he or she is "speaking out of turn." In fact, CRM encourages input and feedback from all members. First-line supervisors and chief officers must embrace the concept as an enhancement of their ability to make decisions and encourage their members to expand their capabilities through training.

CRM has become an industry standard for empowering subordinates and their supervisors to take full advantage of the potential of the combined knowledge, skills, and abilities of an entire crew. Crew resource management concepts should be adopted and implemented in all departments to promote multilevel safety practices and encourage open communication channels. (See Figure 4.5.)

Adopting CRM department-wide normally requires a cultural change in the traditional linear hierarchal chain-of-command system. Generally, fire departments operate according to a top-down management system, which does not normally allow for the consideration of information from lower levels, especially during an emergency situation. Like most skills, mastery is accomplished through training, education, and experience.

crew resource management (CRM) ■ A system of utilizing the collective experience, knowledge, and attentiveness of all members of an emergency crew in the decision-making process to promote safety.

FIGURE 4.5 Crew resource management helps in the decision-making process to promote safety. *Courtesy of Roger B. Conant*

CRM training is a three-step process that may be described as follows:

Step 1:   *Awareness.*   This step is a familiar concept in the fire and emergency services and is built into training in many areas, such as hazardous materials response. Developing awareness is the first step toward implementation in the operational realm. It addresses the five factors necessary to CRM: communications, situational awareness, decision making, teamwork, and barriers.

Step 2:   *Reinforcement.*   This step is designed to implement realistic scenario-based training to support the five conceptual elements introduced in the awareness stage. In fire departments, hands-on training is used to simulate the speed and stress under which decisions must be made.

Step 3:   *Refresher program.*   This final step in CRM training is a basic overview lecture on the five factors, highlighted by classroom role play.

COMMUNICATIONS

Nearly every after-action report ever produced following a serious firefighter line-of-duty death or injury incident includes details on some form of breakdown in communications. A few of those breakdowns are mechanical (radio problems are not uncommon), but more often than not the breakdown is associated with purely human factors. A comprehensive study of the problems associated with the communications issue was prepared by the TriData Corporation of Arlington, Virginia, and published by the National Institute for Occupational Safety and Health (NIOSH) (Frazier, Hooper, Orgen, Hankin, and Williams, 2003).

It was found that often the well-known communications cycle—sender, message, medium, receiver, and feedback—is compromised and critical information is missed or misinterpreted. Other times the information is never conveyed at all. Critical information may not reach the proper source due to an organizational culture that does not encourage feedback from subordinates. Other times the sender may be uncertain of the legitimacy or urgency of what he or she is witnessing and, in an effort to avoid potential embarrassment or ridicule, fails to make the call.

The history of fire and emergency services is filled with examples of critical and sometimes fatal communication breakdowns. The National Institute for Occupational Safety and Health (NIOSH), which investigates many LODD incidents, implicates communication problems as a significant or contributing factor in many of its investigative findings. NIOSH acknowledges two separate communications issues: technical or equipment-based issues and human factor issues (Ridenour et al., 2008). It is those human factors that CRM and situational awareness are meant to address.

SITUATIONAL AWARENESS

Another identified cause in many tragic occurrences is lack of situational awareness. In a general context situational awareness is defined as "the perception of the elements in the environment within a volume of time and space, the comprehension of their meaning and the projection of their status in the near future" (Endlsey and Garland, 2000). Situational awareness for the fire and emergency services can be defined as being acutely alert to what is going on around you, converting that data into meaningful information, and translating the potential effects of these facts to the current situation and beyond. (See Figure 4.6.)

FIGURE 4.6 Situational awareness is a key component of decision making. *Courtesy of Roger B. Conant*

The ability to communicate information usefully is interdependent with situational awareness. The quality of an individual's situational awareness is related to the three factors found in the definition:

- *Perception.* The emergency scene is a tremendously loud, chaotic, and dynamic environment. Efforts to stay focused and prioritize the vast amount of input that a firefighter is inundated with are often compromised by these many distractions. Perception of the vital cues is a key first component to situational awareness.
- *Comprehension.* The individual's ability to sift the data and separate the critical pieces under adverse conditions is a key factor of comprehension. Data are only useful if interpreted correctly. Training, education, experience, and common sense are important to full comprehension.
- *Projection.* The most difficult and important aspect of situational awareness is the ability to translate and transmit the right information to the present and to the near future.

One strategy for maintaining situational awareness consists of the following steps. To anticipate and prevent problems related to on-scene safety, all four steps must be addressed. If one is omitted, safety cannot be maintained.

1. *Maintain control.* This refers to seeing the big picture. Without a true understanding of the big picture, the steps you take to mitigate the situation are likely to make it worse.
2. *Assess the problem in the time available.* Start as soon as you get the call for help. Then when you arrive on scene, size up the situation properly.
3. *Gather information from all sources.* Because the emergency scene continually changes, start asking questions before you respond to the incident, and continue to gather information throughout the call.
4. *Monitor the results.* Check the results of each of your actions. Then be prepared to use alternative plans when the situation calls for it. (Okray and Lubnau, 2004, pp. 86–89)

Decision Making

Situational awareness and decision making are interdependent. The three factors of situational awareness—perception, comprehension, and projection—help form the foundation for decision making. Firefighters are familiar with the process of risk–benefit analysis; that is, weighing the benefits of an action in relation to the risk associated with achieving an objective. To accomplish this analysis, firefighters utilize many facets of their knowledge, skills, and abilities.

TRADITIONAL DECISION MAKING

The traditional decision-making process is not complex. First, the situation must be defined. Next, alternatives must be generated. Then information must be gathered to support or reject the ideas. The most rational or possibly successful solution is selected, and finally the solution is implemented through action.

Not all decisions are made during emergency operations. Instead, most decisions are routine and are made with little cognitive processing. Firefighters often are able to anticipate a situation and provide direction and guidance for both routine and emergency operations. Departments promulgate SOPs/SOGs to address many predictable situations and events. Those procedures are designed to streamline the decision-making process. Limiting the known, less effective possible alternatives ahead of time expedites the process and increases the likelihood that the correct decision will be reached.

An example of a traditional decision-making process is as follows: A company officer arrives at the scene of a working residential structure fire. While completing the size-up of the building (such as occupancy type, fire growth and development, potential life safety concerns, exposures, and so on), the officer is intuitively utilizing the decision-making process. He or she is defining the problem and simultaneously generating alternative actions to implement training, education, experience, SOPs/SOGs, and other inputs. The most viable solution is identified. The action plan is communicated to the crew, and then it is implemented. This process is repeated until the problem has been solved or the incident mitigated.

RECOGNITION PRIMED DECISION MAKING

In 1989 Professor Gary Klein and his associates at MIT began studying how people make decisions and called that process naturalistic decision making (NDM) (Ross, Klein, Thunholm, Schmitt, and Baxter, 2004). NDM focused on the early aspects of the traditional model, including how people make observations and what processes they go through when seeking alternatives. Through observation and research based on the original study, Klein and his associates extracted a new model called **recognition primed decision making (RPD)** (Ross et al., 2004). This model focuses on how people in high-stress situations, such as firefighters, make decisions quickly and accurately.

RPD is predicated on the theory that people use intuitive responses to quickly develop a course of action based on the situation presented. Life-threatening emergencies that firefighters face warrant quick decision making. When confronted with such a problem, firefighters immediately begin to work out a viable solution through a complex mental process wherein less viable alternatives are quickly discarded and a "best" alternative at that time is selected. (See Figure 4.7.) The process is nearly instantaneous, but is conditional to that person's established knowledge, skills, and abilities. People with more training, education, or experiences similar to the one presented are more accurate and effective when choosing the necessary action. The chosen action is then processed through imagination and intuition to quickly extract a possible outcome. If that outcome appears to be acceptable, it is instituted.

Certain variations are acknowledged in the RPD process. Often, the situation presented is well known, or typical, and a quick solution is recognized almost immediately. A simple cause-and-effect process is familiar from past experiences and the correct

recognition primed decision making (RPD) ■ The theory that people use intuitive responses to quickly develop a course of action based on the situation presented.

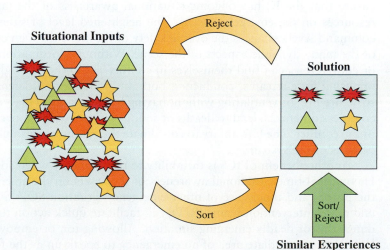

FIGURE 4.7 The decision-making "sifting" process is being able to choose a viable solution through a complex mental process.

solution is already "primed" in the individual's consciousness and needs only to be accessed. In this variation, experienced decision makers are at an advantage in making quick, accurate decisions. Inexperienced decision makers do not have the same cumulative knowledge base and therefore are not as primed to access an accurate solution as quickly.

What happens when the best solution is not readily accessed from our past training, education, and experience? Such a variation is identified when a situation is presented but the solution is not well known. The decision maker immediately begins a process of identifying close or similar situations, and quickly runs possible outcomes through a process of trial and error to find a solution. Past training, education, and experience in this variation of the process becomes an advantage, because such information allows the person to mentally sort and reject many alternatives before even considering them. Inexperienced decision makers will invariably run through more possible trial-and-error scenarios than experienced decision makers before coming to a proper solution.

Can RPD help firefighters make quick and accurate decisions on the emergency scene? Yes, every day. So why do researchers need to study it if it is already happening? Because it is a skill that can be developed through exposure to varied training scenarios and education. Recognition primed decision making has been proven to be invaluable in developing situational awareness skills that help foster safe practices including crew resource management.

Putting It Together with ICS

The first steps toward full implementation of any initiative are obtaining the correct knowledge and developing the ability to apply that knowledge. Many individuals as well as many departments stop at this point because they falsely believe they have completed their task through the initial training programs and development of a few SOPs/SOGs. However, the institutionalization of a principle, such as empowering all firefighters to stop unsafe practices, demands sustained exertion at all levels.

As noted previously, cultural change (and that is what stopping unsafe practices amounts to) and full institutionalization may take generations to achieve. The starting points of initial training and developing SOPs/SOGs must be slowly integrated into the everyday fabric of a department. How is this done? And can it be done in the context of existing SOPs/SOGs such as the incident command system (ICS)? The answer to those questions is more complex.

Every firefighter knows that the overriding benefit of any incident command system is safety through accountability. (See Figure 4.8.) As a management tool, ICS helps to ensure that the IC has ongoing situational awareness of the tasks and location of the resources on the emergency scene. The heightened level of situational awareness at the command level enhances the overall safety of the scene through coordination and control of the many dynamic aspects of the scene. Without coordination, oftentimes individuals or entire crews can find themselves in conflict with other firefighters, such as advancing opposing hose streams, conducting both an offensive and defensive attack on the same structure fire, or ventilating without having hose streams in place. All of these actions and countless others can lead to deaths or serious injuries of those personnel on the losing end of the conflict. The IC's ability to coordinate the resources through ICS is one of the safety benefits of the system.

Another benefit of ICS is the ability to react quickly when something does go wrong. Having real-time situational awareness of the assets on the emergency scene, along with the proper and timely use of other emergency situation SOPs/SOGs such as "Mayday" and rapid intervention teams (RIT), can facilitate quick action to overcome a potentially dangerous or deadly emerging situation. Allowing for, or empowering, personnel at any level in the immediate area of an emergency to speak up at the first signs of trouble will

only enhance those ICS capabilities. It stands to reason that, if firefighters recognize and acknowledge an evolving dangerous condition early, they will be better able to react to, and avoid or avert that danger.

Individual situational awareness must be translated, when appropriate, to the IC in the form of a briefing. Individuals, company officers, and division and group supervisors operating on the incident scene must be empowered and encouraged to speak up and give timely situation status reports to the IC. The reports are essential for the IC to fully comprehend the conditions occurring throughout the emergency scene and react appropriately to those conditions. The IC cannot possibly see and follow every changing condition in every corner of a structure or scene. The IC must rely on others to use their situational awareness and RPD skills, and then communicate that information back to command for integration into the coordinated effort. Empowering all responders to utilize the ability to speak up through the chain of command will complement and enhance the ICS system of safety through accountability.

Recognize and Reward

Most firefighters shun the spotlight. "I was just doing my job" is the common response whenever anyone seeks to praise a firefighter for a job well done. The altruistic sense of personal responsibility to help others is so ingrained that it seems that any external accolades are unnecessary and often embarrassing. Firefighters truly do not believe that the acts they perform in the face of extreme adversity are heroic or laudable, but instead are normal and natural, and therefore not noteworthy. The false sense of humility has put the fire department at a disadvantage when it comes to proactively recognizing personnel for a job well done.

Do you want firefighters to repeat the actions and behaviors of those being recognized and rewarded? Proper recognition can be used to reinforce safety. Ironically, the current mind-set usually is to give awards only for heroic acts that defy safe practices, which embraces and reinforces behavior that does not lend itself to safety. Therefore, the unsafe practices become widely expected as the normal way for other individuals in the department to behave, and so the same actions and behaviors are repeated.

Fire departments should reward positive behavior and outstanding performance when the situation or individual deserves it. Establishment of a full-spectrum departmental recognition program is multifaceted. Often committees are set up to determine the various levels of recognition that will be awarded, including individual and crew citations, years of service, orders of merit, and others.

Recently, fire departments have begun to increase recognition and reward of behaviors, programs, or initiatives that directly impact firefighter safety in a positive way. Any time a positive safety action is performed or safety program is developed and implemented, the department should acknowledge that contribution and encourage others to follow suit by a system of positive reinforcement. This can be as simple as fostering a friendly competition among stations to identify and correct safety hazards in the station, to full development and realization of a new safety policy for the department. Individual and crew citations are a simple and inexpensive way to acknowledge the accomplishment and instill pride and ownership in the safety climate. Remember that verbal recognition does not cost anything and can be quite powerful.

If Firefighter Ramirez (in the scenario described earlier in the chapter) worked in a fire department that had instituted an employee recognition program, he should receive an individual citation for his quick and decisive action, which resulted in averting a potentially life-threatening situation. Once the custom of annual or biennial award programs is entrenched into the fabric of an organization, the opportunities to recommend peers for recognition become apparent and the program grows. Safety is a positive subject and should be treated as such.

Summary

Because the emergency scene can be a dynamic and dangerous environment, the safety of all personnel working on an emergency scene is of paramount importance. (See Figure 4.9.) Ensuring scene safety also is dynamic and the shared responsibility of all firefighters. The IC's ability to coordinate resources and maintain overall scene safety is improved exponentially when every firefighter is trained to develop a capacity for situational awareness. Situational awareness is further enhanced through the development of the model of recognition primed decision making.

When heightened situational awareness is combined with the concepts of crew resource management, empowerment, and challenge and confirm (speak up), through the process of cultural change, then true institutionalization can be achieved. Maintenance of these concepts depends on a department's commitment to the ideal, which is sustained by training, evaluation, and recognition and reward programs.

FIGURE 4.9 Changing culture: firefighter LODDs are not inevitable.
Courtesy of Paul Shea

Review Questions

1. Discuss unsafe practices.
2. List three factors related to the quality of an individual's ability to utilize situational awareness.
3. Describe the term *empowerment* as it relates to the concept of challenge and confirm.
4. Explain how to implement crew resource management principles to enhance overall emergency scene safety.
5. Define how situational awareness capabilities lead to better decision making.

References

Endlsey, M. R., and Garland, D. J. (Eds.). (2000). *Situational Awareness Analysis and Measurement.* Mahwah, NJ: Lawrence Erlbaum.

Frazier, P., Hooper, R., Orgen, B., Hankin, N., and Williams, J. (2003, September). *Current Status, Knowledge Gaps, and Research Needs Pertaining to Firefighter Radio Communication Systems.* Arlington, VA: Prepared by TriData for NIOSH. Retrieved May 19, 2009, from http://www.cdc.gov/niosh/fire/pdfs/FFRCS.pdf

Helmreich, R. L., Merritt, A. C., and Wilhelm, J. A. (1999). University of Texas at Austin Human Factors Research Project 235: "The Evolution of Crew Resource Management Training in Commercial Aviation." *International Journal of Aviation Psychology,* 9(1), 19–32. Retrieved May 20, 2009, from http://homepage.psy.utexas.edu/homepage/group/HelmreichLAB/Publications/pubfiles/Pub235.pdf

International Association of Fire Chiefs. (2003). *Crew Resource Management: A Positive Change for the Fire Service* (3rd ed.). Fairfax, VA: Author. Retrieved May 20, 2009, from http://www.iaff.org/06news/NearMissKit/6.%20Crew%20Resource%20Management/CRM.pdf

International Association of Fire Fighters, AFL-CIO, CLC, and Division of Occupational Health, Safety and Medicine. (2005, 2010). *Improving Apparatus Response and Roadway Operations Safety in the Career Fire Service.* Instructor Guide, p. 2. Retrieved January 27, 2011, from http://www.iaff.org/hs/EVSP/2010%20Instructor%20Guide.pdf

National Fallen Firefighters Foundation. (2004, April 14). *Firefighter Life Safety Summit: Initial Report.* Emmitsburg, MD: Author. Retrieved November 20, 2009, from http://www.firesprinklerassoc.org/ElectedOfficials/FEMA%20Firefighter%20Life%20Safety%20Summit%20Report.pdf

National Fallen Firefighters Foundation. (2005–2009). *Everyone Goes Home: 16 Firefighter Life Safety Initiatives.* Emmitsburg, MD: Author. Retrieved November 26, 2009, from http://www.everyonegoeshome.com/initiatives.html

Okray, R., and Lubnau, Thomas, II. (2004). *Crew Resource Management for the Fire Service.* Tulsa, OK: PennWell.

Ridenour, M., Noe, R. S., Proudfoot, S. L., Jackson, J. S., Hales, T. R., and Baldwin, T. N. (2008, November). *NIOSH Fire Fighter Fatality Investigation and Prevention Program: Leading Recommendations for Preventing Fire Fighter Fatalities, 1998–2005.* Washington, DC: Department of Health and Human Services. Retrieved November 20, 2009, from http://www.cdc.gov/niosh/docs/2009-100/pdfs/2009-100.pdf

Ross, K. G., Klein, G. A., Thunholm, P., Schmitt, J. F., and Baxter, H. C. (2004, July–August). "The Recognition-Primed Decision Model." *Military Review,* pp. 6–10. Retrieved November 20, 2009, from http://www.au.af.mil/au/awc/awcgate/milreview/ross.pdf

Wingspread Conference Report. (1966, February). *Wingspread Conference on Fire Service Administration, Education and Research: Statements of National Significance to the Fire Problem in the United States.* Retrieved November 20, 2009, from http://nationalfireheritagecenter.com/1966Wingspread.pdf

Wingspread Conference Report. (1976, March). *Wingspread II: Statements of National Significance to the Fire Problem in the United States.* Retrieved November 20, 2009, from http://www.nationalfireheritagecenter.com/1976Wingspread.pdf

Wingspread Conference Report. (1986, October). *Wingspread III: Statements of National Significance to the Fire Problem in the United States: Conference on Contemporary Fire Service Problems, Issues and Recommendations.* Retrieved November 20, 2009, from http://nationalfireheritagecenter.com/1986wingspread.pdf

Wingspread Conference Report. (1996, October). *Wingspread IV: Statements of Critical Issues to the Fire and Emergency Services in the United States.* Retrieved November 20, 2009, from http://www.nationalfireheritagecenter.com/1996Wingspread.pdf

Wingspread Conference Report. (2006, April). *Wingspread V: Statements of National Significance to the Fire Service and to Those Served.* Retrieved November 20, 2009, from http://www.nationalfireheritagecenter.com/2006Wingspread.pdf

Ronny J. Coleman

Courtesy of Eric Melcher, Photographer, Volunteer State University

KEY TERMS

accreditation, *p. 108*

certification, *p. 105*

continuing education unit (CEU), *p. 120*

credentialing, *p. 104*

doctrine, *p. 96*

licensure, *p. 107*

lower division, *p. 113*

profession, *p. 97*

professional recognition
(or designation), *p. 108*

upper division, *p. 113*

OBJECTIVES

After reading this chapter, the student should be able to:

- Describe the importance of *training* and *education* in firefighter safety and survival.
- Identify the similarities and differences between *certification* and *accreditation* as they relate to fire and emergency services training.
- Identify at least five National Fire Protection Association professional qualification standards.
- Define *professional* as it relates to the fire and emergency services.
- Distinguish between the Fire and Emergency Services Higher Education (FESHE) conference, the International Fire Service Accreditation Congress (IFSAC), and the National Board on Fire Service Professional Qualifications (Pro Board).

PEARSON
myfirekit™

For practice tests and additional resources, visit **www.bradybooks.com** and follow the **MyBradyKit** link to register for book-specific resources.

Register for **MyFireKit** by following directions on the **MyFireKit** student access card provided with this text. If there is no card, go to **www.bradybooks.com** and follow the **MyBradyKit** link to Buy Access from there.

Meeting the Minimum Training and Education Needs for Professional Development

doctrine ■ A codification of beliefs; a body of teachings, instructions, and taught principles or positions, as the body of teachings in a branch of knowledge or belief system.

Emergency response is a dangerous business. So is being a soldier. The military has adopted a concept called **doctrine**, a codification of beliefs, designed to provide guidance to the leaders who take individuals into harm's way. For example, the U.S. Marine Corps teaches that "all Marines deserve to be properly equipped, properly trained and properly led in combat." The same can be applied to the fire and emergency services: all firefighters deserve to be properly equipped, properly trained, and properly led for the safe performance of their duties on and off the emergency scene. One of the most decorated soldiers from the Vietnam War, Colonel David Hackworth, was quoted as saying "practice does not make perfect— it makes permanent. We must train like we fight and we must fight like we train" (Hackworth and Sherman, 1989). Colonel Hackworth's view can certainly be applied to the role of training, education, certification, and credentialing in firefighter safety.

Because the fire and emergency services is such a hazardous occupation, knowledge and awareness are absolutely essential in preventing firefighters from being killed or injured in various emergency scene situations. Given the best equipment in the world and a courageous capacity to place oneself in harm's way are not nearly as important as being properly qualified and prepared to take action under stress. It is entirely up to the individual firefighter to accept the personal obligation that his or her safety and survival are linked to creating a culture of competency. (See Figure 5.1.)

This chapter is based upon an increasing priority of the fire and emergency services to render firefighters safe under the most difficult conditions. In short, there is no glory

FIGURE 5.1 Professional development provides the knowledge, skills, and abilities in helping prevent firefighter line-of-duty deaths. *Courtesy of Martin Grube*

in dying in the line of duty. The fire and emergency services is moving in the direction of expecting that firefighters will live long enough to enjoy their retirement instead of just being a name on a monument. The tolerance for firefighters engaging in inappropriate and unsafe behaviors is changing. The professional development initiative is a call for action that applies to seasoned and novice firefighters.

Why is it important for all firefighters, regardless of where they are serving, their rank and assignment, size of their department, or any other characteristic of their exposure to emergency services, to be not only trained and certified but also educated to meet a high level of proficiency? What role do training and education play in achieving safety? Training and educating firefighters can help reduce the number of firefighter line-of-duty deaths and serious injuries.

Mechanisms already exist to ensure proper training and education. The notion that fire and emergency response is inherently dangerous and cannot be made safer ignores the fact that almost all other high-risk occupations have established minimum standards and enforce them uniformly. The law enforcement community, for example, would never allow an individual to carry and use a weapon in applying deadly force without meeting minimum qualifications. How can the fire and emergency services justify allowing anyone to fight fire and to enter a burning building without meeting minimum qualifications?

Fire and emergency services structure, practices, and training and education vary widely from nation to nation. However, the nationally recognized system of training and education for firefighters in the United States includes state-sponsored training, regional training academies, colleges and universities, the National Fire Academy, as well as other public and private training and education delivery components. The system, however, is not universally mandated at this time, but is entirely dependent upon individuals, various fire and emergency services departments, fire and emergency services organizations, and other parties adopting and implementing its provisions. Although many departments have resisted accepting professional qualification standards, resistance is eroding rapidly in the face of the need to eliminate the tragic consequences of firefighters being killed and injured due to a lack of knowledge. Those in the fire and emergency services should encourage both the adoption of and the continued commitment to a host of standardized systems that encompass every rank from the entry-level volunteer firefighter to the chief of any fire and emergency services department.

The existing training and education system provides a progression of opportunities for those in the fire and emergency services field to obtain knowledge, skills, and abilities (KSAs) that can be applied throughout their entire career. It is important for all participants in the fire and emergency services field not only to understand the implications of these opportunities but also to have knowledge of the past and present trends in training and education.

Origins and History of Fire and Emergency Services Training

A **profession** is an occupation that requires specialized knowledge accompanied by long and intensive academic preparation. A professional, therefore, may be defined as a person engaged in one of the professions who conforms to its technical or ethical standards. These definitions raise the questions of whether or not a person engaged in the fire and emergency services is, by definition, in possession of specialized knowledge, and whether or not becoming a firefighter involves long and intensive academic preparation. Every firefighter should be a strong career-long supporter of the processes of becoming a professional. (See Figure 5.2.)

profession ■ An occupation that requires specialized knowledge accompanied by long and intensive academic preparation.

FIGURE 5.2 Professional development is a career-long process. *Courtesy of Martin Grube*

ORIGINS OF FIREFIGHTING

The first institutionalized effort to create the term *firefighter* came with the creation of the Roman Fire Brigade in A.D. 64. The brigade identified special task assignments such as hook person, pump operator, and nozzle man. In general, however, for more than two thousand years, firefighting was more a hobby or community obligation than a profession.

A review of the fire service in Europe is important to gain knowledge about the events that led to the creation of the U.S. fire service. Although Benjamin Franklin is often recognized as the father of the U.S. fire service, most of the initial concepts of firefighting in this country were based upon knowledge brought here by European immigrants from the 1600s to the 1800s.

INFLUENCE OF INDUSTRIALIZATION AND JAMES BRAIDWOOD ON THE FIRE SERVICE

As the world became more structured, so did the fire service. Starting about 1850, the idea began that firefighting required a more structured response for those engaging in the practice. In most industrialized nations, firefighting started to become a full-time job opportunity about 1860. Among the first countries to consider firefighting as a job was the United Kingdom, with leadership from Fire Chief James Braidwood, known as the father of the British fire service and the developer of the municipal fire department.

In the wake of several disastrous fires in Edinburgh, Scotland, Braidwood at age 24 was elected fire chief, or master of the fire engines, of Glasgow, Scotland, in October 1824. With his particularly keen interest in the fire service, combined with his inventiveness, initiative, shrewdness, and energy, he soon began to make his mark (Blackstone, 1957). Among other significant contributions, he is credited with developing hiring criteria for firefighters.

One of Braidwood's first actions after achieving the rank of firemaster was to choose a particular type of person to be a firefighter. He selected only slaters, carpenters, masons, plumbers, and smiths, explaining his rationale as follows:

> Men selected from these five trades are also more robust in body, and better able to endure the extremes of heat, cold, wet, and fatigue, to which firemen are so frequently exposed, than men engaged in more sedentary occupations. (Blackstone, 1957, p. 103)

Further, his concept of hiring firemen (a term not considered biased at the time) was limited to selecting individuals who were between the ages of 17 and 25 because he believed that age-group was more able to be readily trained than older individuals who had fixed work habits.

Braidwood was not just an educated theorist. He had witnessed numerous conflagrations and the subsequent failure of firefighters to be equal to the task. As firemaster, Braidwood was known to be a taskmaster. His personal convictions about the training of firefighters were based upon his own experiences. In one fire, he saved nine people, dragging and carrying them to safety. Perhaps for that reason, Braidwood felt that physical fitness was an essential part of the firefighter's needed skills. For the first time in history, he required putting firemen through a course of gymnastics to reach and maintain peak fitness. He drilled his firefighters every Wednesday at four o'clock in the morning and explained his rationale by stating:

> The mornings too, at this early hour, are dark for more than half the year, and thus, the firemen are thus accustomed to work by torch-light, and sometimes without any light whatever, except for the public lamps which are then burning. And, as most fires happen in the night, the advantage of drilling in the dark must be sufficiently obvious. (Blackstone, 1957, p. 104)

Braidwood published a book entitled *On the Construction of Fire Engines and Apparatus, the Training of Firemen and the Method of Proceeding in Cases of Fire* in 1830, which formed the basis for a course of instruction for his firefighters. Braidwood received requests for the book from all over the world.

By the time Braidwood was appointed as the first chief officer of the London Fire Engine Establishment in 1833, he had changed his mind about who made the best firefighters. He stated:

> Seamen are to be preferred, as they are taught to obey orders, and the night and day watches and the uncertainty of the occupation are more similar to their former habits, than those of other men of the same rank in life. (Blackstone, 1957, p. 118)

The end of Braidwood's career in the London Fire Engine Establishment was as remarkable as its beginning. Like many other firefighters who followed in his footsteps, his life was cut short by the very event he had worked so hard to eliminate. James Braidwood was killed at the Great Fire of Toohey Street, Saturday, June 22, 1861, at approximately five o'clock in the evening. A wall collapsed on him while he was inspecting for the safety of one of his crews. He had led the London Fire Engine Establishment for 28 years. Fortunately, the course of action set in motion by Braidwood did not die with him. He was ultimately succeeded by another person of equal mental and physical stamina, Sir Eyre Massey-Shaw.

INFLUENCE OF SIR EYRE MASSEY-SHAW ON FIRE SERVICE TRAINING

Sir Eyre Massey-Shaw was appointed as fire chief of what is now known as the London Fire Brigade in 1873. Massey-Shaw expanded Braidwood's theories on fire protection, including the requirements for entry to the service. He too focused on seamen for his recruits:

> A smart man, who has served at sea for a few years, and has a taste for the work of a fireman, can be brought forward for duty within an average period of about six or eight weeks; a man equally smart, but without the advantage of a seaman's training, may possibly be brought forward within about as many months, but even at the end of that time he would hardly be as expert as a seaman in climbing and the use of ropes. (Massey-Shaw, 1876, p. 305)

Massey-Shaw's criteria for entrance into the brigade included that the candidates be under age 25, have a chest measurement of over 37 inches, be at least five feet five inches in height, pass a physical examination by a physician, and be able to read and write. Further, candidates had to be able to raise a fire escape ladder single-handedly. All appointments were tentative, and Massey-Shaw imposed a three-month probation period on all new candidates.

After their selection, the candidates were required to complete a basic training program. Massey-Shaw thoroughly documented his efforts at providing a "curriculum" for the professional firefighter. In his textbook, published in 1876, Massey-Shaw stated:

> From the remotest periods of antiquity to the present time the business of extinguishing fires has attracted a certain amount of attention: but it is a curious fact, that, even now, there is so little method in it, that it is a very rare circumstance to find any two countries, or even two cities in one country, adopting the same means or calling their appliances by the same names. (Massey-Shaw, 1876, p. v)

Under Massey-Shaw's administration, the London Fire Brigade had also moved from a partially volunteer operation to one of full-time paid firefighters. The transition was created to a large degree by the need to have skilled firefighters combating large fires in their first few moments of origin, instead of waiting until they had reached block-wide conflagrations. Sir Massey-Shaw expressed his thinking clearly in statements such as the following:

> The importance which I attach to a sound system of training will probably be understood when I state my conviction, founded on what appears to me the clearest and most positive evidence, that some of the greatest losses by fire, which the world has ever experienced, have been owing to want of skill on the part of firemen. It is true that want of discipline may justly be credited with a considerable portion of the blame, but, as a practical man, I do not hesitate to assert that, where there is no skill, discipline becomes almost impossible and is, at least under such circumstances, of very little use, so far as the extinguishing of the fire is concerned. (Massey-Shaw, 1876, p. 306)

Massey-Shaw's belief in formal training was proven when he established London's first fire training academy. Its requirements were as follows:

> Each on appointment joins this class and learns the use and manipulation of all the appliances, as explained in the foregoing pages. At the same time, he lives in the state, and by degrees is taught the general working of the brigade; but during this period never attends a fire, except on an emergency, and then only under the personal charge of his instructors. Nothing is so destructive of sound education in this way as permitting men to attend fires before they know how to handle the appliances properly, and the youngest hands are therefore brought out as little as possible. (Massey-Shaw, 1876, p. 305)

The seminal concept of training replacing sheer bravery and stress was continued in Massey-Shaw's observation that:

> It may perhaps be said that great numerical strength will make up for deficiency of skill and knowledge; and this may, no doubt, be to some extent correct; at least it appears to be the theory established in many places; but I am inclined to believe that, for dealing with great emergencies, no amount of numerical strength, even when combined with discipline, can compensate for the absence of skill and knowledge, and on this account, I consider a proper system of training, before attending fires, the only true method for making men real firemen. (Massey-Shaw, 1876, p. 306)

Massey-Shaw's system worked. His fire academy was originally established using a lead instructor and two assistant instructors. Candidates were constantly evaluated during their initial training, and those that were unfit were released. Massey-Shaw's system also depended upon practical reinforcement of the knowledge gained in the academy. He directed that:

> When a man is pronounced competent by the instructor, he is removed from the drill class, and is posted to a station, where he receives further training and instruction from the officer in charge, who entrusts him, at first, with work of the simplest kind, and by degrees, as he gains experience, with all the duties of his position. (Massey-Shaw, 1876, p. 305)

In accomplishing all this organizational work, Massey-Shaw made it quite clear that he hoped that his efforts would not have to be "reinvented" by someone else. In the closing of the introduction to his textbook, he stated:

> I therefore have every confidence that [this book] will be of service to all who are interested in the preservation of life and property, and especially to those who have devoted themselves to the practical work of extinguishing fires and who, whether their claim be conceited or not, consider with my fellow laborers and myself, that the business, if properly studied and understood, is work being regarded as a profession. (Massey-Shaw, 1876, p. 305)

Clearly, Massey-Shaw recognized that firefighting was a dangerous occupation. At the time he favored hiring sailors or former military personnel because he wanted to have personnel accustomed to working under stressful conditions for long periods of time, the process in the United States was just starting to become as formal. In 1873, Massey-Shaw wrote the following after visiting several U.S. fire departments:

> When I was in America it struck me forcibly that although most of the chiefs were intelligent and zealous in their work, not one that I met even made the pretension to the kind of professional knowledge, which I consider so essential. Indeed one went so far as to say that the only way to learn the business of a fireman was to go to fires . . . a statement about as monstrous and as contrary to reason as if he said that the only way to become a surgeon would be to commence cutting off limbs, without any knowledge of anatomy or of the implements required.
>
> There is no such short cut to proficiency in any profession, and the day will come when your fellow countrymen will be obliged to open their eyes to the fact that if a man learns the business of a fireman only by attending fires, he must of necessity learn it badly, and that even what he does pick up and may seem to know, he will know imperfectly, and be incapable of imparting to others.
>
> I consider the business of a fireman a regular profession requiring previous study and training as other professions do; and I am convinced that where study and training are omitted, and men are pitch forked into the practical work without preparation, the fire department will never be capable of dealing satisfactorily with great emergencies. (Massey-Shaw, 1876, pp. 110–111)

CREATION OF PAID FIRE DEPARTMENTS IN THE UNITED STATES

The path of converting firefighting from a random opportunistic vocation into a full-time job was then paralleled in the United States. The first paid fire departments were created about the time of the Civil War. One of the motivating factors in their creation was the acceptance of an advanced technology: the steam-operated fire pumper. Powerful, but dangerous, the steamer introduced an increasing emphasis on being properly trained and evaluated, which resulted in the creation of training schools and academies. As early as 1880, recruit academies emerged at numerous locations around the United States. Although a tremendous amount of energy was devoted to exchanging information on basic and advanced skill sets, no "standard" for the performance of a firefighter existed. (See Figure 5.3.)

For over 50 years, recruit academies adopted both text material and curricula as technology continued to progress. The same five decades saw significant changes in fire protection technology: automatic sprinklers were developed and the first fire-alarm systems created.

Despite the advances in fire protection, the people of the United States experienced several catastrophic fires during this era. On the same day in 1871, both the Great Chicago Fire and the wildland fire in Peshtigo, Wisconsin, resulted in large loss of life. These events provided more justification for the creation of paid fire departments.

The fire season of 1910 finally put a spotlight on the scope and breadth of both fire protection and firefighting. A huge wildland fire that destroyed much of the northwest United States resulted in the creation of the U.S. Forest Service. The year 1910 was declared "The Year of the Fires" because of the deaths of so many civilians and firefighters and huge property losses. Just one year later, the infamous Triangle Shirtwaist Factory fire occurred in New York.

CREATION OF *TRADE ANALYSIS* FOR FIRE SERVICE OCCUPATION

Although no record exists of his personal response, there is evidence that a young fire officer in the Los Angeles Fire Department observed all of those tragedies. The fire officer, Ralph J. Scott, later became fire chief of Los Angeles and president of the International Association of Fire Chiefs. He wrote to the Office of Vocational Education in Washington, DC, offering his opinion that there should be a set of published standards for firefighter training.

As a result of that letter, the Los Angeles Fire Department developed a document called *Trade Analysis of Fire Engineering* (Gowell, Tebbetts, and Baker, 1932). The analysis identified a set of specialized skills-based textbooks that were needed in training to become a firefighter. Adopted by the federal government, the *Trade Analysis* quickly became a reference source for the development of training materials. In the 1930s, this document began to be distributed through the network of those advocating improvements in firefighter training. The International Fire Service Training Association (IFSTA) began to publish an increasingly diverse number of textbooks in order to meet some of the demands suggested by the *Trade Analysis of Fire Engineering*.

Origins of Fire Service Certification

ROLE OF THE WINGSPREAD CONFERENCE

In 1966, the Johnson Foundation brought together a group of fire service leaders for a brainstorming session about the future of the fire service. The group has continued to meet every 10 years to monitor and assess the accomplishment of a variety of improvements in this profession. At the very first meeting, the group clearly recognized the need for a widespread acceptance of an improvement in fire service training and education.

Such recognition had been suggested by the statements made in the Wingspread statements, a series of visionary declarations (Wingspread Conference Report, 1966). The first was the statement that "Professional status begins with education"; the second was "the scope, degree and depth of the educational requirements for efficient functioning of the fire service must be examined."

A profession:

- Should rest upon a systemic body of knowledge of substantial intellectual content and on the development of personal skill in the application of this knowledge to specific cases
- Must set up standards of professional conduct that take precedence over the goal of personal gain
- Should have an association of members, among whose functions are the enforcement of standards and the advancement and dissemination of knowledge
- Should prescribe ways—controlled to some degree by the members of the professional association—of entering the profession by meeting certain minimum standards of training and competence (Silk, 1960, p. 9)

At the last Wingspread Conference in 2006, participants encouraged acceptance of a universal credentialing system and recognized the need for fire-related higher education degree programs in every state. The participants agreed that to reach the necessary level of competence, the fire and emergency services should consider both the credentialing system and higher education degree programs. The need for professional development in the fire and emergency services remains a concern today.

ROLE OF THE IAFF IN CREATING RECRUITING AND TRAINING STANDARDS

Also in 1966, the International Association of Fire Fighters passed Resolution 111-1966. The resolution, authored by Texans Ernest A. Emerson (later to become Texas state fire marshal) and Alcus Greer, called for setting up standards for areas such as recruiting and training in the fire service. The IAFF staff was instructed to prepare materials that included a recruiting program, a training program, and the recommendation for associate in arts and bachelor's degree programs. The call was out for improved training and educational opportunities and, more importantly, for the establishment of some standard criteria to measure a firefighter's competency.

Emmett Cox, an eminent member of the fire service, was called out of retirement and asked to direct this important IAFF study committee. He did so with a number of other committee members, including Carl McCoy and the International Fire Service Training Association located at Oklahoma State University.

What is important now is what this all means for the current generation of firefighters. The consequence of the historical evolution was simple: firefighting had gone from a

FIGURE 5.4 One of the many skills included in firefighter training standards. *Courtesy of Martin Grube*

social function of neighbors helping neighbors to a specific set of skills and knowledge basic to a person's being able to operate efficiently, effectively, and safely. The evolution introduced the need for specific terms to describe how the training and education process was to be utilized by the fire service. (See Figure 5.4.)

DEFINING THE TERMS

Adopting a minimum set of professional standards was fundamental to establishing a competency. There are levels to provide recognition of professional competence, not only in the fire and emergency services but also in other professions. The levels include:

- Credentialing
- Certification
- Licensure
- Professional recognition (or designation)
- Accreditation

All five terms have been adopted by the fire and emergency services to recognize various levels of achievement and ability to perform. More importantly, the process of achieving these forms of recognition for a person meeting specific standards for job performance is an integrated part of assuring firefighter safety and survival.

The relationship between how a person can be certified or licensed to perform provides the basis for the contention that separate terms are needed. Although anyone can read a book or attend a class to become familiar with a set of standards, that person needs to be evaluated to achieve any level of recognition.

Credentialing

Credentialing is the process of formal recognition or technical competency and performance by evaluating and monitoring adherence to an applicable professional standard for direct or peer review. In addition, credentialing verifies an individual by investigation and observation. It defines a scope of practice that he or she may provide. The criteria must

credentialing ■ The process of formal recognition or technical competency and performance by evaluating and monitoring adherence to an applicable professional standard for direct or peer review. In addition, it verifies an individual by investigation and observation and defines a scope of practice that he or she may provide.

FIGURE 5.5 IAFF Fire Ground Survival Training. *Courtesy of Travis Ford, Nashville Fire Department*

be directly related to quality of care. In the fire and emergency services, credentialing is often associated with the level of training an instructor has received or is qualified to give. (See Figure 5.5.)

One of the more critical aspects of protecting the integrity of the training system is the training, certification, and credentialing of instructors. Although no one nationwide approach encompasses all instructor credentials, the following two elements are fairly consistent:

- The use of a standard for instructor qualifications
- A mechanism for the state (through the fire marshal's office or through the colleges and universities) to recognize an individual who meets these instructor qualification standards

The credentialing of instructors is important because it ensures accountability for those who are conducting the learning experience for individuals attempting to become certified or licensed. No compromise can be accepted in ensuring that instructors stick to the standards and teach to the designated curriculum.

In some states instructors are given designations that are tiered; that is, basic instructor, advanced instructor, and so forth. Some states require very specific credentials for a person who is a teacher of teachers.

Certification

Certification refers to the confirmation of certain characteristics of a person, that is often, but not always, achieved after some form of external review, education, or assessment by a third party. One of the most common types of certification is that linked to a specific occupation, which means a person is certified as being competent to complete a job or task, usually by the passing of an examination.

Advocates of certification point to the fact that training and education of individual firefighters is a critical aspect of firefighter safety and should be a part of minimum

certification ■ Confirmation of certain characteristics of a person, that is often, but not always, achieved after some form of external review, education, or assessment by a third party.

requirements of all fire and emergency services departments. However, it should be recognized that each state is responsible for its own certification process.

There are two general types of certification: (1) those that are valid for an entire lifetime, once the exam is passed, and (2) those that require a person to be recertified after a certain period of time. A good example of lifetime versus renewable certification is the difference between a firefighter certification (some of which last a lifetime) and an emergency medical technician certification (which requires periodic recertification). Other examples include the continuing education requirements in the field of hazardous materials or the very specialized physical skill sets required for special hazards such as dive rescue or technical rescue. (See Figure 5.6.)

Certifications can differ within a profession by the level or specific area of expertise to which they refer. In the fire and emergency services, *being certified* usually relates to basic firefighting or emergency medical service skills. Using certification as a means of discriminating between those that should be allowed to be in harm's way versus those who should be retrained is not limited to strictly career-based departments, but is being used by volunteer departments also. Fire, a mass casualty incident, communicable disease, and the adverse effects of hazardous materials do not recognize the distinction between career and volunteer firefighters. (See Figure 5.7.)

Levels of certification usually reflect levels of complexity and increased skill requirements. Being certified at the Fire Fighter I level is not the same as being certified at the Fire Officer III Level. More and more fire and emergency services departments are requiring basic entry-level firefighters to be certified. Many job flyers and position descriptions now include requirements for both minimal training and educational requirements, along with those for promotion. Including such certification requirements in recruitment material will likely continue as more and more departments participate in the certification process.

Note that *certification* does not refer to being legally able to practice or work in a profession. The legal "ticket" to practice is called *licensure,* as discussed next.

FIGURE 5.7 There are various levels of hazardous materials certification for emergency response. *Courtesy of Travis Ford, Nashville Fire Department*

Licensure

Licensure refers to the granting of a license by a state, which gives an individual permission to practice a profession legally. Such licenses are usually issued to regulate an activity that is deemed to be dangerous or a threat to the person or the public or that requires a high level of specialized skill. Physicians, attorneys, engineers, surveyors, and architects are among the professionals who are typically licensed. Certification and licensure are similar in that they both require the demonstration of a certain level of knowledge, skill, or ability. However, important basic differences exist, as seen in Table 5.1.

In some states, a person must be licensed to engage in EMS activity, whereas other states require certification. In both cases, these terms define the level of activity permitted by the department.

In the case of certain occupations and professions, licensing is granted by a state licensing board composed of advanced practitioners who oversee the applications for licenses. Such a board often operates in cooperation with a professional body. (For example, the Society of Fire Protection Engineers develops the professional examination in fire protection engineering; individual states administer the examination and issue licenses to successful applicants.) Licensing often involves accredited training or examinations, but licensing practices vary a great deal from profession to profession and from state to state.

licensure ■ The granting of a license by a state, which gives an individual permission to practice a profession legally.

TABLE 5.1	Certification and Licensure Compared	
CERTIFICATION	**LICENSURE**
An employment qualification	A legal requirement to practice a profession
Generally issued by a professional association	Generally issued by a government agency

Both certification and licensure are conducted as part of a state's legal authority rather than that of the federal government. Sometimes the term *registration* is used to indicate that a person has the skill set to practice a particular profession (such as engineering or surveying) or to obtain a special privilege (such as driving an emergency vehicle or treating an injured person). One of the best reasons for licensing (or registration) is to ensure that the public will not be harmed by the incompetence of the practitioners. Moreover, licensing might serve to restrict parties who are not qualified to engage in dangerous tasks or assignments.

Professional Recognition (or Designation)

professional recognition (or designation) ■ Process based upon examination of an individual's life experience that includes both certification and licensure, but is based upon the notion of providing a capstone to the training and education system for an individual serving in top management positions.

The term **professional recognition (or designation)** is a recent addition to the pedigree of career development processes. It is the process based upon examination of an individual's life experience that includes both certification and licensure, but is based upon the notion of providing a capstone to the training and education system for an individual serving in top management positions.

One example of such a system is the professional designation (recognition) system operated by the Center for Public Safety Excellence for fire chief officers. The system is also closely linked to the idea of a professional development program that was developed and implemented by the International Association of Fire Chiefs Professional Development Committee. Obtaining a chief fire officer (CFO) or chief medical officer (CMO) designation means that the individual has proven through education and leadership or management skills that he or she has obtained a certain status recognized by others within the same profession.

Accreditation

accreditation ■ Official authorization by a third party to deliver training and education programs that conform to a set of standards; given to organizations that then have the ability to certify individuals.

The last term that requires definition is **accreditation**, the official authorization by a third party to deliver training and education programs. In the context of the fire and emergency services, accreditation is based upon the organization's testing and evaluation processes to ensure they conform to a set of standards. An accredited training program receives that recognition only after allowing its testing and evaluation processes to be reviewed in their entirety. *Accredit* is not the same as *certify*: organizations are accredited and individuals are certified, licensed, or recognized. When an organization is accredited, the institution itself has the ability to certify individuals. (See Figure 5.8.)

In the context of the definition, there are standards, courses of instruction, testing processes, issuance of certificates, and a host of other activities. An accredited organization has been authorized by another organization to deliver training programs. In the fire and emergency services training world, the accrediting organizations that have been created to provide oversight are Pro Board and International Fire Service Accreditation Congress (IFSAC), as discussed later. In the cases of colleges and universities, the accrediting organizations depend on the courses being offered and the area of the country in which the program is taught. The six regional bodies that accredit colleges and universities are:

- Middle States Association of Colleges and Schools (MSA)
- New England Association of Schools and Colleges (NEASC)
- North Central Association of Colleges and Schools (NCA)
- Northwest Commission on Colleges and Universities (NCCU)
- Southern Association of Colleges and Schools (SACS)
- Western Association of Schools and Colleges (WASC)

Check the U.S. Fire Administration Web site (http://www.usfa.dhs.gov/nfa/higher_ed/resources/resources_schools.shtm) for the most up-to-date lists of colleges and universities offering fire and emergency services–related degrees with accreditation.

Gaining Recognition of the Professional Qualification Standards

In many fields, including the fire and emergency services, a person must overcome a series of barriers or achieve a certain status before gaining recognition as a professional. In much the same way, an *occupation* needs to overcome barriers in order to become a *profession*. So how has the fire and emergency services continued its transition, through training, education, certification, and credentialing, toward professionalism in ensuring firefighter safety and survival?

BIRTH OF THE NATIONAL PROFESSIONAL QUALIFICATION SYSTEM

As professional qualification standards and the development of recognition systems for the fire and emergency services have evolved, numerous efforts have been made to maintain a sense of national uniformity in professional qualification standards. These efforts included the creation of organizations such as the Joint Council of Fire Service Organizations.

In 1972, the Joint Council of Fire Service Organizations founded the National Professional Qualifications System (NPQS). The Joint Council established a nine-member board, soon called the Pro Board, to oversee the more formalized process of getting professional qualification standards published and distributed to all fire and emergency services departments.

Notably, until that time the term *professional* had not been used in describing most fire and emergency services activities. Although the fire and emergency services considered itself professional, a subtle difference exists between being called professional and being recognized as such. That recognition had not yet been achieved when the Joint Council and Pro Board began their work.

To have a set of current standards, the Pro Board asked the National Fire Protection Association (NFPA) to create a set of professional qualification standards through its technical committee process. The development of the professional qualifications to a large degree continued the evolution of the work of Braidwood, Massey-Shaw, and others. More importantly, the standards were vetted through a contemporary process involving individuals actually doing the jobs.

NFPA 1001 AND OTHER PROFESSIONAL QUALIFICATION STANDARDS

The consensus process resulted in the first contemporary fire service professional qualification standard being made available in 1974: NFPA 1001, *Standard for Fire Fighter Professional Qualifications*. Since that time, an entire series of professional qualification standards have been developed to align with the fire and emergency services' evolving career development processes. The current NFPA professional qualification standards include the following:

- NFPA 1000, *Standard for Fire Service Professional Qualifications Accreditation and Certification Systems*
- NFPA 1001, *Standard for Fire Fighter Professional Qualifications*
- NFPA 1002, *Standard for Fire Apparatus Driver/Operator Professional Qualifications*
- NFPA 1003, *Standard for Airport Fire Fighter Professional Qualifications*
- NFPA 1005, *Standard for Professional Qualifications for Marine Fire Fighting for Land-Based Fire Fighters*
- NFPA 1006, *Standard for Technical Rescuer Professional Qualifications*
- NFPA 1021, *Standard for Fire Officer Professional Qualifications*
- NFPA 1026, *Standard for Incident Management Personnel Professional Qualifications*
- NFPA 1031, *Standard for Professional Qualifications for Fire Inspector and Plan Examiner*
- NFPA 1033, *Standard for Professional Qualifications for Fire Investigator*
- NFPA 1035, *Standard for Professional Qualifications for Public Fire and Life Safety Educator*
- NFPA 1037, *Standard for Professional Qualifications for Fire Marshal*
- NFPA 1041, *Standard for Fire Service Instructor Professional Qualifications*
- NFPA 1051, *Standard for Wildland Fire Fighter Professional Qualifications*
- NFPA 1061, *Standard for Professional Qualifications for Public Safety Telecommunicator*
- NFPA 1071, *Standard for Emergency Vehicle Technician Professional Qualifications*
- NFPA 1081, *Standard for Industrial Fire Brigade Member Professional Qualifications*

Note that these professional qualification standards deal, directly or indirectly, with firefighter safety and survival. For example, NFPA 1001, NFPA 1003, NFPA 1005, NFPA 1051, and NFPA 1081 apply to personnel operating under emergency (and often unsafe) conditions, thus directly affecting safety. Other standards (NFPA 1071, for example) have a more indirect but nonetheless significant impact on firefighter safety.

PARTICIPATION IN THE STANDARDS-MAKING PROCESS

Individuals have been used to identify important issues that arise between the annual updating processes and bring them before the committee for review and justification. This demonstrates the ability of how one individual can make a difference. Furthermore, a focused group of individuals working together can make a world of difference. The standards-making process requires participation to keep it relevant and viable. Individual commitment includes making the effort to identify needed changes in the standards. The suggestion does not imply that all fire and emergency services members need to be on an NFPA technical committee. But everyone in the fire and emergency services—particularly those responsible for safety and for training and education—should review the standards and submit recommendations through the participative process to make sure that external influences do not weaken the standards process.

The Joint Council of Fire Service Organizations was eventually dissolved and, in July 1990, its successor, the National Board on Fire Service Professional Qualifications (NBFSPQ), was incorporated as a freestanding organization. Today the NBFSPQ, often called the Pro Board, provides 72 levels of certification using 16 separate NFPA standards (Pro Board, 2010). It should be noted that the NFPA standards are updated approximately every four years.

PEARSON
myfirekit

Please visit MyFireKit Chapter 5 for a copy of the application to serve on an NFPA technical committee.

INTERNATIONAL FIRE SERVICE ACCREDITATION CONGRESS

A second organization that has played a major role in the development of fire and emergency services training, education, certification, and credentialing is the International Fire Service Accreditation Congress (IFSAC). Established in 1990, the entity accredits not only fire and emergency services training systems but also higher education institutions. IFSAC is organized in a fashion similar to that of the Pro Board in that it has a board of directors and uses a process whereby the organization seeking accreditation must submit its processes and structure for evaluation.

State training systems use these two different systems (i.e., the Pro Board and IFSAC) depending upon background and commitment to the process. Both the Pro Board and IFSAC accredit entities that provide fire and emergency services certification. They neither publish standards nor deliver courses of instruction. They conduct assessments and provide third-party validation to state and other delivery systems (such as private companies). Basically they want to verify that the system complies with the appropriate standards. In other words, the Pro Board and IFSAC validate the process that allows an individual receiving a certificate or a degree to do so with credibility, by confirming that course offerings, institutional support, and qualified faculty are committed to the proper educational needs of the students. IFSAC also accredits higher education degree-granting institutions offering fire science–related two-year and four-year degrees.

NEED FOR CHANGES IN STANDARDS TO ENSURE FIREFIGHTER SAFETY

What continues to make the standards-setting process and the certification of training and education programs relevant to firefighter safety is the ongoing effort to keep the professional qualification standards current. Another objective is to keep certification and education matched to job and safety requirements in the fire and emergency services.

Emergency medical services and hazardous materials (HazMat) response provide two examples of the increasing levels of sophistication needed in the field of fire and emergency services. The development and publication of standards for EMS, Hazmat response, and operations are as much for the safety of the practitioner as for the benefit of a patient or property owner. The "safety stakes" keep escalating for those who respond to these increasingly complicated emergencies daily. The need for firefighters to perform EMS, HazMat, and various specialized technical rescue services (such as confined space, high-angle, or water rescue) safely requires updated standards and training. (See Figure 5.9.)

There is almost always need for new technology in modern society. Good examples might include coping with alternative-fueled vehicles, responding to solar energy installations, and dealing with outbreaks of communicable diseases. There simply is

(a)

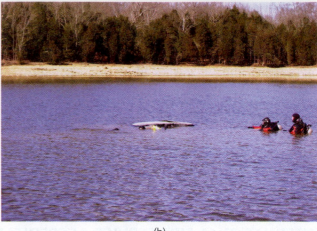
(b)

FIGURE 5.9 Hazardous materials response and water rescue have increased the level of training and education needed to operate safely, effectively, and efficiently on the scene. *(a) Courtesy of Buddy Byers, District Chief, Nashville Fire Department, (b) Courtesy of Walter Demonbreun, Jr.*

no room for an inadequately trained person to confront these types of emergencies. Standards must almost always be modified and improved to cope with changes in technology and the demands placed upon the fire and emergency services.

That does not mean that everyone will be adequately prepared. Not every fire and emergency services department uses minimum standards. Not every department requires certification or licensure. Many fire and emergency services departments are under the impression that whatever training or certification is offered is sufficient for every responder. The fire and emergency services, in general, lacks a sense of appreciation for the importance of the overall professional development process.

COLLEGES AND UNIVERSITIES

The creation of credentialing systems has not resolved the debate about whether or not the fire and emergency services has achieved professionalism. Instead the debate has continued based upon another aspect of achieving a high level of competency: education through colleges and universities.

Two factors have motivated colleges and universities to become more engaged in firefighter training and education. The first factor was the GI Benefit program that was expanded right after the Korean War, which ended in 1953. The GI Benefit program made service members eligible for benefits for education compensation through the 1960s and 1970s. The development of recruit academies that were linked to colleges and universities provided a natural channel to guide firefighters into degree-granting institutions. The second factor was research by Don Favreau (1971) to examine the number of community colleges and their vocational education training programs providing instruction to fire departments in the 1960s. Many community colleges began to embrace the idea of using their funds mechanism to conduct basic recruit academies and develop two-year associate of arts (AA), associate of science (AS), or associate of applied science (AAS) degree programs as a foundation for officer development.

In 2000, Dr. Thomas Sturtevant conducted a national survey of fire service degree programs. The effort, *A Study of Undergraduate Fire Service Degree Programs in the United States* (2001), provided a snapshot of the status of and current and future issues affecting colleges and universities offering fire-related degree programs.

Doctor Denis Onieal, superintendent of the National Fire Academy, has questioned why fire and emergency services training and education has not been given the same consideration as education of other professionals such as doctors and lawyers. He expressed it as follows:

> Then why aren't we given the professional status of physicians and nurses, architects and engineers, and attorneys and accountants? Well, those professions have some things that the Fire and Emergency Services do not yet have; there are a few more steps. (NFA, 2001)

The process of increasing the educational requirements for the fire and emergency services is still under development. For example, in-residence bachelor's degree programs have been developed, and the Degrees at a Distance Program (DDP) has been created by the National Fire Academy. In some parts of the country, four-year degree and graduate-level degree programs are available to expand the body of knowledge in the fire and emergency services. The four-year programs—also called **upper division** programs, as compared to two-year **lower division** programs—are primarily aimed at those who aspire to management and leadership roles. The body of knowledge encompasses the type of topics that would parallel the same level of education required for a public administration degree.

Upper division and graduate-level programs for the fire and emergency services are continuing to develop. The expansion has raised the question of how to coordinate programs at the national level to bring about a more stable and uniformly recognized fire and emergency services higher education system.

upper division ▪ An academic environment in which individuals receive college credit aimed at courses of instruction that are no more than a four-year level degree; i.e., university.

lower division ▪ An academic environment in which individuals receive college credit aimed at courses of instruction that are no more than a two-year level degree; i.e., community college.

Fire and Emergency Services Higher Education (FESHE)

The U.S. Fire Administration's (USFA) National Fire Academy (NFA) has recognized colleges and state training systems as among the critical elements of professional education and has been working to coordinate the two- and four-year academic fire and emergency services degree programs. In June 2002, more than one hundred representatives from colleges and universities nationwide convened at the National Fire Academy in Emmitsburg, Maryland, to approve the work of previous attendees from the first conference in 1999 in adopting a model curriculum for fire-related programs. The result is an organization of postsecondary institutions (called Fire and Emergency Services Higher Education, or FESHE), which now meets annually to promote higher education and enhance the recognition of fire and emergency services as a profession. FESHE strategic goals include working collaboratively with these schools of higher education to encourage them to:

- Develop a national model for an integrated, competency-based system of fire and emergency services professional development.
- Develop a national model for an integrated system of higher education from associate to doctoral degrees in fire and emergency services.
- Prepare the nation for all hazards by developing well-trained and academically educated fire and emergency services responders.

By the end of 2009, FESHE had developed model curricula for 15 fire service courses on the associate's and bachelor's degree levels. Currently the following six courses constitute the theoretical core for all fire-related associate-level degree programs in the model curriculum: (See Figure 5.10.)

- Building Construction for Fire Protection
- Fire Behavior and Combustion
- Fire Prevention
- Fire Protection Systems
- Principles of Emergency Services
- Principles of Fire and Emergency Services Safety and Survival

FIGURE 5.10 The FESHE program mark represents the ideal that within the ivory towers of higher education, firefighters and fire officers, armed with the knowledge and a college degree, can reduce the human and economic impact of fires on their communities.

It is noteworthy that one of the core courses includes *Principles of Fire and Emergency Services Safety and Survival* and an advanced course on the same topic. A complete list of FESHE model courses for both the associate and bachelor's degree curriculum, with course objectives, course outlines, and related material, is found on the USFA Web site (see http://www.usfa.dhs.gov/nfa/higher_ed/feshe/feshe_model.shtm).

Participation in FESHE does not change what member educational organizations do; in fact, the process actually strengthens their standing. With a national, reciprocal system of training and education, these organizations could become the outside agency that assures standards competency—the *de facto* medical, law, engineering, or nursing board for the fire and emergency services.

> Professional standards are particularly important in high-risk industries such as the fire service. We must share the same values if we are to evolve further as a profession. The widespread adoption of the accreditation and certification movement . . . will go far in ensuring that this trend continues to the benefit of each one of us. (Pro Board, 2010)

Since the beginning of FESHE, participants have helped coordinate and promote the standardization of higher education in fire and emergency services as a profession with the standardization of curriculum.

State Professional Development Collaboration

The U.S. fire and emergency services has 50 state systems of professional development. Relationships between training, certification, and higher education vary from state to state because no single national system for professional development exists. Cooperation varies between them. Agreeing to standard documentation that each entity will accept for appropriate credit can be done by seeking partnership for state training, state firefighting, and emergency training associations and fire-related degree programs collaborating to do the following:

- Seek ways, when and where appropriate, to award college credit for training certification.
- Seek partnerships with state training systems to explore ways to include college and university courses as part of certification requirements.

- Design coordination to build a smooth transition from the certification environment, which is basically skills based, and the education environment, which is knowledge based.

Virginia's lead fire agency developed its fire program, the first state professional development collaboration to occur after a two-year effort. In 2006 the professional development model for Virginia was completed. This resulted in every fire science associate degree program in the state, regardless of location or institution, offering the same theoretical core FESHE model courses.

Fire and Emergency Services Trends and Patterns

Throughout most of the history of professional development in the fire and emergency services, participation in both training and education has been voluntary. Early on, leaders from both labor and management invested in the training and education systems without having any real expectation of a direct benefit. However, the trends and patterns in society overall and in the fire and emergency services are specifically to continue to raise the bar on setting minimum requirements that will become mandatory. Society in general has recognized the need for minimum standards for many agencies that are part of the fire and emergency services world. Specifically, the medical field has maintained a very tight rein on the requirements needed to provide emergency medical services. Law enforcement has seen a significant increase in its emphasis on mandatory competency requirements.

The impact of human resource decisions, job requirements, educational incentive programs, and continuing education units (CEUs) for specialized training and education is already required for both competency and safety.

EMERGENCY MEDICAL SERVICES (EMS)

Although the fire and emergency services has almost always been a first responder for medical emergencies, an increased emphasis on the level of emergency medical services is relatively new. At the end of the Vietnam War, the medical profession recognized that military medics provided an improved level of intervention in the field. Certain people suggested that the fire and emergency services might be best positioned to provide paramedicine, or advanced life support (ALS), in civilian and local government. The pilot program that initiated the delivery of EMS at this advanced level was originally created in Los Angeles County, California. Many individuals can trace their interest in the fire and emergency services to a television series entitled *Emergency!* in which two young paramedics, Gage and DeSoto, saved lives by responding to medical emergencies. The image of the firefighter as a medical health professional was initiated at that time.

The health profession is generally one in which a person exercises skill or judgment or provides a service related to:

- Preservation or improvement of the health of individuals
- Treatment or care of individuals who are injured, sick, disabled, or infirmed

The transition of EMS at the advanced level into the fire and emergency services was not as seamless as many people seem to recall. In the first place, certain people did not believe that firefighters in general had the knowledge, skills, and abilities to deliver medical care. That belief was soon overcome, and the fire and emergency services began to work side by side with physicians and other emergency department staff to raise the possibility of surviving an accident or illness in the field. (See Figure 5.11.)

EMS highlighted the importance of being certified to perform a skill effectively and efficiently. Few would argue that certification is needed to perform CPR, but would not necessarily be required to enter a burning building wearing protective clothing. Becoming authorized to provide a discrete level of health care was a linchpin in the professionalization

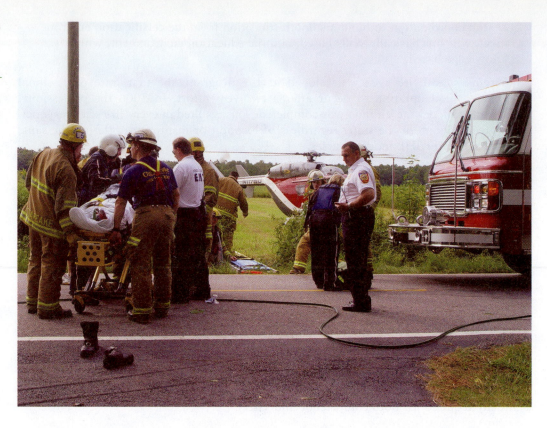

FIGURE 5.11 Firefighters
working with both EMS
and life flight nurses.
Courtesy of Martin Grube

of the fire and emergency services. EMS was the first component of the fire and emergency services in which individuals across the country are certified or licensed according to recognized national standards.

Providing EMS, of course, exposed firefighters to new hazards, including needlesticks, bloodborne and airborne pathogens, and various communicable diseases. Exposure to such hazards emphasized the need for new techniques of firefighter safety and survival (proper disposal of sharps and the use of gloves and masks, for example).

HAZARDOUS MATERIALS

The fire and emergency services has continuously been confronted with problems associated with hazardous materials. In fact, it was not uncommon for hazardous materials to be involved in many events resulting in firefighter fatalities. Over the last two decades, it has not been acceptable to respond to hazardous materials incidents without minimum levels of training. Recognition has grown that specialists need to be better qualified to perform higher levels of sophisticated action at the scene of an emergency. The recognition has resulted in the concept of "levels of training," which often fall outside of standard curriculum for basic firefighting.

The field of hazardous materials was perhaps the first area of specialization to recognize different levels of response, such as:

- Awareness
- Operations
- Technician
- Specialization (U.S. DOT, 2010) (See Figure 5.12.)

Hazardous materials response has started to incorporate these levels that ensure that only qualified personnel perform certain sophisticated HazMat activities. Specific regulations

FIGURE 5.12 Only qualified personnel can perform certain hazardous materials tasks on the emergency scene. *Courtesy of Travis Ford, Nashville Fire Department*

authorized by OSHA and other governmental agencies spell out the training and experience requirement for each level. These standards are changed from time to time and have an impact on the ability of the fire and emergency services department to maintain these areas of specialization (U.S. DOT, 2010). For example, those individuals who wear the highest level of protective clothing and enter a hazardous materials hot zone must have a much higher level of training than those who respond to a small spill or leak in which a firefighter's normal personal protective ensemble (PPE) provides adequate protection.

As the concept of levels of training has developed, areas of specialization have begun to emerge that are directly related to the danger of performing actions at the emergency scene involving hazardous materials.

OTHER HIGH-RISK SPECIALTIES

Position-related training (for firefighters, emergency vehicle drivers/operators, and officers, for example) has been the development of training and education for high-risk specialties. Examples of those high-risk specialties include high-angle rescue, swift water, or confined space rescue, along with various other technical rescue specialties. The basis for certification in these high-risk specialties is to ensure that individuals are adequately trained to use the tools and technology effectively, efficiently, and safely on the emergency scene. As specialties become more complicated and the risk increases, the need for specialized training and education is likely to expand. (See Figure 5.13.)

FIRE PREVENTION QUALIFICATION SYSTEMS

Parallel to the emphasis on qualifications for fire operations personnel is an increased emphasis on the qualification of fire prevention personnel. With increased emphasis on complex fire code enforcement, fire prevention personnel need very specific skills. Fire

(a)

(b)

FIGURE 5.13 Certification is required in high-risk areas such as technical rescue and fireboat operations, to help firefighters perform safely, effectively, and efficiently. *(a) Courtesy of Martin Grube, (b) Courtesy of Captain Terry L. Secrest*

prevention organizations, specifically the International Code Council and the National Fire Protection Association, have created certification systems for fire inspectors. Although those systems are based on NFPA 1031, *Standard for Professional Qualifications for Fire Inspector and Plan Examiner;* NFPA 1033, *Standard for Professional Qualifications for Fire Investigator;* and NFPA 1037, *Standard for Professional Qualifications for Fire Marshal,* they tend to be less visible and less widespread than the operations-oriented certification systems. One can easily argue that plan reviewers and fire inspectors help to create a safer building or environment and so enhance firefighter safety and survival. Plan reviewers and inspectors may be contributing to firefighter safety in very profound ways, and the certification process helps ensure that competency. (See Figure 5.14.)

Certification programs based on NFPA 1035, *Standard for Professional Qualifications for Public Fire and Life Safety Educator,* tend to be state based (rather than provided by an organization) and are currently available in Alabama, Alaska, Arizona, Connecticut, Florida, Georgia, Massachusetts, Minnesota, North Carolina, Oregon, Pennsylvania, South Carolina, Virginia, and Washington. It can be argued that certification to NFPA 1035 or knowledge of the standard improves firefighter safety and survival by preparing the public to be more fire safe; thus the number of responses to an emergency incident is decreased.

QUALIFICATION SYSTEMS FOR SUPERVISORS

An emerging component of fire and emergency services training and education is the notion that minimum qualifications exist to supervise individuals in the field. The National Incident Management System (NIMS) is an example of the concept that supervisors need special training and education in the area of incident management. In the past, it was assumed that those holding the rank of an officer or chief in a specific department knew what they were doing. A series of lessons learned from major fires and other emergency incidents has resulted in the fire and emergency services now recognizing the need to establish minimum qualifications to allow a person to supervise emergency scene operations (National Wildfire Coordinating Group, 2009).

At this time, the incident qualification system is focused on wildland. But the influence of natural hazards or human-made disasters on major mobilization of resources and the increased emphasis on interoperability (such as using rescue teams from one state to assist in another and the impact of terrorism) are placing more weight on the importance of officers and chiefs being properly credentialed. Major mobilization of resources and interoperability may involve other agencies such as law enforcement, public works, and

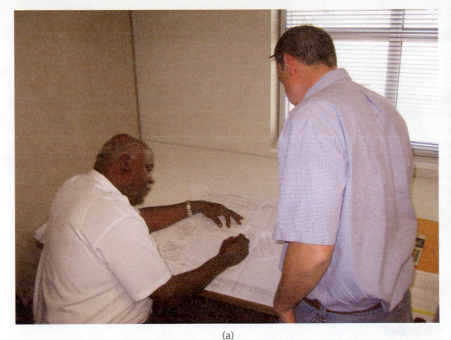

(a)

(b)

(c)

FIGURE 5.14 Plan reviewers, fire inspectors, and fire investigators all contribute significantly to firefighter safety. *(a) Courtesy of Travis Ford, Nashville Fire Department, (b, c) Courtesy of Eric Melcher, Photographer, Volunteer State University*

the health department, especially when an incident may involve terrorism. The training, education, and credentialing of officers can have a profoundly positive effect on firefighter safety and survival.

HUMAN RESOURCE DECISIONS

The human resources (HR) departments for fire departments are responsible for implementing requirements for training and education. As the availability of certification courses has increased, more communities have established certification as the criterion for job selection. The same phenomenon is taking place in upward mobility within departments: HR departments and some fire departments have incorporated certification and education already for emergency vehicle driver/operators and officers of all ranks as part of the minimum requirements for promotion. Some cities require a two- or four-year degree just to be considered for hiring or promotion.

The net result of human resources department or fire department commitment to the use of certification or degrees as a hiring or promotion prerequisite has lead to the requirement in many departments to provide career development guides and develop succession plans that will help guide individuals as they move forward with their own careers. Rather than taking random courses, individuals with access to succession planning or career development guides have an opportunity to plan their career by selecting appropriate courses of instruction. In this way individuals with a certification or degree can be evaluated in a process that ensures their preparation for promotion.

There are two professional development programs that help officers keep responders and their communities safer: the executive fire officer program and the professional credentialing for chief fire officer and chief medical officer.

The executive fire officer (EFO) program offered at the National Fire Academy helps officers enhance professional development through a series of four courses that include the following:

- Executive Development
- Executive Analysis of Community Risk Reduction
- Executive Analysis of Fire Service Operations and Emergency Management
- Executive Leadership

After 25 years of the EFO program, there are almost 3,000 alumni of the four-year program. In addition, many applied research projects archived at the National Emergency Training Center/Learning Resource Center cover issues that affect the overall performance of fire and emergency services, including topics that specifically address safety and survival.

Another program now considered across the United States is the professional credentialing designation by the Center for Public Safety Excellence for chief fire officer (CFO) and chief medical officer (CMO). This professional designation specifies minimum eligibility requirements by validating experience, training, and education competencies against a standard of excellence. The CFO and CMO designations create a level of professionalism that can be recognized by community leaders. By the end of 2009, there are approximately 630 CFOs, 36 CMOs, and 19 individuals who have earned both designations. The Center for Public Safety Excellence has also announced the availability of professional credentialing for the fire officer designation.

JOB REQUIREMENTS

Those fire departments with access to comprehensive training and education systems have tended to adopt minimum job requirements at the recruit level faster than those that lack access to such delivery systems. The result is that it is becoming increasingly difficult to get hired or promoted without meeting these minimum requirements.

EDUCATIONAL INCENTIVE PROGRAMS

Both the training and education systems have been supported by some departments through educational incentive programs. Under incentive programs, individuals are financially rewarded for maintaining certain levels of certification. The phenomenon of educational incentive programs is reflected in labor–management agreements. The specifics in many cases may be contained in a memorandum of understanding (MOU). Tuition reimbursement for courses taken and completed has also become an important part of the educational incentive for many fire and emergency services departments.

CONTINUING EDUCATION UNITS (CEUs)

Educational incentive programs can include requirements for a specific number of **continuing education units (CEUs)**. A CEU is a measure of training or education required

continuing education unit (CEU) ■ A measure of training or education required to maintain a certification or license; often equivalent to 10 hours of classroom time.

FIGURE 5.15 Acquiring continuing education requires that a certain set of skills be demonstrated.
Courtesy of Dave Brasells

to maintain a certification or license. Essentially a CEU recognizes that an individual has spent a specific number of hours in class or performed a certain set of skills to demonstrate currency in his or her certification. Generally, 1 CEU represents 10 hours of participation in a training or education program. Good examples of the use of the CEU can be found in the field of emergency medicine or hazardous materials. (See Figure 5.15.)

Training providers often grant CEUs on the basis of a variety of delivery systems. For example, some CEUs require an instructor to actually observe the candidate performing a task, such as CPR. In other situations, a person can obtain CEUs through long-distance learning, which may include the use of self-paced text and testing, either online or through correspondence. Regardless of whether obtained through face-to-face exposure, seat time in a class, or completing a checkoff list, the CEU assures that skills do not degrade when those skills must be critically honed to perform in the field. In the fire and emergency services, a CEU recognizes recurrent training that provides effectiveness, efficiency, and safety.

Individual Commitment to the Training and Education Process

The fire and emergency services is well respected by society, but that respect is not necessarily based on job performance. Sometimes it is based on the fact that society recognizes the danger of firefighting. The manner in which this danger is sometimes measured is in the loss of life. Those are individuals, not just numbers. Every year we go back to the National Fallen Firefighters Memorial and pay a form of respect that is a Pyrrhic victory. The modern motto being promoted is "Everyone Goes Home." Everyone—labor, management, family, and friends—wants to see the fire and emergency services become a safer occupation. Establishing and maintaining professional credentials should be one of the fire and emergency services' highest priorities and deepest values. Such professionalism requires a commitment on the part of all decision makers in the fire and emergency services.

There are those who disagree with the need for professionalism and credentials. They do not believe that standards are for them or that there is a consequence of failure for not adopting standards. The individual commitment begins with you. It is relayed through those you have influence over and will ultimately be measured by how your department and your peers support and implement training and education.

We compared firefighter safety to the safety of those who go into armed combat. This is not an accidental comparison but a relevant one. The people who pay the price for failing to recognize the importance of meeting standards go far beyond the casualties themselves. Our community pays the price in terms of liability. Our families pay the price in terms of emotional impact, and our profession pays the price with lack of credibility in certain circumstances.

A department with a high commitment to setting and keeping standards is going to be a more credible department than one that does not have that level of commitment. Everyone from the rookie to the chief should have a career development plan that incorporates exposure to the appropriate training and education, testing and measurement of the professional standards, and official recognition by an accredited entity that each person has achieved a level of competency. (See Figure 5.16.)

IMPACT OF TECHNOLOGY

In the early days of the fire and emergency services, firefighters learned their jobs by observing other firefighters. Instruction was limited to face-to-face training, literature, and physical application. As the body of knowledge in the fire and emergency services has expanded, that has become inadequate. As a result, textbooks have been published, audiovisual materials have been presented, simulations have been conducted, all in the interest of transferring knowledge to the firefighter. The technology of receiving information is now undergoing even further change. Specifically the introduction of the computer

FIGURE 5.16 Every firefighter needs a professional development plan throughout his or her entire career to remain competent. *Courtesy of Martin Grube*

as a delivery mechanism has provided an entirely new way of obtaining information. It is now recognized that computer-based training and long-distance learning are increasingly becoming part of the delivery system for professional development.

Distance education primarily takes place on the Internet. Utilizing a variety of software packages, the student is exposed to text, graphics, audio, video, and feedback mechanisms. This type of learning is called asynchronous learning. Asynchronous communication occurs when someone transmits an electronic message to which an individual can respond at a later time. The value of asynchronous learning is that the student is able to work at his or her own pace and complete the program when time best suits the student. Students are given their own passwords to reach the course site on the Internet. Once a student is enrolled in the program, he or she can complete modules and return to the site as time allows. At the end of each module there is usually some form of assessment procedure that tests the student on knowledge or skills obtained during the online experience. There are advantages and disadvantages to this type of program depending upon personal learning style, comfort with technology, the support level of one's environment, discipline, and the time frame. For example, many colleges and universities along with the National Fire Academy are now offering online courses of instruction.

Looking to the Future

If the past is prologue, then the present must be evidence of its impact. Trends and patterns that have been developing over a period of one hundred years are not likely to reverse themselves. The quest for professionalism continues as the emphasis on increasing the level of intensity for meeting standards has not diminished over time. On the issue of professional development, the 2006 version of the ongoing Wingspread series states:

> Significant strides have been made in fire service professional development, but improvement is still needed. The fire service needs to continue to evolve as a profession as have other governmental entities that operate in the environments where we work as well as other governmental organizations and the private sector. These skills are as important in the volunteer and combination fire services as they are in the career fire service.
>
> *Universal acceptance and use of a credentialing system will help in professional development, but the availability of degree programs in fire science and fire department management are necessary to reach the level of competence needed for firefighters and fire executives. Each state should have at least one two-year degree program available in the community college system, to provide basic knowledge and skills. Bachelor degree programs should also be available to firefighters who wish to pursue them, at reasonable cost.* (Wingspread Conference Report, 2006, p. 8)

No one can accurately predict the future, but if the last 50 years of development have created any sense of how chiefs will proceed in the future, increasing demands on the fire and emergency services will definitely increase the need to be properly trained, properly equipped, and properly educated before an individual is considered qualified to go into harm's way.

Summary

Having stated that the trends and patterns have taken the fire and emergency services to where it is today, it is important to recognize that the system also can erode over time. The individual and organizational commitment required to get us to where we are today should be regarded as fundamental. The level of commitment in the future by both leaders and labor in the fire and emergency services needs to be equally intensive. Every year a fire chief must prepare and defend a budget. Every year a department must make choices regarding its priorities. Providing justification regarding keeping a training program at a high level of competency and providing support for education incentive programs are crucial, because these programs can be easy targets during tough economic times. The fire chief's role is to continually be assessing the skill set of the department, which includes preserving the integrity of the system for certification and qualification. One cannot assume that this will happen. To the contrary, it is not uncommon for the first casualty of a budget battle to be a training system. The military would not tolerate removing its ability to train and neither should the fire and emergency services.

Although it once might have been true that anyone who could pick up a bucket and throw water on a burning building would be called a firefighter, that phenomenon has long since passed into history. The term *firefighter* has a series of requirements attached that define the person as having been trained, tested, and certified to be part of a department that responds to all types of emergencies. Notably, no distinctions have been drawn about whether or not the firefighter receives compensation for this activity; compensation is a feature of employment and not one of competency. In fact, certain individuals are not certified or licensed, let alone trained or educated, to perform in this field, but are nonetheless employed. Other individuals are certified, licensed, and educated to perform in this field, and they are not compensated.

The difference is it that a person that is trained to meet specific standards, has the knowledge of the entire spectrum of performance, and has been tested and evaluated is much more likely to be able to perform effectively, efficiently, and most importantly, safely.

Review Questions

1. Describe how the capabilities of individuals who become firefighters should be primarily measured.
2. Discuss the difference between credentialing, certification, licensure, professional recognition (or designation), and accreditation.
3. List the organizations that provide accreditation to fire departments authorizing them to issue certificates to demonstrate competency.
4. Describe the reasons for revisiting the NFPA professional qualification standards over time.
5. List the six recommended FESHE core courses.

References

Blackstone, G. V. (1957). *A History of the British Fire Service*. London: Routledge and Kegan Paul.

Favreau, D. (1971, June). *Fire Service Education 1971: A Survey and Historical Developments of Fire Service Education in the United States*. Albany, NY: International Fire Administration Institute.

Gowell, J. L., Tebbetts, W. W., and Baker, J. F. (1932). *Trade Analysis of Fire Engineering, Los Angeles Fire College*. Los Angeles: Los Angeles Fire Department.

Hackworth, D. H., and Sherman, J. (1989). *About Face: The Odyssey of an American Warrior*. New York: Touchstone.

Massey-Shaw, E. (1876). *Fire Protection: A Complete Manual of the Organization, Machinery, Discipline, and General Working of the Fire Brigade of London*. London: C. and E. Leighton.

Massey-Shaw, E. (1945). *The Business of a Fireman*. London: Lomax, Erskine & Company.

National Fire Academy (Producer). (2001). *Highlights from the Second Fire and Emergency Services Higher Education (FESHE) Conference* [Motion picture]. Available from the United States Fire Administration, 16825 South Seton Avenue, Emmitsburg, MD 21727.

National Fire Protection Association. (2008). *NFPA 1001: Standard for Fire Fighter Professional Qualifications.* Retrieved April 3, 2009, from http://www.nfpa.org/aboutthecodes/AboutTheCodes.asp?DocNum=1001

National Wildfire Coordinating Group. (2009, June). *National Interagency Incident Management System Wildland Fire Qualification System Guide.* Retrieved October 13, 2009, from http://www.nwcg.gov/pms/docs/pms310-1.pdf

Pro Board. (2010). *National Board on Fire Service Professional Qualifications: History and Overview.* Retrieved October 13, 2010, from http://www.theproboard.org/history_and_overview.htm

Silk, L. S. (1960). *The Education of Businessmen.* New York: Committee for Economic Development, No. 11, p. 9.

Sturtevant, T. B. (2001). *A Study of Undergraduate Fire Service Degree Programs in the United States.* Retrieved April 5, 2009, from http:www.dissertation.com/library/112130xa.htm

U.S. Department of Transportation. (2010). *49 Code of Federal Regulations Part 172: Subpart H—Training.* Retrieved October 13, 2010, from http://ecfr.gpoaccess.gov/cgi/t/text/text-idx?c=ecfr&rgn=div5&view=text&node=49:2.1.1.3.7&idno=49#49:2.1.1.3.7.8

U.S. Fire Administration. (2010). *Fire and Emergency Services Higher Education (FESHE) Model Curriculum.* Retrieved October 13, 2010, from http://www.usfa.dhs.gov/nfa/higher_ed/feshe/feshe_model.shtm

Wingspread Conference Report. (1966, February). *Wingspread Conference on Fire Service Administration, Education, and Research: Statements of National Significance to the Fire Problem in the United States.* Retrieved November 20, 2009, from http://nationalfireheritagecenter.com/1966Wingspread.pdf

Wingspread Conference Report. (2006, April). *Wingspread V: Statements of National Significance to the Fire Service and to Those Served.* Retrieved November 20, 2009, from http://www.nationalfireheritagecenter.com/2006Wingspread.pdf

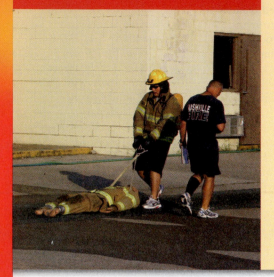

Martha Ellis

Courtesy of Travis Ford, Nashville Fire Department

OBJECTIVES

After reading this chapter, the student should be able to:

- Identify the role of fitness in the fire and emergency services.
- Explain the need to commit fully to emergency responder fitness and proper nutrition.
- Recognize the need for a comprehensive fitness program for every fire department.
- Recognize the legal considerations in implementing performance testing.
- Identify the elements of performance test development.
- Discuss the importance of having an annual physical ability assessment.
- Explain the need for a rehabilitation program in every fire department.

PEARSON

myfirekit™

Understanding All of the Medical, Fitness, Rehab, and Performance Standards

In the United States there are approximately 800,000 volunteer and 300,000 career fire-fighters. Each one can be completely resting one moment and working at full capacity the next, because an emergency incident never provides time to prepare. The "time to prepare" is truly before the incident occurs. To endure such demand on the body, firefighters must be in excellent physical condition. (See Figure 6.1.)

Yet, over the past 20 years, cardiovascular-related issues have consistently accounted for 40% to 50% of all firefighter fatalities annually. In fact, according to the U.S. Centers for Disease Control and Prevention (CDC), firefighters suffer the highest mortality rates due to cardiovascular events on the job (NIOSH, 2007). Such rates suggest that they may be responding to emergency incidents while unfit to perform their duties.

A significant percentage of the nation's firefighters are not in the best physical condition. Boston University's School of Medicine conducted a study from 2004 to 2007 on area recruits under the age of 35 (Tsismenakis et al., 2009). The study found that, based on **body mass index (BMI)**, which is the measure of body fat based on height and weight, 43.8% were overweight and 33% were obese. **Obesity** is a condition in which an individual has a very high amount of body fat in relation to lean body mass, or a body mass index of 30 or higher. A similar study of volunteer firefighters reported that 35% were overweight and 41% were obese (Yoo and Warren, 2009). Although many fire departments do not rely solely on BMI in determining poor conditioning, the poor conditioning (extra weight) in such a large segment of the fire and emergency services strongly suggests an increased risk of injuries and heart disease.

Departments have an obligation to all stakeholders to provide for personnel safety and health. A great deal of time, research, and money goes into protective clothing, high-tech equipment, training, and emergency response vehicles, as is proper. Yet the idea of seriously addressing health-and-fitness criteria with that same quantity and quality of effort and resources often has a lower priority. The lesser priority given to health means chances are good that the very purpose of the fire services—"to serve and protect"—is jeopardized.

body mass index (BMI) ■ Measure of body fat based on height and weight.

obesity ■ A condition in which an individual has a very high amount of body fat in relation to lean body mass, or a BMI of 30 or higher.

FIGURE 6.1 Firefighting is physically demanding. *Courtesy of Martin Grube*

Physically fit and healthy personnel are requirements in business. Yet, the USFA reported an estimated 737,000 firefighters are not offered any type of a wellness or fitness program (USFA, 2006, p. v). If any other business were to neglect such a critical component of its operational needs, it would risk going under. In the business of emergency response, such neglect risks lives.

We live in a time when obesity and poor health are becoming the norm within our communities—the very communities from which the fire and emergency services draws firefighting candidates. According to the CDC, in 2008 Colorado was the only state that had a prevalence of obesity less than 20%, 32 states had a prevalence greater than 25%, and six had a prevalence equal to or greater than 30% (CDC, 2009). So, how can the fire department ensure that firefighters are fit and healthy throughout their careers?

First and foremost, the needs of the community should dictate the caliber of employee hired. The challenge lies in selecting and retaining employees by way of a fair, objective, valid, and accessible process to all.

Second, and most importantly, in conjunction with departmental efforts, each individual firefighter must take personal responsibility for his or her own physical and mental condition. Most people who are not as fit as they should be fall into one of two categories: those who know they are out of condition and those who are in denial. Those in the second and more dangerous category can literally kill themselves and those around them trying to prove that they still can keep up.

Third, it is important for all firefighters to be able to meet a fitness standard and pass a medical exam at any given point in their career. The fireground is consistently demanding, the tones unfailingly go off, and the expectation that individuals can and will perform their duties is ever present. The human element is the only variable in the equation! It is poor planning to wait until a high-rise fire to find out whether crews can get the necessary gear to the floor below the fire and still have the reserves to make an attack. Monitoring the health of personnel as closely as the status of tools and emergency response vehicles seems only proper. (See Figure 6.2.)

FIGURE 6.2 Any high-rise fire response may be one of the most physically demanding incidents on the body.

Current Recommended Standards

There are several standards that specifically deal with firefighter health issues. These standards offer guidelines that provide important practices that should be adopted and followed in the fire and emergency services.

NFPA STANDARDS

The NFPA provides much needed guidelines and recommendations for all aspects of fire prevention, suppression, and training. Its standards are usually well founded in history and necessity. The following four NFPA standards pertain specifically to the health, safety, and fitness of firefighters:

> NFPA 1500, *Standard on Fire Department Occupational Safety and Health Program*
> NFPA 1582, *Standard on Comprehensive Occupational Medical Program for Fire Departments*
> NFPA 1583, *Standard on Health-Related Fitness Programs for Fire Department Members*
> NFPA 1584, *Standard on the Rehabilitation Process for Members During Emergency Operations and Training Exercises*

NFPA 1500

The NFPA 1500 standard addresses many aspects of firefighter safety, including medical and fitness requirements (2007a). For example, NFPA 1500, Section 10.1, references NFPA 1582 and offers considerable detail on the medical examination, including a definition of who should be allowed to conduct the physical exam (a licensed physician who has a solid understanding of the demands of the emergency services). NFPA 1500 clearly states that "candidates and members who will engage in fire suppression shall meet the medical requirements specified in NFPA 1582. . . ." This is not a practice reserved for new hires, but applies to "any member who will engage in fire suppression." NFPA 1500, Section 10.2, states that "The fire department shall develop physical performance requirements for candidates" and that "candidates shall be qualified as meeting the physical performance requirements established by the fire department prior to entering into a training program to become a firefighter." Meeting this requirement is a critical element because a lifestyle of "fit for duty" should be established prior to the hiring process. Some individuals may get in shape only to be hired and quickly fall back into their poor habits after they are hired.

What may be surprising to those already in the fire department is NFPA 1500's reference to incumbent firefighters. It reads, "Members who engage in emergency operations shall be annually qualified as meeting the physical performance requirements established by the fire department. Members who do not meet the required level of physical performance requirements shall not be permitted to engage in emergency operations" (NFPA 1500, Chapter 10.2). The only way to apply the standard is to have objective and measurable physical performance requirements. Because the relationship between work completed and time is directly related to success or failure on the emergency scene, the most objective means of measuring job-related physical ability is with a validated, time-based performance test.

In NFPA 1500, Section 10.2.5, states, "Members who are unable to meet the physical performance requirements shall enter a physical performance rehabilitation program to facilitate progress in attaining a level of performance commensurate with the individual's assigned duties and responsibilities." Skill proficiency is an important factor for incumbents.

In NFPA 1500, Section 10.3.1, goes on to say, "The fire department shall establish and provide a health and fitness program that meets the requirements of NFPA 1583 *Standard on Health-Related Fitness Programs for Firefighters* [in order] to enable members to develop and maintain a level of fitness that allows them to safely perform their assigned functions."

NFPA 1582

The NFPA 1582 standard is very specific about many aspects of the medical examination. Primarily for new hires, Category A Medical Condition (Section 3.3.13.1) is severe enough to "preclude a person from performing as a member in a training or emergency operational environment by presenting a significant risk to the safety and health of the person or others." Primarily for incumbents, Category B Medical Condition (Section 3.3.13.2), "based on its severity or degree, could preclude a person from performing as a member in a training or emergency operational environment by presenting a significant risk to the safety and health of the person or others" (Section 5.1.1). To be able to identify these potential debilitating medical conditions could save both the life of the individual and the lives of countless others.

In order to determine what could be detrimental to performing firefighting functions, NFPA 1582 (2007b) has identified 13 essential job tasks and descriptions:

1. Performing firefighting tasks (e.g., hose line operations, extensive crawling, lifting and carrying heavy objects, ventilating roofs or walls using power or hand tools, forcible entry), rescue operations, and other emergency response actions under stressful conditions while wearing personal protective ensembles and self-contained breathing apparatus (SCBA), including working in extremely hot or cold environments for prolonged time periods.
2. Wearing SCBA, which includes a demand valve–type positive-pressure face piece or HEPA-filter mask, and requires the ability to tolerate increased respiratory workloads.
3. Exposure to toxic fumes, irritants, particulates, biological (infectious), and nonbiological hazards, or heated gases, despite the use of personal protective ensembles and SCBA.
4. Depending on the local organization, climbing six or more flights of stairs while wearing fire protective ensemble weighing at least 50 lb (22.6 kg) or more and carrying equipment/tools weighing an additional 20 to 40 lb (9 to 18 kg).
5. Wearing fire protective ensemble that is encapsulating and insulated, which will result in significant fluid loss that frequently progresses to clinical dehydration and can elevate core temperatures to levels exceeding 102.2°F (39°C).
6. Searching, finding, and rescue-dragging or carrying victims ranging from newborns up to adults weighing over 200 lb (90 kg) to safety despite hazardous conditions and low visibility.
7. Advancing water-filled hose lines up to 2-1/2 in. (65 mm) in diameter from fire apparatus to occupancy (approximately 150 ft [50 m]), which can involve negotiating multiple flights of stair, ladders, and other obstacles.
8. Climbing ladders, operating from heights, walking or crawling in the dark along narrow and uneven surfaces, and operating in proximity to electrical power lines or other hazards.
9. Unpredictable emergency requirements for prolonged periods of extreme physical exertion without benefit of warm-up, scheduled rest periods, meals, access to medication, or hydration.
10. Operating fire apparatus or other vehicles in an emergency mode with emergency lights and sirens.
11. Critical, time-sensitive, complex problem solving during physical exertion in stressful, hazardous environments, including hot, dark, and tightly enclosed spaces further aggravated by fatigue, flashing lights, sirens, and other distractions.
12. Ability to communicate (give and comprehend verbal orders) while wearing personal protective ensemble and SCBA under conditions of high background noise, poor visibility, and drenching from hose lines or fixed protection systems (sprinklers).
13. Functioning as an integral component of a team, where sudden incapacitation of a member can result in mission failure or in risk of injury or death to civilians or other team members.

FIGURE 6.3 Performance of the 13 essential job tasks of NFPA 1582 requires firefighters to be medically fit. *Courtesy of Captain Terry L. Secrest*

Based on the essential job tasks identified by NFPA 1582, it is reasonable to expect candidates and incumbents to willingly participate in medical exams. It is especially reasonable for all stakeholders to expect fire departments to properly provide the medical exams at the time of hiring and *every year* thereafter. NFPA 1582 should help determine an individual's medical ability to perform the duties required without presenting a significant risk to the individual's or others' safety and health. (See Figure 6.3.)

NFPA 1583

The NFPA 1583 standard addresses the fitness requirements described in NFPA 1500. To better understand NFPA 1583 (2008a), clarification of the terms *fitness assessment* and *fitness standard* along with *physical abilities* is necessary:

- *Fitness assessment.* The term refers to the process of evaluating an individual's overall level of fitness. A person's fitness level can be assessed by measuring five different components of fitness by way of a fitness assessment: aerobic/anaerobic capacity, muscular strength, muscular endurance, flexibility, and body composition (i.e., body-fat percentage). Fitness assessments are a great way to identify the strengths and weaknesses of an individual, making exercise prescription a more precise and effective means to improvement.
- *Fitness standard.* A fitness standard for the fire and emergency services is one that empirically matches the demands of actual firefighting tasks to specific fitness measurements. The standard is determined by testing the physical abilities of a firefighter using content-valid test measurements with which the skills performed match actual firefighting tasks. The fire service typically refers to these tests as incumbent performance standards.
- *Physical abilities.* As the name implies, this term refers to a benchmark that represents an acceptable level of fitness as demonstrated through a physical ability test. The test must be developed to specifically reflect the physical ability required to accomplish essential job functions. Note, however, that because there can be perceived significant differences among departments for what is required of local firefighters, the fitness standard could be completely different, depending on where it is being administered. Validation of fitness standards has been developed to prevent departments from creating arbitrary standards that may be punitive or discriminatory.

Unless specific statistical correlations have been drawn between fitness assessment scores and a valid fitness standard for any given department, the fitness assessment may be used only to evaluate a candidate or member's fitness status. It may not be used to assess physical ability to perform on the emergency scene.

The program defined in NFPA 1583 includes the following: the assignment of a qualified health and fitness coordinator, periodic fitness assessments for all members, an exercise training program that is available to all members, education and counseling regarding health promotion for all members, and a process for collection and maintenance of health-related fitness program data (Section 4.2).

NFPA 1584

The NFPA 1584 standard is an integral part of maintaining the health and safety of fit, medically qualified firefighters when on the emergency scene or during training activities. It sets the standard on the rehabilitation process for members during emergency operations and training exercises.

In 2008 NFPA 1584 became a standard that required every fire department to have a plan in place for how it would provide **rehabilitation**, the restoration of an individual to a state of preparedness, on the emergency scene or during training exercises that last more than one hour. That rehabilitation process includes the following:

rehabilitation ■ Restoration of an individual to a state of preparedness.

- Rest
- Hydration to replace body fluids
- Cooling (passive or active)
- Warming
- Medical monitoring
- Emergency medical care, if required
- Relief from climatic conditions (heat, cold, wind, rain)
- Calorie and electrolyte replacement
- Accountability
- Release

By evaluating vital signs of individuals while operating on the emergency scene or conducting training activities, it is possible to identify those individuals who are in need of rest and rehydration. (See Figure 6.4.)

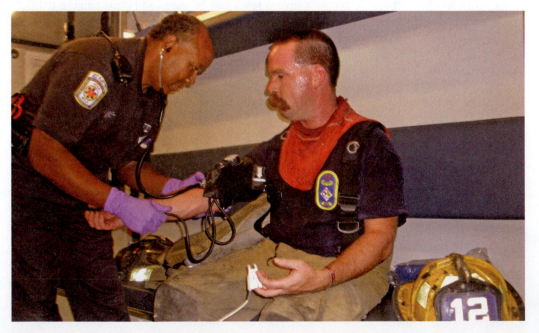

FIGURE 6.4 Every fire department must have a policy and procedure in place for evaluating vital signs on the emergency scene or during training activities. *Courtesy of Travis Ford, Nashville Fire Department*

The intent of NFPA 1584 is threefold, aiming to (McEvoy, 2007):

- Provide ongoing education on when and how to rehabilitate
- Provide the supplies, shelter, equipment, and medical expertise to firefighters where and when they need it
- Create a safety net for members unable or unwilling to recognize when they are fatigued

The overall approach of NFPA 1584 is to provide proper education in nutrition, hydration, and rest starting in the recruit academy. Done properly, rehab can increase an individual's work capacity, improve personnel resources on the emergency scene, and allow firefighters to properly rehab themselves using the equipment, supplies, and medical expertise provided on the scene by the department.

Where to Start?

LEADERSHIP

Leadership at any level must lead by example. Although the likelihood of a chief needing to actually perform physical tasks at an emergency scene is usually remote, that is irrelevant. All chief officers of the department should be among the first in line to perform physical performance testing. It is important for them to make an effort to step up and train with the crew. They should be as thoroughly engaged in the requirements and process as the newest probationary emergency responder. The leaders in the fire service must be willing to do whatever it is they are asking of their membership.

Chief officers must be willing to make the tough decisions for the good of the department, but that may directly impact an individual. In all likelihood the individual will be someone who has come up through the ranks with the chief, toiled through recruit academy 20-odd years ago, and fought side by side with the chief during some of the biggest fires in their municipality. However, leaders cannot afford to make professional decisions based on personal feelings. If everyone is to meet the same standard, by definition there can be no exceptions.

Leaders within the fire department do not necessarily come loaded down with brass. Leadership is possible from any rank. Fire departments are led from the bottom up sometimes more effectively than from the top down. Leading by example is everyone's responsibility. Subordinates value a committed administration, just as any officer appreciates an employee who is eager to contribute to the department.

It is up to the department's leaders to do everything they can for the program to achieve success. Time, equipment, and a positive attitude are all areas that cannot be shortchanged.

DEVELOPING POLICY

Chief officers can commit to their decisions by making them into policy. New programs especially need solid direction, which well-written policies and procedures can provide. Policies that support a health and fitness program should address:

- Implementation timelines
- Logistics of all aspects of the program
- Definitions of criteria
- Clearly spelled out steps to be taken in the event someone is unable to successfully complete the assessment in the determined time frame

The chief's decisions are unlikely to be made in a vacuum. There should be a multitude of individuals involved in the development of policy for the health and fitness program. The decision process needs to have just as many opinions on and perspectives of what the final product should look like. The chief must consider all angles and possibilities and, if necessary, arbitrate any heated discussions or conflicts within the policy development process. The chief should manage this from a position of knowledge and a sense of how he or she would like to see the program develop.

CURRENT PROGRAMS

Several organizations have worked tirelessly in an effort to reduce firefighter line-of-duty deaths by developing programs that target specific wellness and fitness issues. The programs have been developed to help fire departments better understand the need for implementing wellness and fitness programs.

THE IAFF/IAFC WELLNESS/FITNESS INITIATIVE

In 1996 the International Association of Fire Fighters (IAFF) and the International Association of Fire Chiefs (IAFC) joined forces to create the Fire Service Joint Labor Management Wellness/Fitness Initiative (IAFF, 2009a). This collaboration was truly an unprecedented step in the right direction. Ten fire departments contributed to the development of the program: Austin, Calgary, Charlotte, Fairfax County, Indianapolis, Los Angeles County, Miami-Dade, New York, Phoenix, and Seattle.

At the foundation of the IAFF/IAFC Wellness/Fitness Initiative is the belief that the overall wellness and fitness system must be holistic, positive, rehabilitating, and educational. The current WFI includes the following major components: medical, fitness, injury and medical rehabilitation, behavioral health, cost justification, and data collection and implementation.

NATIONAL VOLUNTEER FIRE COUNCIL

In the *Health and Wellness Guide for Volunteer Fire and Emergency Services,* the National Volunteer Fire Council (NVFC) identifies five key components for success (USFA and NVFC, 2008, p. v):

■ Regular health and fitness screenings and medical evaluations
■ Fitness program (cardiovascular, strength, and flexibility training)
■ Behavioral modification, addressing smoking, hypertension, diet, cholesterol, and diabetes
■ Volunteer education
■ Screening of volunteer applicants

Since its original publication in 1992, the guide has been updated twice, most recently in 2008. It presents case studies to demonstrate how volunteer departments can implement their own health and wellness programs and maintain their success over time.

As part of its Heart-Healthy Firefighter Program, the NVFC has created an interactive guide for individuals to develop and implement a fitness program called the Fired Up for Fitness Challenge (NVFC, 2008).

In 2008 the NVFC compiled a report called *Emerging Health and Safety Issues in the Volunteer Fire Service* (USFA, 2008c) that addresses the emerging issues of health and wellness program development and implementation.

Organizations such as the IAFF, IAFC, and NVFC have made great progress toward developing health and fitness programs with the Wellness-Fitness Initiative (WFI), and yet additional work is still needed in the fire and emergency services to realize any marked improvements in health-related firefighter fatalities and injuries.

Fatality statistics still reveal that nearly 50% of all annual firefighter fatalities that are health related fall into the category of "stress and over exertion" (Moore-Merrell, Zhou, McDonald, Fisher, and Moore, 2008; USFA, 2009). The follow-up question is why.

■ Is it due to lack of departmental attention to death and injuries?
■ Is there an advantage to firefighters being off the job when injured?

- Is there less incentive to be on unrestricted duty?
- Are there state laws that give tax breaks that act as incentives to emergency responders who become injured on the job?

Although that fatality statistic includes both career and volunteer firefighters, the job environment is the same for everybody. The challenge is that more volunteer departments without any health and fitness programs respond to emergency incidents every day; they may not be fit enough to actually do the job. Typically, lack of such programs may be because of a lack of interest or a budget issue to maintain any type of program. However, professional firefighters should be expected to hold higher standards with financial support that will hopefully lead and guide volunteer firefighters.

MEDICAL STANDARDS

Firefighters serve in an occupation that is unpredictable and physically demanding, and perform various tasks on the emergency scene. Therefore, being medically fit to perform the variety of tasks should be a requirement for every fire department.

Factors Threatening Firefighters

In 2008, over 40% of all firefighter line-of-duty deaths were attributed to cardiovascular system failures (USFA, 2009). Although in some years the numbers have varied slightly, cardiovascular-related issues are the number-one killer of firefighters annually (Hales, 2008, p. 1). This statistic does not have to be so, if the fire and emergency services does everything it can to prevent those deaths. (See Figure 6.5.)

RISK FACTORS

Why are emergency responders so susceptible to cardiovascular disease? To answer the question, it is important to examine both intrinsic and extrinsic risk factors.

FIGURE 6.5 There are several intrinsic and extrinsic risk factors that affect cardiovascular-related issues. *Courtesy of Wayne Haley*

Intrinsic Risk Factors

Intrinsic risk factors are circumstances that can increase a person's likelihood of experiencing a cardiac episode. They come from internal and oftentimes uncontrollable sources. Such factors lie in genetic predisposition.

Coronary heart disease (CHD) does not happen overnight. Unless there is an acute congenital disorder, the actual development of CHD happens over an extended period of time. Family history has been shown to be a high-risk factor for development of CHD in an individual. If an individual's father or brother experienced a heart attack, bypass surgery, or sudden death prior to age 55, or the individual's mother or sister experienced similar episodes prior to age 65, then that individual is at a higher risk for developing heart disease than the general population (American College of Sports Medicine, 2006, p. 22).

Extrinsic Risk Factors

Extrinsic risk factors are circumstances that can increase a person's likelihood of experiencing a cardiac episode from external and oftentimes controllable sources. They include the firefighting environment and its impact on the human body, as well as what emergency responders do to themselves, such as smoking and chewing tobacco, unhealthy eating habits, using alcohol, and so on.

Firefighters are often forced to expose themselves to rapid increases in physiological demands. Heart rate, stroke volume, and blood pressure can increase dramatically during a response. Depending on the circumstances, those demands are taxing even to a healthy heart.

A cardiovascular system that is constantly being asked to go from a resting status to maximum effort is a system under stress. Such stress is met by the body's "fight-or-flight response," which includes a mix of blood shunting and adrenaline surging and a host of neurological impulses. It is the body's way of providing adequate energy to accommodate the need to run or put up a fight, much like starting a cold vehicle, dropping it into gear, and stomping on the accelerator. It will respond if it is a finely tuned vehicle. However, if the vehicle has not been well maintained, it will likely sputter black smoke, backfire a couple of times, and shut down completely.

A recent study of Indianapolis firefighters (Brown and Stickford, n.d.) concluded that "Some heart rates observed in rescue operations were in excess of 100% of predicted heart rate maximums and sustained for 20–40 minutes. Heart rates of this magnitude are extreme and may be responsible for triggering catastrophic cardiovascular events seen in firefighters possessing underlying heart rate" (p. 50). The study further states that "firefighters with pre-existing cardiovascular disease exposed to the physical and emotional stress of a fire scene are in extreme risk of experiencing a myocardial infarction, stroke, or other cardiovascular system collapse" (p. 72). (See Figure 6.6.)

The Harvard School of Public Health studied data from 1994 to 2004 to examine firefighter fatalities due to coronary heart disease. Researchers aimed to identify specifically which firefighting tasks appeared to increase the risk of dying from a coronary event. They stated:

> In parallel with our finding of a significantly increased risk of death from coronary heart disease during fire suppression, as compared with nonemergency duties, the risk was significantly elevated during physical training. This finding is consistent with investigations implicating intense physical activity as a strong triggering factor, especially among physically inactive persons. (Kales, Soteriades, Christophi, and Christiani, 2007)

The Harvard study further indicated that "firefighters were generally fit going into the service, but over the course of a number of years—because of not exercising regularly and not eating right—are becoming obese" (Kales et al., 2007). The cruelest statistic revealed through the study is that 20% of the deaths from on-duty heart attacks occurred in individuals who had been previously diagnosed with coronary heart disease.

intrinsic risk factors ■ Circumstances that can increase a person's likelihood of experiencing a cardiac episode. They come from internal and oftentimes uncontrollable sources.

extrinsic risk factors ■ Circumstances that can increase a person's likelihood of experiencing a cardiac episode from external and oftentimes controllable sources.

PEARSON
myfirekit™

Please visit MyFireKit Chapter 6 for a list of heart and lung presumption legislation by state.

FIGURE 6.6 Firefighters are frequently required to respond and perform at emergency scenes shortly after waking up. *Courtesy of Travis Ford, Nashville Fire Department*

The Harvard study proposed various environmental factors (extrinsic risk factors) as plausible explanations for the high mortality from cardiovascular events among emergency responders. They include, but are not limited to, smoke and chemical exposure, heat stress, shift work, psychological stressors, and a high prevalence of cardiovascular risk factors (Kales et al., 2007).

Specific chemical exposures common to firefighting include carbon monoxide, hydrogen cyanide, and particulate matter. All are known to lead to or exacerbate cardiovascular disease. Even moderate exposure to carbon monoxide can be very detrimental to cellular respirations when bound with hemoglobin. Carbon monoxide binds more readily to oxygen receptors on hemoglobin molecules, which dramatically decreases the amount of oxygen delivered to the cardiac muscle cells. (See Figure 6.7.)

Cardiac muscle cells that do not receive sufficient amounts of oxygen (i.e., ischemia) will fatigue prematurely, leading to decreased cardiac output and possibly angina. Even low levels of exposure to carbon monoxide have been shown to more than double the risk of death and nearly quadruple the incidence of cardiovascular events in otherwise healthy people (Hedblad, Engström, Janzon, Berglund, and Janzon, 2006).

Heat stress can be very taxing on the cardiovascular system. When a firefighter overheats, the heart rate increases in an effort to facilitate cooling. A weak cardiovascular system is not going to be able to keep up with the cooling demands on an extended incident (firefighting campaign) or when operating in the summer months on any emergency scene. (See Figure 6.8.)

More ambiguous but worth consideration is the emergency responder's shift schedule and other daily responsibilities such as psychological stressors. Firefighters operate not only in urgent circumstances but also when sleep deprived and potentially stressed by the nature of the call.

Other Risk Factors

Some risk factors can be controlled by the individual, such as smoking/tobacco use, alcohol use, and drug use. Once identified, such risk factors may require lifestyle changes as well as treatment. Diseases such as diabetes and cancer also are considered risk factors.

Smoking/Tobacco Use. In 2008, an estimated 70.9 million people in the United States aged 12 or older were current (past month) users of a tobacco product (SAMHSA, 2009), even though virtually all recent studies support the notion that cigarette smoking is detrimental to circulatory health. The American Heart Association considers cigarette smoking the most important risk factor for men and women under the age of 50 (AHA, 2010). Cigarette smoking increases the risk of developing several chronic disorders such as atherosclerosis, cancer, and chronic obstructive pulmonary disease (AHA, 2009).

Obviously, smoking is detrimental for a person expected to perform at an emergency scene. As a consequence, some fire departments already prohibit any form of tobacco use on or off duty. Never start smoking. If you already have, do whatever it takes to quit.

Note that the effects of smokeless tobacco are detrimental to the user. They include an increased risk of oral cancer from 28 cancer-causing agents and gum disease (NCI, n.d.).

Alcohol Use. The Centers for Disease Control and Prevention (CDC) reports that certain health problems, such as cirrhosis of the liver, pancreatitis, various cancers, high blood pressure, and psychological disorders, are associated with excessive alcohol use (CDC, 2010a).

A study of the Cincinnati Fire Department showed 29% of the personnel surveyed had scores on the Michigan Alcoholism screen test that suggested problems with alcohol use (Boxer and Wild, 1993). Its statistics are in line with a recent alcohol survey of firefighters from across the country in which 37% reported that a member either missed duty at the fire station or missed a call due to alcohol consumption (Kirk, 2005).

Recently the International Association of Fire Chiefs (IAFC) introduced a "zero tolerance" policy statement regarding the use of alcohol. The policy bans alcohol consumption "by any members of any fire or emergency services agency/organization at any time when they may be called upon to act or respond as a member" (West, 2003).

Drug Use. Unfortunately, it appears that drug abuse has found its way into the fire and emergency services in both career and volunteer departments of all sizes. Drugs impact judgment, reasoning, and physical ability, which can have a direct effect on safety. In 2008 an estimated 22.2 million people were classified as having substance dependence or abuse in the preceding year (SAMHSA, 2009). The study *Drugs in the Workplace* provides data on the relationship between drug use and behaviors that have an impact on the workplace (NIDA, 1989).

Diabetes. In 2003 the NFPA adopted a blanket ban on employment as a firefighter of anyone who takes insulin (Crawford, 2004). Almost 11% of people over the age of 20 have diabetes, which leads to the complications of heart disease and stroke, high blood pressure, and other health-related issues. Unfortunately, almost 2 million individuals are diagnosed with diabetes each year (American Diabetes Association, 2011).

There are two types of diabetes, Type I and Type II. Type I diabetes, which usually occurs in children and young adults, does not allow the body to produce insulin. Type II diabetes is the most common, and either the body does not produce enough insulin or the cells ignore the insulin.

Cancer. The CDC ranks cancer as the second leading cause of death (CDC, 2009). Twenty-six states have firefighter cancer presumption legislation (pending in eight more states) that provides workers with compensation benefits (National League of Cities, 2009). However, coverage is not the same from state to state.

The IAFC Firefighter Cancer Support Network provides information about cancer prevention and screening. The Fire Fighter Cancer Foundation International offers outreach, support, and resource assistance programs.

Research shows that environment, including diet and lifestyle, causes up to 90% of all cancers (Kang, David, Hunt, and Kriebel, 2008). One study (LeMasters et al., 2006) suggests three cancers that are significantly associated with firefighting and designated as a probable risk: multiple myeloma (cancer of bone marrow), non-Hodgkins lymphoma (affecting immune system), and prostate and testicular cancer. The study suggests another

PEARSON
myfirekit

Please visit MyFireKit Chapter 6 for a list of the states with cancer presumption legislation.

six cancers that are possibly associated with firefighting. Another area of concern is breast and cervical cancer, because more women are joining the fire and emergency services every year.

REDUCING RISK FACTORS

Although genetics has been shown to be a significant intrinsic risk factor, it does not mean an individual is doomed if he or she has a family history of a particular condition, nor that an individual is off the hook if he or she does not. Risk factors such as smoking, hypertension, dangerous HDL to LDL cholesterol ratios, impaired fasting glucose, obesity, and a sedentary lifestyle all can be positively impacted with some conscious effort.

The mere act of moving will increase one's chances of survival immensely. Studies have shown that regular physical activity has a positive impact on several of the intrinsic risk factors leading to cardiovascular disorders, including hypertension, serum cholesterol level, fasting glucose, and least surprising of all, obesity (Sharkey, 1997). The IAFF has created the Fit to Survive program (IAFF, n.d., *Fit to Survive*), which helps individuals reduce the chances of developing heart disease. It includes menu planning, portion distortion, stop-drop-control high blood pressure, and smoking cessation.

Nutrition Training and Counseling

Another critical component of good health and fitness is a healthy diet. Eating in the fire station can be challenging if everyone is not on board with a healthier approach to satisfying dietary needs, especially when there is a fire station cook. If one thinks one's fellow crew members hold one's life in their hands on the fireground, consider the power they hold with one's health and well-being in the kitchen of the fire station. (See Figure 6.9.)

Many feel lost when it comes to nutrition. It seems as though every day brings is a new fad diet or opinion on the latest dietary recommendations. Here again is another advantage of having properly trained certified fitness coordinators (CFCs)/peer fitness trainers (PFTs)

FIGURE 6.9 Healthy cooking at the station plays an important role in a firefighter's overall health and wellness. *Courtesy of Travis Ford, Nashville Fire Department*

accessible to the department's membership. **Fitness coordinators** are individuals who are trained in exercise prescription and nutrition management. Although some states may require individuals to be certified nutritionists in order to provide detailed nutritional information, CFCs/PFTs will be able to provide nutritional recommendations and guidance to anyone who needs it. Individuals with extreme nutritional needs, such as those with diabetes, should first consult a registered dietitian or physician.

fitness coordinators ■
Individuals who are
trained in exercise pre-
scription and nutrition
management.

No Smoking Program

Countless studies have proven that smoking kills, both the smokers themselves as well as their families, friends, and coworkers who are exposed to them. It is tragic to see so many fire and emergency services personnel succumb to the addiction to nicotine. To remove an expensive piece of equipment designed to protect your respiratory system during an incident only to light up a cigarette is as incongruous as taking off one's helmet and then hitting oneself on the head with a hammer.

Some fire departments have barred smokers and users of any tobacco product from becoming firefighters. Some departments have implemented policies that can result in termination if an individual is caught smoking. More should do the same. Individuals who smoke should be disqualified from employment, and incumbents who smoke should be given a timeline and assistance to quit with the help of some type of smoking cessation program. Although allowing smoking in the fire station is probably commonplace, if smoking is allowed it should be done only outside of the fire station, including the emergency response vehicle bay areas.

FITNESS STANDARDS

According to FEMA, the fire service has seen a gradual decrease in the number of firefighter injuries both on the fireground and off (USFA, 2008a). Injuries have decreased 16% between the years of 1995 and 2004 overall and 23% on the fireground. That may sound like great news. However, the fire service has seen a significant decrease in the number of fires. When the injury occurrences were tracked on a per-fire basis, the decrease was only 3% (USFA, 2008a).

The National Institute of Standards and Technology (NIST), under the U.S. Department of Commerce, conducted a study in 2002, *The Economic Consequences of Firefighter Injuries and Their Prevention* (NIST, 2005). It reviewed data from the NFPA and categorized the nature of all injuries reported in 2002. Sprains, strains, and muscle pain accounted for almost 50% of all reported injuries. Nearly 35% of all injuries resulted in time off from work.

NIST noted that the costs of firefighter injuries could be staggering. Consider time off with pay; increased worker's compensation insurance premiums; the cost of staffing for the vacancy; and the cost of personnel conducting investigations, data collection and documentation, potential medical retirement, and training of replacement personnel, to name just a few expenses. In comparison to these expenses, the cost of prevention may not be as expensive as initially thought.

In addition to examining the cost of firefighter injuries, NIST discovered that only about 20% of over 26,000 fire departments surveyed maintained basic firefighter fitness and health programs in the United States (NIST, 2005). No one should be satisfied until every career, paid-on-call, and volunteer fire department has a comprehensive, measurable fitness and wellness program.

The University of Illinois in Chicago conducted a study that reviewed over 1,300 worker's compensation claims from fire departments in northern Illinois between 1992 and 1999 (Walton, Conrad, Furner, and Samo, 2003). The results identified overexertion as the cause of injury in 33% of the claims. Of those claims, 49% resulted from lifting. The mean of per-claim costs for those overexertion injuries was $9,715. The cost for overexertion injuries would be much greater today. It is important to note that *overexertion is* a relative term. What may be overexertion to the out-of-condition individual requires minimal effort of the physically prepared. However, on any given day, even the fittest and healthiest firefighter can get punished by performing task on a structure fire.

FITNESS COORDINATORS

Often the membership can feel a bit lost if physical fitness has not been an everyday part of their lives. They can be frustrated, scared, and uncertain about where to start. They need guidance, encouragement, and motivation. Taking the time and money to train several of a department's emergency personnel to be CFCs is a great way to kick off a fitness program.

Having CFCs embedded throughout the department, accessible to all, will provide the needed lifeline to which many of the more concerned can turn. The fitness coordinators can provide needed fitness and wellness information to the department. Granted, the most likely to volunteer to become a fitness coordinator will be the athletes within the department. The key is to provide a diverse pool of trainers so there will be someone whom everyone feels comfortable with turning to for help; but to ensure the credibility of the program, all should be able to pass any test the department is considering.

There are several options for how to certify fitness coordinators. NFPA 1583 provides information in Table A.5.2.1 about professional organizations providing training. Among them is the IAFF/IAFC, which has a course called Peer Fitness Training Certification Program (PFT) (IAFF, 2010) through the American Council on Exercise (ACE). Another option is training through nationally recognized personal trainer certification programs, which provide accredited CFC training through the American Council of Sports Medicine (ACSM). An agency's administrators or fitness program development committee can best determine the particular department's needs.

FITNESS TRAINING AND EQUIPMENT

Once you have the map, you need the means to "get there." Given that CFCs/PFCs are in place, a plan can be developed for each member. All plans, however, must consider the necessity of the time and tools to accomplish the goal. Some fire departments have chosen to implement alternative programs such as education on self-assessments. Fitness challenges, group fitness classes, cross-fit teams, and so on are some excellent alternatives when resources are limited. (See Figure 6.10.)

FITNESS TRAINING SCHEDULES

The best most personnel can hope for is to have some time during a shift to complete a workout before the tones go off. If there is a pattern in a station's run volume, schedule workout time appropriately. Getting a workout completed first thing in the morning is usually most effective. As the day wanes, often so does the desire to start one's workout. Check the vehicle and equipment, clean the station, and get right to the workout.

Fitness Equipment

Rarely does a department have unlimited funds. In most cases the norm is quite the contrary. Money for fitness equipment, however, is money well spent. It does not take a fortune to get the essentials. Start with a rack of dumbbells, a couple of kettle bells, an adjustable bench, a fit ball, and a piece of aerobic equipment, and the department is good to go until next year's budget goes into effect. Yes, the ideal would be to outfit a workout room with all the trick equipment on the market, but that really is not necessary. Where money is lacking, creative solutions can take over. As long as there is gravity, resistance training is available.

One way to ease the burden of providing equipment is to utilize the local YMCA, community centers, private gyms, and collegiate facilities. Some fire departments allow

FIGURE 6.10 Cross-fit training is becoming a popular workout routine. *Courtesy of Martin Grube*

for an exchange of sick time pay to pay for membership fees. Such an exchange can be beneficial on several fronts. The obvious benefit is that crews have access to good-quality equipment and facilities. The benefit to the department is there are no maintenance issues with the equipment, which can add up. (See Figure 6.11.)

Another benefit that may not provide an immediate financial return but can pay off in the long run is public exposure. When emergency response personnel are out in the

FIGURE 6.11 Local facilities provide crews with good-quality equipment. *Courtesy of Martin Grube*

public, striving to improve themselves, working elbow to elbow with the very people they protect, it provides a fabulous opportunity to connect and relate with their customers, and to promote the professional manner of the fire service. It showcases their dedication to career and community. Equally as important to remember, however, is that the exact opposite can be true. Professionalism, courtesy, and consideration need to be demonstrated at all times, just as in any public setting, or more harm than good will be the result.

Responsibility lies with each member to contribute to the effort of health and physical fitness on his or her off time. To rely solely on the time available at work for achieving the level of fitness required to meet the demands of the job could be foolhardy. Members need to take personal responsibility, even if it means utilizing personal time to help maintain their fitness level. With today's fire and emergency services' work schedules, exercising just a few days per month may not provide an adequate level of fitness.

PERFORMANCE STANDARDS

What does support for a department's physical performance test look like? One way to support the program is to rename it physical performance assessment instead. Looking at the whole program should provide a measure of how well the department is doing and recommendations on how to improve. As noted earlier, adhering to the parameters of NFPA 1582 and NFPA 1583 is a great place to start to provide for the physical requirements necessary for departmental personnel. It demonstrates a level of caring for your membership's health and well-being that comes from a different angle than the physical ability test or standard (PAT/PAS). The standard is objective, meaning there is little room for guesswork because it is black and white. It helps keep denial and avoidance out of the missed diagnosis equation. It is potentially lifesaving. Being fit is a component of the assessment, but not necessarily the main focal point.

Medical assessments provide an opportunity to measure potential job-related effects on a member, in addition to screening for any disorders that would limit the ability to function on an emergency scene. Getting baseline measurements for hearing, vision, residual lung capacity, ECG, or diseases detectable with blood or urine sampling gives firefighters a record of their physical status at hiring. Such data could be relevant if a person were to fall victim to a career-ending injury or illness.

Having a measurable means to assess the readiness of a department's membership is just the beginning. Think of the PAT/PAS as the underpinning of the program. Standing alone, it will serve only a fraction of its potential purpose. Build on it with all necessary components to make that test a mere annual formality, and you have success.

A Question/Consideration

Although the WFI program has many good components, some say it is missing the key element to success—time-based performance standards. The controversy lies in the idea that such a test should be given to both incumbents and candidates for the job of firefighter. Should an older experienced firefighter be judged by the same physical standards as a young recruit? Some have suggested that the same test for all could lead to unfair punitive action against incumbents.

Indeed differences exist between an incumbent and a candidate, including training, on-duty injuries, work-related accumulation of stress, and so on. However, what leaders within individual departments must realize is that the rewards of the successful completion of a validated annual performance test are continued employment, improved health, and potentially a longer and more productive life for personnel, even after retirement. That is what makes incorporating *both* entry and incumbent PATs/PASs into the culture of the fire service a necessity. When firefighters really watch each other's backs, they are working to maintain high standards, which in turn breeds dignity, pride, and respect, of both self and the department.

The implementation process is key because if it is too vague, then nobody will apply it. Testing should even be considered for a particular fire department's needs, such as wildland versus structure firefighting, or a combination of both.

Legal Considerations

A discussion of performance test development commonly leads to the topic of discrimination. Doing the right thing and staying out of court are worthy goals for any administrator. In general, the biggest gap between groups of individuals is likely to be between men and women. Age is an issue in incumbent testing. Test results are considered to have disparate impact if 80% of an open class can pass a given test but less than 80% of a protected class cannot pass the same test (U.S. EEOC, Title VII, 1964).

When a PAT/PAS is developed with the intent of filtering candidates based solely on the essential functions of the job and the physical constructs required to execute those functions, chances are that test will withstand a discrimination challenge. The key is to determine the level of physical performance necessary to perform the essential functions of the job and then accurately test for that level of physical ability. Such consideration should protect any department from error and from most lawsuits rooted in Title VII or the Americans with Disabilities Act.

When developing a performance standard for the fire department, evaluate the practical, legal, and scientific realities associated with job-related physical performance testing. It is important to become educated in case law that documents favorable and unfavorable court decisions. Just as one would not go into a medical procedure ignorant of its intricacies, it is vital to approach the departmental PAT/PAS with the same level of ownership. Yours is not the first ship to sail these waters.

The more scientific or statistical justification that can be gathered to support the selected test development model will work in everyone's favor. Consider the logistics of testing when developing departmental standards. The less complicated they are the better, and utilizing existing gear and facilities will save money and heartache. Considering the essential functional requirements of fire suppression and emergency response will help keep the test sound and responsible.

If a test simply discriminates between those physically capable of executing the essential functions of the job and those incapable, the department is on the right track. Will that mean that more men will pass your test than women? Without a doubt. That is just physiology. It will be up to the candidate to do what he or she can to overcome those physical challenges.

In addition, examining the physical fitness constructs that support job performance is very important. What physical abilities will demonstrate that a person can successfully complete the necessary emergency scene functions? Such scrutiny likely translates into actually incorporating non-skill-based fireground activities into the PAT/PAS. Non-skill-based activities need to be considered because, unless the department is requiring previous fire training as a condition of hiring, its officers cannot ask untrained candidates to demonstrate proficiency in skills they have not learned.

To be able to demonstrate the physical ability to complete a given task is only one part of the equation. In order to be effective on the emergency scene, tasks must be performed in a timely manner. Therefore, some time component to the test development process must exist.

UNIVERSAL APPLICABILITY

A department's PAT/PAS could possibly serve it in three ways: as an initial screening for entry-level candidates, as an annual time-based performance standard, and as a return-to-duty performance ability standard. Each one of the tests should vary somewhat

FIGURE 6.12 Physical ability standards help in the initial screening process of entry-level candidates. *Courtesy of Travis Ford, Nashville Fire Department*

so as not to be confused, perhaps by creating certain tasks to test incumbents on critical skills:

- *Initial physical ability standard.* The first use of the PAT/PAS, naturally, is as an initial screening for entry-level candidates. The department must seek out the best of the best or it will be paying for it for years to come. Qualified candidates are out there. Do not dismiss the right to hold candidates to a high physical ability standard. (See Figure 6.12.)
- *Annual performance standard.* Just as firefighters recertify for EMS readiness, hazardous materials, and other job-related certifications, they should be required to re-demonstrate physical proficiency to do the job. Firefighters are expected to perform on the emergency scene in a timely manner; therefore, it makes sense to evaluate that ability with a time-based performance standard on an annual basis to evaluate the proficiency of critical basic skills.
- *Return-to-duty standard.* Just as the fire service runs tests on an engine to ensure that it is suitable for service—and if it is not, it is taken out of service until it is repaired and passes all proper tests—so it should test members who have been out of service for some time. An agency should not release members back into service until a release from the department's physician/medical director (this may not be the member's physician) is obtained and those members pass the PAT/PAS. In addition, when an emergency service provider is put on light duty due to an injury or illness, that individual must demonstrate physical proficiency prior to being allowed back into the field. Trained incumbents should be held to a different and higher standard than entry-level candidates.

Test Development Basics

Typically, tests include gear, tools, or other pieces of equipment. It is wise to weigh departmental equipment and assess the fire department to support decisions to include specific weight substitutions, or height or distance requirements within the test. The more the test is personalized, the more realistic and defensible it will be.

JUSTIFICATION

Collecting data to support test development is critical. Specific tasks incorporated into the final test design need to be validated. Who knows the job better than the individuals who do it every day? Survey 60% to 75% of the membership to ascertain a list of tasks they consider necessary for the successful intervention of any emergency, or just use the IAFF PAT/PAS template that is already a required part of the process in order to gain a license. Such action will produce an extensive list of essential functions. Allowing the membership to rank those functions based on frequency, criticality, and arduousness is a great way to establish a list of tasks for which candidates and incumbents should demonstrate physical proficiency.

Once the department has determined what it is going to test and how much weight is to be carried for the determined distances or heights, it is time to establish a cutoff time for the test. To run incumbents through the test and take the average time is a technique that can be defended only if your membership represents an acceptable level of readiness. Because the test is designed to determine that very thing, allowing the membership to determine the cutoff score is like grading on a curve. The cutoff time needs to reflect acceptable performance, not validate what is already in existence. For that information, Dr. Paul Davis established a double-blind study to help determine the cutoff score objectively through a pacing study (Sharkey and Davis, 2008). Several video demonstrations are recorded to provide a broad sampling of anonymous individuals moving through the course at various speeds. The idea is to get samples equally spaced and dispersed through a reasonable time span.

CURRENTLY ACCEPTED PERFORMANCE STANDARDS

Several excellent publications go into great detail on work-related physical requirements, and an administrator would be well advised to research all options before selecting a test development program for his or her department. Requesting references from the various service providers will allow the department to check on the effectiveness and usability of different programs.

Candidate Physical Ability Test

The WFI's candidate physical ability test (CPAT) is part of the IAFF/IAFC Joint Labor Management Initiative (2009b). The CPAT tests entry-level applicants, who are usually in their prime, to measure their physical ability to enter recruit school. More than 900 fire departments are currently licensed to use it (IAFF, n.d., *CPAT Licensing*).

CPAT is the only internationally recognized test created by both labor and management and overseen by the Department of Justice during development. No other candidate physical ability test has withstood the extent of its review. The U.S. Equal Employment Opportunity Commission has agreed not to bring a lawsuit through April 2011 based upon any claim that the CPAT has an adverse impact for women candidates against any fire department that utilizes CPAT (IAFF/IAFC, n.d., p. 18). The test has the backing of the IAFF and IAFC, which gives a level of assurance to fire departments that it will be maintained by the IAFF and IAFC, its originators.

The CPAT requires an individual to wear a 50-pound weighted vest during which the individual simulates the following:

- *Stair climb.* Climbing stairs with a 25-pound simulated hose pack.
- *Hose drag.* Stretching and advancing hose lines, charged and uncharged.
- *Equipment carry.* Removing and carrying equipment from fire apparatus to fireground.
- *Ladder raise and extension.* Placing and raising a ground ladder to the desired floor or window.
- *Forcible entry.* Penetrating a locked door and breaching a wall.

- **Search.**   Crawling through dark areas to find victims.
- **Rescue drag.**   Removing a victim from a fire building.
- **Ceiling pull.**   Checking a ceiling to locate fire extension.

Each individual component has individual failure points because speed causes mistakes, and mistakes on the emergency scene can cause firefighter death and injuries.

Other Testing Models

Another recognized test development model, for both candidates and incumbents, was designed by Dr. Paul Davis. His model has been challenged and prevailed in courtrooms at nearly every level. The key to the model's success is that it evaluates several key elements to ensure a test has relevance to a specific department's work requirements and environment. (See Figure 6.13.)

Dr. Paul Davis has defined 10 employment axioms (Sharkey and Davis, 2008), which hold great relevance in the fire and emergency services:

1. Employment in emergency services is not a right; it is a privilege.
2. Although largely sedentary, the job does have, at times, profoundly demanding physical requirements.
3. The physical demands of the emergency scene define the activities and the response of the emergency responder.
4. The physically fit firefighter has a greater probability of success than his less fit counterpart (i.e., more is better).
5. Physical fitness is an attribute that, if not maintained, degrades over time.
6. There are no prohibitions against applicants or incumbents improving their physical fitness commensurate with the requirements of the job.
7. No physical ability test is perfect.
8. There is no such thing as a physical ability test that does not have adverse impact.
9. The fact that a test has an adverse impact does not make it illegal.
10. There can be no physical fitness without a physical training program.

REHAB STANDARDS

Remember, even healthy, fit firefighters can get injured or become exhausted on the emergency scene. It is critical to monitor fire crews during and after an emergency operation. Even in the controlled environment of the drill ground, emergency personnel may experience everything from mild dehydration to cardiac arrest. If the potential for these types of problems is present in a controlled environment of the drill ground, imagine adding the chaotic, unpredictable, and unforgiving elements of the emergency scene. (See Figure 6.14.)

Comprehensive Emergency Rehabilitation Program

Establishing emergency personnel rehab on the emergency scene can save lives. It can be as simple as a shady or cool place to go for a few minutes to grab a snack and a drink to the full rehabilitation area supported by Red Cross or local community agencies, such as community fire department volunteers known as fire buffs. Agencies should make efforts to provide healthy snacks and drinks when they respond to the emergency scene. A key element of rehab is a couple of qualified individuals who can monitor and record vitals on all members involved in the emergency operation. Individuals returning to the hazard zone with undetected abnormalities in their vital signs represent missed opportunities to mitigate a potential disaster. However, NFPA 1584 does not mandate vital sign measurement, and some departments may not use it as a tool for rehab.

Again, take into consideration that given half a chance an individual will go so far as to deny chest pain to avoid being embarrassed or humiliated in front of peers. Individuals working in rehab must be strong enough to be able to make the tough decision to take someone out of service if abnormalities are found. The responsibility to remove an individual from the emergency scene should rest with the safety officer or incident commander. However, everyone is ultimately responsible for being properly rehabbed, and

FIGURE 6.14 A key element of rehab is having local fire buffs provide a much needed service on the emergency scene.
Courtesy of Daniel A. Nelms, Emergency Service Training

FIGURE 6.15 Taking vital signs on the emergency scene is an important part of being properly rehabbed. *Courtesy of Martin Grube*

only under certain situations should the incident commander get involved in requiring an individual to be rehabbed. (See Figure 6.15.)

Doing this is much more difficult than one may think. Because on-scene rescuers often are friends, one having to send another to the hospital from the emergency scene for further testing can be uncomfortable. Could you send a responder to the hospital after he begged you to let him finish up the incident and promised to get to his doctor as soon as possible? Such an occurrence would reveal a firefighter's true toughness, because only a very tough individual, grounded in the principle of doing the right thing, would follow through and keep that responder from returning to the scene.

IMPLEMENTATION

All in all, current programs provide a great platform from which to work. The two most powerful organizations in the fire service—the IAFF and IAFC—have moved fitness and health-related issues in the fire service into the spotlight. Working toward monitoring medical, fitness, performance, and rehab standards as the norm is in its incipient phase. However, the effectiveness of future fitness programs will rely on the support of the IAFF and IAFC.

Making the Commitment

The fire and emergency services needs all of its governing agencies to make a commitment that goes beyond voluntary fitness programs. It requires a valid means to make those very difficult human resources decisions when safety takes precedence over sparing someone's feelings. However, individuals should know the expectations and consequences beforehand and be provided adequate retraining and rehabilitation to meet the minimum standard. The fire and emergency services needs to take that final step and commit to NFPA 1500 completely, and NFPA 1582 and 1583 must include clear direction and criteria for the medical and fitness testing process. Although fire departments should recognize that the standards exist for them, it is still their responsibility to implement those standards.

Every fire department in the United States should be adopting the fitness and medical criteria defined in NFPA 1500, NFPA 1582, and NFPA 1583. To many, such adoption seems obvious, because the membership have become occupational athletes. As occupational athletes, both the lives of the public that the fire and emergency services is sworn to protect and firefighter and emergency responder lives require the membership to be ready, willing, and able to meet the next challenge and make a difference.

Setting Up for Success

Ideally, the importance of implementing a proper fitness program is apparent to all stakeholders. Unfortunately, to actually arrive at the moment when standard operating procedures/guidelines (SOPs/SOGs) are in place and implementation of a program ensues is usually a long fought battle.

Support for such a program can be earned. It will just take time. The success of the program, however, will depend on the buy-in, support, and participation of all department personnel. Although success may be difficult to accomplish, never lose sight of the fact that the battle is worth fighting because it will save lives.

Listen, reason, discuss, and apply all possible aspects of logic to make a case. The battle will be significantly more difficult if the leadership cannot get the membership to understand the far-reaching significance of a shift in thinking. Know that the position is sound, but not inflexible. There is a lot of room for flexibility during the planning and implementation phase, which is a great time to gather suggestions and for both labor and management to develop ownership in the project.

In order to increase the odds of success for a fitness program, full transparency and inclusiveness of all stakeholders or their representatives are recommended. That means hashing out the details of a program with all parties who either will be impacted by its application or could potentially have a say in how it is executed or developed.

The human resources division will be among the stakeholders, and it should agree that the uniqueness of fire and emergency services must be reflected in the process of filling and maintaining the operational positions on the force with healthy, capable individuals. From a risk management perspective, anything short of a multistage evaluation of all candidates with an emphasis on cognitive skills, basic academic aptitude, judgment, written and oral communication skills, and physical ability would be creating a potentially dangerous situation for any fire department. The logic can be extended to the testing of the physical ability of incumbents as well.

City attorneys and risk managers are among the stakeholders who should be brought into the loop. They will recognize the gravity of getting it right the first time. A good risk manager understands the relationship between prevention and possible loss through injury or death. His or her job is to save the municipality money. Most would say that is an easier task if approached from a prevention perspective or from the offensive side as opposed to the defensive side.

Prevention of opposition is critical, because the department could be met with opposition from a union or labor association. A proactive approach is a good idea to get the union or labor association involved early as well. Win its support first, before proceeding to invest time and effort into the program. Without the support of labor, the department is likely wasting its time.

Have a plan on how to promote the program to nonbelievers. In order to sell a plan, have one. Bar none, the most skeptical and toughest sell will be to the incumbent firefighters. They will experience the most profound personal impact of a wellness-fitness program, as well as feel threatened and, in a way, helpless. Change will be happening to them unless department leaders intervene and make them a part of the change. First and foremost, listen. Not so much to the irrational justifications of why not to implement the program, but more to incumbent firefighters' genuine concerns about its implementation. Make it sound as though the program is intended for their health and well-being, not just to weed out those who are not physically fit.

Typically the fear lies in perceived, imminent failure. Incumbent firefighters are certain that whatever the test ends up being, it will be some grueling event, designed for the young and perfectly fit. They will be justifying their defiance of the program with statements such as, "They can't expect us to be able to do what the young kids can do." The testing process should be directed to the minimum standard, and the physicals should address the underlying health issues only. If an individual cannot meet that minimum standard, then that person is a danger to him or herself, his or her peers, and the public. However, a perfectly fit person that lacks experience to properly perform and make decisions is a danger. The public should expect nothing less than someone who is healthy, fit, and capable to perform from the fire departments they support through tax dollars.

Timelines will be critical. The process of implementation should take upward of two to three years. It is important to ensure that the first time the membership sees the department's performance standard the only expectation will be to walk through it. Just having the opportunity to touch the equipment and walk through the course without an expectation to perform can take the bite out of the fear of the unknown.

The following year, the expectation could be completing the course within 30 to 45 seconds of the established cutoff time, for example. Such completion should be attainable for most. There can be a great deal of improvement realized within a year's time as long as each member is doing his or her part as well. Fitness training can be considered for those not meeting the standard. Peer fitness trainers can assist with exercise regiments, diet, and so on.

Not until the third run-through would the expectation be that all members whose job descriptions include operating on the emergency scene complete the test in the allotted amount of time. It is an administrative decision about how the process unfolds, but even at this point, if there are members who are missing the mark, but have made measurable improvements over the implementation period, there may be consideration for that improvement. Eventually, and sooner rather than later, all members engaging in emergency scene activities will have to meet the standard.

Naturally, incumbent firefighters are fearful of losing their jobs, a reasonable fear. Once a department establishes a standard, when individuals are not meeting it, it has a duty to act. The liability is too great not to. Does that mean firing someone the first time out? Absolutely not. There should be very little job impact in the implementation process. Be looking for individuals' willingness and proof of participation, along with improvement. Assuring people that implementation will be gradual and well supported should help alleviate some of the anxiety. Departments can offer remediation to those who are not meeting the mark.

CHAPTER **REVIEW**

Summary

The health and well-being of the nation's firefighters depends on addressing the problem of line-of-duty deaths and injuries at the local fire department level. Everyone in the fire and emergency services should strive to develop and implement a total health and wellness program.

The expectation to perform is ever present in the fire and emergency services. The time has come to decide whether to evaluate an individual's ability to meet those expectations in a controlled environment or leave them to chance. Many look at fitness standards in the fire service as something that the department does "to" them. In fact, a health and fitness program is the greatest gift an administration can give to its membership. It is the gift of health, longevity, support, understanding, education, productivity, and life.

There is only one way to ensure that every member of the department is in that ready state, and that is through an annual physical ability test supported by a health and fitness program. After all, it is the same job with the same safety issues for all members, regardless of age or gender, with the same expectations, and the same demands.

Review Questions

1. According to NFPA 1500, how often should firefighters be tested for physical competency?
2. What are the five components of fitness that should be evaluated on an annual basis?
3. What are four intrinsic cardiac risk factors that can be positively impacted by exercise?
4. What are the five main components of a complete fitness program?
5. Of all the intrinsic cardiac risk factors, which one can be eliminated immediately and have a significant impact on the health and well-being of an individual?

References

American College of Sports Medicine. (2006). *ACSM's Guidelines for Exercise Testing and Prescription* (7th ed.). Philadelphia: Lippincott Williams & Wilkins.

American Diabetes Association. (2011). *Diabetes Statistics*. Retrieved February 1, 2011, from http://www.diabetes.org/diabetes-basics/diabetes-statistics/?utm_source=WWW&utm_medium=DropDownDB&utm_content=Statistics&utm_campaign=CON

American Heart Association. (2009). *Learn and Live: Cigarette Smoking and Cardiovascular Diseases*. Dallas, TX: Author. Retrieved June 8, 2009, from http://www.americanheart.org/presenter.jhtml?identifier=4545

American Heart Association. (2010). *Cigarette Smoking and Cardiovascular Diseases*. Dallas, TX: Author. Retrieved March 15, 2010, from http://www.americanheart.org/presenter.jhtml?identifier=4545

Boxer, P. A., and Wild, D. (1993, April). "Psychological Distress and Alcohol Use Among Firefighters." *Scandinavian Journal of Work, Environment, and Health*, 19(2), 121–125.

Brown, J., and Stickford, J. (n.d.). *Physiological Stress Associated with Structural Firefighting Observed in Professional Firefighters*. Retrieved June 10, 2010, from http://www.indiana.edu/~firefit/pdf/Final%20Report.pdf

Centers for Disease Control and Prevention. (2009, March 6). *Leading Causes of Death*. Retrieved October 15, 2010, from http://www.cdc.gov/nchs/fastats/lcod.htm

Centers for Disease Control and Prevention. (2010a, July 20). *Alcohol and Public Health: Frequently Asked Questions*. Atlanta, GA: Author. Retrieved October 15, 2010, from http://www.cdc.gov/alcohol/faqs.htm#16

Centers for Disease Control and Prevention. (2010b, September). *Overweight and Obesity: U.S. Obesity Trends, Trends by State, 1985–2009*. Retrieved October 12, 2010, from http://www.cdc.gov/obesity/data/trends.html

Crawford, B. A. (2004, May 1). "Sorry, Sugar." *Fire Chief*. Chicago, IL: Penton Media Inc. Retrieved March 15, 2010, from http://firechief.com/health-safety/fitness/firefighting_sorry_sugar/index.html

Hales, T. (2008, October 2). "Heart Disease in the Fire Service: Cause for Concern." Paper presented at the PERI virtual conference. Retrieved March 14, 2010, from http://www.riskinstitute.org/peri/images/file/S908-D9-Hales.pdf

Hedblad, B. O., Engström, G., Janzon, E., Berglund, G., and Janzon, L. (2006). "COHb% as a Marker of Cardiovascular Risk in Never Smokers: Results from a Population-Based Cohort Study." *Scandinavian Journal of Public Health, 34*(6), 609–615. Retrieved March 11, 2010, from http://sjp.sagepub.com/cgi/content/abstract/34/6/609

International Association of Fire Fighters. (n.d.). *CPAT Licensing.* Washington, DC: Author. Retrieved May 13, 2010, from http://iaff.org/hs/well/cpatlicensestory.htm

International Association of Fire Fighters. (n.d.). *Fit to Survive: The Fire Fighter's Guide to Health and Nutrition.* Washington, DC: Author. Retrieved March 17, 2010, from http://www.iaff.org/hs/FTS/ftsdefault.asp

International Association of Fire Fighters (IAFF). (2009a). *Fire Service Joint Labor Management Wellness-Fitness Initiative.* Retrieved September 15, 2009, from http://www.iaff.org/HS/Well/wellness.html

International Association of Fire Fighters (IAFF). (2009b). *Wellness/Fitness Initiative (WFI): Fire Service Joint Labor Management Wellness-Fitness Task Force Candidate Physical Ability Test Program Summary.* Retrieved June 8, 2009, from http://www.iaff.org/HS/CPAT/cpat_index.html

International Association of Fire Fighters. (2010). *PTF Candidate Information Guide.* Retrieved February 1, 2011, from http://www.iaff.org/hs/PFT/PFT%20Candidate%20Guide.htm

International Association of Fire Fighters and International Association of Fire Chiefs. (n.d.). *Candidate Physical Ability Test (CPAT)* (2nd ed.). Washington, DC: Author.

Kales, S. N., Soteriades, E. S., Christophi, C. A., and Christiani, D. C. (2007). "Emergency Duties and Deaths from Heart Disease Among Firefighters in the United States." *The New England Journal of Medicine, 356*(12), 1207–1215. Retrieved August 10, 2009, from http://content.nejm.org/cgi/content/full/356/12/1207

Kang, D., David, L. K., Hunt, P., and Kriebel, D. (2008, February 27). "Cancer Incidence Among Male Massachusetts Firefighters, 1987–2003." *American Journal of Industrial Medicine, 51*, 329–335.

Kirk, J. (2005, June 1). *Results of Online Alcohol Survey Released.* Reston, VA: International Association of Fire Chiefs. Retrieved March 15, 2010, from http://www.iafc.org/displayindustryarticle.cfm?articlenbr=27233

LeMasters, G. K., Genaidy, A. M., Succop, P., Deddens, J., Sobeih, T., Barriera-Viruet, H., et al. (2006). "Cancer Risk Among Firefighters: A Review and Meta-analysis of 32 Studies." *Journal of Occupational and Environmental Medicine, 48*(11), 1189–1202. Retrieved March 15, 2010, from http://www.firerehab.com/Columnists/Mike-McEvoy/articles/327047-Making-Rehab-a-Requirement-NFPA-1584/

McEvoy, M. (2007, March 3). *Making Rehab a Requirement: NFPA 1584.* Firerescue1.com. Retrieved May 13, 2010, from http://www.firerehab.com/Columnists/Mike-McEvoy/articles/327047-Making-Rehab-a-Requirement-NFPA-1584/

Moore-Merrell, L., Zhou, A., McDonald, S., Fisher, E., and Moore, J. (2008, August 7). "Contributing Factors to Firefighter Line-of-Duty Deaths in the United States." Washington, DC: IAFF. Retrieved May 17, 2010, from http://www.iaff.org/08News/PDF/InjuryReport.pdf

National Cancer Institute. (n.d.). *Smokeless Tobacco and Cancer.* Retrieved November 15, 2009, from http://www.cancer.gov/cancertopics/factsheet/Tobacco/smokeless#r1

National Fire Protection Association. (2007a). *NFPA 1500: Standard on Fire Department Occupational Safety and Health Program.* Retrieved May 17, 2010, from http://www.nfpa.org/aboutthecodes/AboutTheCodes.asp?DocNum=1500

National Fire Protection Association. (2007b). *NFPA 1582: Standard on Comprehensive Occupational Medical Program for Fire Departments.* Retrieved May 17, 2010, from http://www.nfpa.org/aboutthecodes/AboutTheCodes.asp?DocNum=1582&cookie%5Ftest=1

National Fire Protection Association. (2008a). *NFPA 1583: Standard on Health-Related Fitness Programs for Fire Department Members.* Retrieved May 17, 2010, from http://www.nfpa.org/AboutTheCodes/AboutTheCodes.asp?DocNum=1583

National Fire Protection Association. (2008b). *NFPA 1584: Standard on the Rehabilitation Process for Members During Emergency Operations and Training Exercises.* Retrieved May 17, 2010, from http://www.nfpa.org/AboutTheCodes/AboutTheCodes.asp?DocNum=1584&cookie%5Ftest=1

National Institute for Occupational Safety and Health. (2007, June). *Preventing Fire Fighter Fatalities Due to Heart Attacks and Other Sudden Cardiovascular Events.* NIOSH Publication No. 2007-133. Cincinnati, OH: NIOSH—Publications Dissemination. Retrieved August 1, 2009, from http://cdc.gov/niosh/docs/2007-133/#background

National Institute of Standards and Technology. (2005, March). *The Economic Consequences of Firefighter Injuries and Their Prevention: Final Report* (NIST GCR 05-874). Arlington, VA: TriData Corporation. Retrieved August 10, 2009, from http://www.fire.nist.gov/bfrlpubs/NIST_GCR_05_874.pdf

National Institute on Drug Abuse. (1989). *Drugs in the Workplace: Research and Evaluation Data.* NIDA Research Monograph, Number 91. Bethesda, MD: Author. Retrieved March 15, 2010, from http://www.drugabuse.gov/pdf/monographs/download91.html

National Institute on Drug Abuse. (2009, December 3). *Tobacco/Nicotine*. Bethesda, MD: Author. Retrieved March 15, 2010, from http://www.drugabuse.gov/DrugPages/Nicotine.html

National League of Cities. (2009, April). *Assessing State Firefighter Cancer Presumption Laws and Current Firefighter Cancer Research*. Retrieved March 15, 2010, from http://www.nlc.org/ASSETS/CEB0ADD3BBD646CF917926525FA87FD7/FFPresumptionReport.pdf

National Volunteer Fire Council. (2007). *Keep It Strong: NVFC Heart-Healthy Firefighter Resource Guide*. Retrieved September 15, 2009, from http://healthy-firefighter.org/files/documents/2005resourceguide.pdf

National Volunteer Fire Council. (2008). *Fired Up for Fitness Challenge*. Retrieved September 15, 2009, from http://www.healthy-firefighter.org/atp/

Sharkey, B. (1997). *Fitness and Health* (4th ed.). Champaign, IL: Human Kinetics Publishing.

Sharkey, B. J., and Davis, P. O. III. (2008). *Hard Work: Defining Physical Work Performance Requirements*. Champaign, IL: Human Kinetics Publishing.

Substance Abuse and Mental Health Services Administration. (2009). *Results from the 2008 National Survey on Drug Use and Health: National Findings*. Office of Applied Studies, NSDUH Series H-36, HHS Publication No. SMA 09-4434. Rockville, MD: Author. Retrieved March 15, 2010, from http://www.oas.samhsa.gov/nsduh/2k8nsduh/2k8Results.cfm

Tsismenakis, A. J., Christophi, C. A., Burress, J. W., Kinney, A. M., Kim, M., and Kales, S. N. (2009, March 19). "Epidemiology. The Obesity Epidemic and Future Emergency Responders." *Obesity: A Research Journal*. Retrieved March 14, 2010, from http://www.nature.com/oby/journal/vaop/ncurrent/full/oby200963a.html

U.S. Equal Employment Opportunity Commission. (1964, July 2). Title VII of the Civil Rights Act of 1964. Public Law 88-352, 78 Stat. 241. Retrieved June 11, 2009, from http://www.eeoc.gov/laws/statutes/titlevii.cfm

U.S. Fire Administration. (2006). *Four Years Later: A Second Needs Assessment of the U.S. Fire Service*. FA-303. Emmitsburg, MD: Department of Homeland Security/USFA. Retrieved March 14, 2010, from http://www.usfa.dhs.gov/fireservice/grants//needs-assess2.shtm

U.S. Fire Administration. (2008a, February). *Fire-Related Firefighter Injuries in 2004*, p. 3. Retrieved August 10, 2009, from http://www.usfa.dhs.gov/downloads/pdf/publications/2004_ff_injuries.pdf

U.S. Fire Administration. (2008b, February). *A Profile of Fire in the United States 1995–2004* (14th ed.). FA-315. Emmitsburg, MD: Department of Homeland Security/USFA. Retrieved August 10, 2009, from http://www.usfa.dhs.gov/downloads/pdf/publications/fa_315.pdf

U.S. Fire Administration (2008c, March). *Emerging Health and Safety Issues in the Volunteer Fire Service*. FA-317. Emmitsburg, MD: Department of Homeland Security/USFA. Retrieved September 15, 2009, from www.usfa.dhs.gov/downloads/pdf/publications/fa_317.pdf

U.S. Fire Administration. (2009). *A Provisional Report: On-Duty Firefighter Fatalities in the United States*. Retrieved August 10, 2009, from http://www.usfa.dhs.gov/downloads/txt/08-fatality-summary.txt

U.S. Fire Administration and the National Volunteer Fire Council. (2008). *Health and Wellness Guide for the Volunteer Fire and Emergency Services*. Emmitsburg, MD: Author.

Walton, S. M., Conrad, K. M., Furner, S. E., and Samo, D. G. (2003, April). "Cause Type, and Workers' Compensation Costs of Injury to Firefighters." *American Journal of Industrial Medicine, 43*(4), 454–458.

West, P. (2003, September 4). "IAFC Adopts 'Zero-Tolerance' Alcohol Policy." *Fire Chief*, Chicago, IL: Penton Media Inc. Retrieved March 15, 2010, from http://firechief.com/news/firefighting_iafc_adopts_zerotolerance/

Yoo, L., and Warren, F. (2009, August). "Prevalance of Cardiovascular Disease Risk Factors in Volunteer Firefighters." *Journal of Occupational and Environmental Medicine, 51*(8), 958–962. Retrieved November 15, 2009, from http://journals.lww.com/joem/Abstract/2009/08000/Prevalence_of_Cardiovascular_Disease_Risk_Factors.14.aspx

7

Data Collection and Research

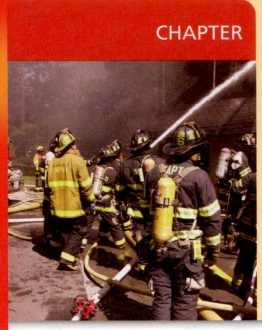

Charles R. Jennings

Courtesy of Anthony Drago, White Plains Department of Public Safety

KEY TERMS

market failure, *p. 163*

peer review, *p. 157*

policy, *p. 168*

research, *p. 157*

OBJECTIVES

After reading this chapter, the student should be able to:

- Understand the history of the fire and emergency services safety problem.
- Identify institutional factors that make fire and emergency services research difficult.
- Recognize the importance of historical research to current levels of firefighter safety.
- Know the most common causes of fire and emergency services fatalities, and the systems and definitions used to track them.
- Identify policy research and what distinguishes it from fire protection engineering or fire science.
- Explain the role of research and data in improving firefighter safety.
- Know the sources and uses of national data on firefighter safety.
- Identify fire and emergency services–academic partnerships and collaborations and their funding sources.
- Identify emerging topics for future research.

PEARSON

myfirekit™

For practice tests and additional resources, visit **www.bradybooks.com** and follow the **MyBradyKit** link to register for book-specific resources.

Register for **MyFireKit** by following directions on the **MyFireKit** student access card provided with this text. If there is no card, go to **www.bradybooks.com** and follow the **MyBradyKit** link to Buy Access from there.

What Is Research?

The term *research* is perceived by many firefighters as something remote and irrelevant to the daily concerns of the fire and emergency services. But that is not true. There is a strong relationship between research and firefighter safety. But to describe it, one must first understand what research is and know about the requirements for the development of a research agenda.

Research is defined in the *Merriam-Webster Online Dictionary* as "investigation or experimentation aimed at the discovery and interpretation of facts, revision of accepted theories or laws in the light of new facts, or practical application of such new or revised theories or laws." It is a systematic activity that includes documentation of results. Although images of laboratory coats and test tubes might come to mind, research can be done in the field. It may involve the hard sciences, such as medicine or chemistry, as well as the social sciences, including policy analysis, organizational behavior, human factors, and even interview-based techniques. Data sources such as incident or injury reports and census data can be used as the basis for research, particularly to reveal national trends or problems.

Singleton and Straits (2004) list the four primary approaches to research:

- Experiments
- Surveys
- Field research
- Analysis of secondary data (such as census information or other agency reports)

Each of these approaches may be appropriate for different aspects of a research question. For example, if someone wanted information on emergency vehicle driver behavior, he or she could run experiments to simulate conditions in a laboratory setting, conduct a survey of emergency vehicle drivers, observe emergency vehicles being driven to actual responses, or analyze accident statistics and training records. Ideally, the same question may be addressed by multiple researchers using differing approaches, which can lend greater certainty to findings.

What distinguishes research from more casual inquiries are the attention to a well-defined methodology, a formal presentation of results, and an unbiased evaluation of both the methodology and the resulting conclusions. To achieve an unbiased evaluation, the research report must be submitted before publication for **peer review** by recognized experts who were not directly involved in the research.

Research is a neglected topic in the fire and emergency services, but it should not be. Learning everything about the history of line-of-duty deaths and the ways firefighters work both in the station house and in the field can play an important role in keeping today's responders safe. However, because that information is not often looked at from a research point of view, the lack of high-quality information about firefighter deaths and injuries is a major barrier to improving fire and emergency services safety today.

Yes, research is a neglected topic in the fire and emergency services. Yet when actions that may be as simple as filling out an incident report are critical to ensuring the safety of all personnel, is it not time to change? However, before anything can change, all firefighters—novices as well as experienced personnel—must first become aware of the need for research. Without the support of fire and emergency services personnel, there will be little chance of more and higher quality data being recorded, collected, and drawn upon for the answers to the safety problems associated with the job.

research ■ Investigation or experimentation aimed at the discovery and interpretation of facts, revision of accepted theories or laws in the light of new facts, or practical application of such new or revised theories or laws.

peer review ■ A process in which research is subjected to review by outside experts before publication.

The Safety Problem

According to the U.S. Fire Administration (USFA, 2009c, p. 6), there was an average of 118 firefighter fatalities annually in the United States from 1977 to 2008. The number drops to 106 if the September 11, 2001, attacks are excluded. The number of fatalities

has declined from the early years when reliable records were kept, but has tended to plateau at around 100 per year since 1990. (See Figure 7.1.)

Fireground operations is the most common activity at the time of death (USFA, 2009c, p. 13), followed closely by responding to or returning from alarms, other on-duty activities (such as inspections or administrative work), training, non-fire emergencies (EMS or other service calls), and the period immediately after return from an alarm. The activity at the time of death for 2008 is shown in Table 7.1.

The types of activity and causes of injury at time of death have remained fairly unchanged over time. However, when deaths per fire are examined, a troubling finding emerges. Deaths per fire have actually trended upward in the last several years. Firefighter deaths per fire have declined in the last two years, but are still higher than they were a few years ago. There has been no consistent pattern of decline. (See Figure 7.2.)

Why should there be a *consistent decrease* in deaths? Because over the last 30 years, tremendous investments have been made in equipment, training, and policy to make firefighting safer. Despite better equipment, legal mandates, new national standards, and an increased emphasis on safety, the number of fatalities per fire has not decreased consistently.

The National Firefighter Near-Miss Reporting System summarizes the same concern by observing that the number of deaths and lost-time injuries has not decreased in spite

TABLE 7.1	Activity at Time of Death in 2008
TYPE OF DUTY	**NUMBER OF DEATHS**
Fireground operations	28
Responding/returning	24
Other on-duty activities	30
Training	12
Non-fire emergencies	11
After an incident	13

Source: USFA, *Firefighter Fatalities in the United States in 2008* (Emmitsburg, MD: U.S. Fire Administration/FEMA, September 2009), p. 13.

FIGURE 7.2 There has been no consistent pattern of decline based on the current number of structure fires versus line-of-duty deaths and injuries.
Courtesy of Captain Terry L. Secrest

of significant advances in protective equipment and apparatus (National Firefighter Near-Miss Reporting System, n.d.).

CHANGING DEFINITIONS OF LINE-OF-DUTY DEATH

Differences in how words are defined—such as the terms *line-of-duty death* and *on-duty death*—can complicate research analysis and must always be taken into consideration. For example, two organizations that track fire and emergency services deaths are the U.S. Fire Administration (USFA) and the National Fire Protection Association (NFPA). However, the National Institute for Occupational Health and Safety investigates and gathers data based on its own definitions. The information is used by other organizations, including the National Fallen Firefighters Foundation (NFFF), the International Association of Fire Chiefs (IAFC), and the International Association of Fire Fighters (IAFF).

The U.S. Department of Justice (DOJ) collects information related to fire and emergency services deaths when a death benefit is claimed. For the DOJ, the term *on-duty death* is more accurate than *line-of-duty death*. That is because the Hometown Heroes Survivors Benefits Act of 2003 (2003) uses the term *on-duty death* to ensure that a firefighter who suffers a fatal heart attack or stroke while on duty qualifies for survivor benefits, just like the firefighter who suffers a fatal injury.

> Previous to implementation of the Hometown Hero's Act in, December 15, 2003, [fire-fighters] who became ill as the result of a heart attack or stroke after going off duty needed to register some complaint of not feeling well while still on duty in order to be included in this study. For [firefighter] fatalities after December 15, 2003, [firefighters] will be included if they become ill as the result of a heart attack or stroke within 24 hours of a training activity or emergency response. For [firefighters] who become ill after going off duty—if the activities while on duty were limited to tasks that did not involve physical or mental stress—will not be included in this study. (USFA, 2009c)

Hence, the total number of fire and emergency services deaths that have been counted since the act's implementation has increased, not because there have been more deaths but because of the changed definition. The USFA is currently using the Hometown Heroes Act definition, although it notes the numbers of deaths under its old and expanded criteria. The NFPA continues to use its methodology, which requires a complaint to be lodged before going off duty.

Both the NPFA and USFA count every death that occurs while on duty. These are categorized as deaths that occur after an incident or as deaths that follow a release from duty, the latter usually as a result of cardiovascular causes. Criteria the NFPA and USFA use for counting a death may be summarized as follows:

- *Hometown Heroes Act.* Any cardiovascular-related death within 24 hours of being "on-duty" if any nonroutine stressful or strenuous activity was part of that duty.
- *NFPA.* Deaths must be clearly linked to on-duty status and directly related to an incident or other activity that occurred on duty. Each case is reviewed for consistency.

Physical fitness has been identified as a contributing factor to firefighter on-duty deaths. The National Volunteer Fire Council (NVFC) Heart Healthy Initiative screened career and volunteer firefighters at several national events in 2006. Its screenings found that 30% had hypertension, average cholesterol levels were near or above 200, and obesity was a concern as well. Thirty percent of those firefighters screened had hypertension, and another 47% were considered to be in the early stages of developing high blood pressure (NVFC, 2010). Those limited data and other studies suggest that the health and physical fitness of a significant proportion of the fire and emergency services community is poor. An NFPA study found that well over half of those victims for which an autopsy or other data were available had prior heart-related conditions or had arteriosclerosis occluding at least 50% of arterial capacity (Fahy, 2005, p. 2).

The NVFC Heart Healthy Initiative has been credited by many firefighters as the key to understanding their personal level of risk and moved them to adopt a more heart-healthy lifestyle. It is hoped that as the program and its aims are more widely adopted in the fire and emergency services, a decrease in injuries and deaths related to cardiac-related factors will be seen.

CHRONIC THREATS TO HEALTH

In addition to the acute health threats facing the fire and emergency services are a number of chronic conditions that may pose a threat to health. These threats include diesel exhaust exposure, exposure to fire contaminants at scenes and on protective gear, medical exposures from EMS calls, chronic stress, critical incident stress, and sudden waking.

Stress

Psychological stress has received limited study among the fire and emergency services community. Effects of exposure to incident-related stress have been shown to contribute to depression, substance abuse, and anxiety (Bacharach, Bamberger, and Doveh, 2008; Boxer and Wild, 1993; and Mendez, Molloy, and Magaldi, 2006). No systematic data are available to show whether or not the possibility of a greater than average number of suicides exists in the fire and emergency services, but such concerns should be examined in light of anecdotal reports from allied professions. Psychological screening—which could identify personnel who are at greater risk of experiencing difficulties—is not a routine part of the hiring process in many organizations. (See Figure 7.3.)

Sudden Waking

The International Association of Fire Chiefs (IAFC) is currently conducting a study on sudden waking and its effects on the health of firefighters. Waking suddenly from sleep is

FIGURE 7.3 Fire and emergency services can be stressful both physically and mentally. *Courtesy of Anthony Drago, White Plains Department of Public Safety*

believed to have the potential to cause long-term damage to the heart. Several studies have shown that firefighters can experience high levels of physical and psychological stress during performance of fire or emergency scene duties (Angerer, Kadlez-Gebhardt, Delius, Raluca, and Nowak, 2008; Kales, Soteriades, Christophi, and Christiani, 2007; and Womack, Green, and Crouse, 2000).

DIFFERENT DEPARTMENTAL SETTINGS

According to the USFA, there are over 30,000 fire and emergency services departments in the United States (USFA, 2009b). Those departments vary from rural single stations to large urban departments with thousands of personnel. There are over 1.1 million career, volunteer, paid-on-call, and combination firefighters (USFA, 2009b). Estimating the number of firefighters is difficult because local standards vary so greatly, as do standards for membership. Local government expenditures for fire and emergency services were $25.9 billion in 2002 (U.S. Census Bureau, 2005a).

There is no legal requirement for departments to report statistics on operations, casualties, or incidents to the federal level. Neither are there any mandatory minimum national standards for training, education, or experience for appointment, retention, or promotion. As a consequence of the highly decentralized and mostly unregulated structure, very little definitive information is available on the nature of fire and emergency services problems. (See Figure 7.4.)

Most departments do not routinely meet with neighboring departments specifically for safety purposes. Data on incidents that could be valuable learning tools for other departments are sometimes not shared out of embarrassment, pride, or simple lack of a forum to do so. In other cases, fear of liability or lack of awareness or resources may cause significant incidents to be poorly documented. Such circumstances, in addition to

FIGURE 7.4 Fire and
emergency services issues
cannot be properly
addressed and supported
without statistical data.
Courtesy of Bill Ketchum

the diversity in departmental structures, make identifying the causes of problems and their
solutions a challenge. But without good data on deaths, injuries, and their causes, solutions for all firefighters will remain elusive.

If there is to be a means of improving firefighter safety nationally, fire and emergency
services departments must start by embracing full participation in data reporting through
the National Fire Incident Reporting System (NFIRS).

History of Fire and Emergency Services Research

The long history of research and its contributions to emergency responder safety is important to understanding why the fire and emergency services must commit to it. Departments
and organizations working together have achieved great gains in firefighter safety in the
past, and much can be learned from these efforts for the future.

NECESSARY RESOURCES

To understand the history of fire and emergency services research, it is necessary to understand the essential resources that sustain research. Research in a field depends on the
presence of two types of resources:

- A supply of researchers with advanced degrees, who usually are assigned to an
  academic institution
- Money for funding those researchers, which may come from the federal government via grants or from academic institutions, foundations, nonprofits, and the
  private sector

Neither of these two requirements exists in today's environment. One of the problems
created by too few qualified researchers and the lack of funding is "critical mass," or the
minimum resources needed to start and maintain a venture. For example, if there is not
enough funding, then it is impossible to build and support a scholarly field devoted to the
needs of the fire and emergency services. As a consequence, academic institutions do not

hire the faculty necessary to devote time to the necessary research and publication. And promising students, who may have completed major graduate research on fire and emergency services topics, abandon the area of inquiry for lack of opportunities after graduation.

At present, existing research efforts are sporadic and diffuse, with different disciplines contributing research on an irregular basis and without consistency. A scholarly field of policy research would define research needs and serve to bridge the gap between research in specialized disciplines and the fire and emergency services itself.

Currently, the executive fire officer (EFO) program at the National Fire Academy requires attendees to complete an applied research project that relates to issues of fire and emergency services. The recently started Ph.D. program at Oklahoma State University will require a similar research effort.

Another problem created by lack of funding affects the research that is actually being performed. Fire and emergency services safety problems may call for research designs that are too broad for the funding available. Political pressures and lack of research capability result in overly ambitious and underfunded projects that fail to address basic knowledge gaps. Such conditions—a lack of funding to adequately support an ongoing research project—can lead to projects that are poorly defined and insufficiently coherent. In addition, if the findings of such projects cannot be duplicated or evaluated properly, as they must be to be considered valid, "stretching" the meaning of findings could lead to error in their application.

THE CALL FOR RESEARCH: 1947 ONWARD

The sorry state of fire and emergency services research is well documented. The federal government has produced numerous studies that acknowledge the problem (Austin et al., 1998; National Commission on Fire Prevention and Control, 1973; NRC, 1961; USFA, 1987). However, in most cases the studies have inspired little meaningful change. The reasons are numerous, but the lack of recognition by the fire and emergency services of the role of research and its importance in addressing real-world issues must be one cause.

The first known federal report calling on investments in fire service–oriented research came by way of Harry S. Truman with what is now called the President's Fire Prevention Conference of 1947 (USFA, 2002b). The report, which was visionary for its grasp of emergency response needs, suggested forming state fire colleges under the control of leading colleges or universities. The report saw the need to attract competent chief officers and called for lateral movement between departments and consolidation of smaller departments. Interestingly, it did not call for creation of a federal fire agency. Most notably, the 1947 report recognized the lack of research-based information to guide fire and emergency services management and called for collaboration between academic institutions and the fire and emergency services organizations.

The next major study was done by the National Research Council in the early 1960s (NRC, 1961). The study recognized something called **market failure**, or the inability of normal economic markets to allocate resources efficiently (*Economist*, 2009). It recognized the failure of the market to address the then national priorities of fire research and called for greater emphasis on research into the fire problem (deaths, injuries, and other losses). Market failure is due to nonexcludability (for example, a streetlight) and non-rival consumption, meaning that the use of the good by one person does not diminish its usefulness for others (Pearce, 1992).

The state of fire research at that time was summarized as follows:

> . . . fire suppression and fire prevention have been under study for a great number of years by a wide variety of private and governmental organizations. . . . However most of this effort is applied work, a good deal of which is directed toward the problems of satisfying code requirements and finding remedies for very specific problems . . . areas of economic interest to the whole nation are often of insufficient interest, to any one group, to produce a desirable overall level of attention. (NRC, 1961)

market failure ■ The inability of normal economic markets to allocate resources efficiently due to nonexcludability (for example, a streetlight) and non-rival consumption, meaning that the use of the good by one person does not diminish its usefulness for others.

The NRC report essentially means that the private sector will fund research only when it has a direct and immediate benefit to its business. In the absence of government funding, private sector priorities explain why so much of the research that does take place is directed at code compliance, approval of specific products, or parochial issues such as evacuation of air traffic control towers, while basic questions such as the most efficient way to attack a fire and the effectiveness of officer training go unstudied.

The absence of a healthy fire and emergency services research field means that problems are ill defined and research efforts are directed by organizations and interest groups that may have limited funds, or that may be more responsive to narrow interests or political expediency than would be the case with a rigorous scholarly research community.

It can be argued that the problem of market failure is applicable to the fire and emergency services problem because the service is a public good, meaning that its benefits are not limited to those who pay and that there is no way to limit its provision to those who pay for the service. However, because the fire and emergency services is characterized by incomplete information, meaning that local fire and emergency services may not be aware of research and its potential contributions, it is not economically attractive to spend resources on any problem the fire and emergency services may have, either because organizations are not willing to pay or because the benefits would go to so many organizations that it would not be practical to solicit funding from each of them.

The best-known federal report on fire and emergency services issues was a 1970s report called *America Burning* (National Commission on Fire Prevention and Control, 1973). Among its conclusions was a call for research. The report gave four reasons: it would save more lives, reduce deaths and injuries to firefighters, reduce property losses, and protect the public at lower cost. It is perhaps the most influential report thus far because it triggered the strongest federal response to issues of fire service research ever.

The *America Burning* report proposed a program for research into safer equipment and called for an expanded federal role in fire research overseen by a newly formed federal fire agency. As a result, *America Burning* was responsible for the formation of the first federal fire agency and the first systematic and well-funded fire service policy research undertaken in the United States. However, after a brief period of activity, federal involvement in fire and emergency services research declined sharply around 1980.

In 1987, a new report called *America Burning Revisited,* sponsored by the USFA, again looked at the lack of research. It found that although research is part of its mission, "the USFA . . . [has] the legislative authority to do these studies, but lacks money and personnel" (USFA, 1987, p. 23).

About 10 years later a Blue Ribbon Panel report (Austin et al., 1998) was commissioned by the Federal Emergency Management Agency (FEMA) to examine the withered state of federal fire and emergency services activity. The panel found that research was almost at the point of "extinction" and faulted the failure to defend research programs during the budget process and management of those efforts when projects were undertaken. The report recommended a $12 million annual research budget for the USFA.

The 2002 report, *America at Risk,* found that "research on the science of fire, fire behavior, the suppression and extinguishing of fire, and fire service operations is inadequate" (USFA, 2002a, p. 21).

A 2003 report by the National Academy of Sciences (National Research Council of the National Academies Committee, 2003, p. vii) examined the state of fire research and found that the *America Burning* report recommendation for research funding was $26 million in 1973, or $113 million in 2003 dollars. Overall, federal funding for fire and emergency services research had decreased by 85% since 1973. The report found that public policy as it applies to the fire and emergency services was not receiving adequate attention. The last three Wingspread Conferences (Wingspread Conference Reports, 1986, 1996, 2006) have all focused attention on the need for data collection and research. Since 1986 data collection and research has remained a significant issue with conference attendees.

In 2005, as a result of the renewed focus on firefighter safety, the National Fire Service Research Agenda Symposium (National Fallen Firefighters Foundation, 2005) set out a number of research issues directly and indirectly related to the problems of firefighter safety. The report did not address funding requirements. Even with that missing vital component, the 2005 report was a critical first step in again raising awareness of the need for research.

FIRE AND EMERGENCY SERVICES RESEARCH IN ACTION

To understand the contribution that research can make in reducing line-of-duty death and injury among firefighters, it is useful to revisit the sole period in U.S. history of significant and sustained federal funding for fire and emergency services research. The "golden era" was a result of the National Science Foundation's program entitled Research Applied to National Needs (RANN). The program lasted from 1971 to 1977 and coincided with the formation of the National Fire Prevention and Control Administration.

Funding for RANN, which came through the National Science Foundation, totaled $500 million over the program's life and supported all elements (fire and non-fire-related) of the program. The major focus on fire and emergency services included breathing apparatus development and deployment analysis research. It was the largest and last major federal effort in applied fire research for the fire and emergency services.

New SCBA Developed

In 1971 NASA teamed with the National Bureau of Standards (later renamed the National Institute of Standards and Technology) to redesign the self-contained breathing apparatus (SCBA) to make it lighter and easier to use. Changes included an improved face mask, lighter cylinder, higher pressure, and more ergonomic harness. In 1974, the new SCBA was field-tested in New York, Houston, and Los Angeles (National Bureau of Standards/NASA, 1986). It later became the industry standard. (See Figure 7.5.)

FIGURE 7.5 Modern SCBA was developed from federally funded research in the 1970s.
Courtesy of Anthony Drago, White Plains Department of Public Safety

Redesigned SCBA improved usage of the technology and greatly reduced smoke inhalation injury in the fire and emergency services. Smoke inhalation, gas inhalation, or respiratory distress declined from over 12,000 cases (18.5%) in 1981 to just under 3,000 in 2004 (8% of injuries), according to NFPA statistics (NFPA, 2010). It is probably the single biggest advancement in firefighter safety. NASA continues to adapt technology for use by the fire and emergency services.

National Institute of Standards and Technology

The foremost locale for fire science and engineering research related to firefighter safety has been at the National Institute of Standards and Technology (NIST), located in Gaithersburg, Maryland. The federal government laboratory, part of the U.S. Department of Commerce, has a Building and Fire Research Laboratory. NIST has been involved in fire research from its inception and continues to pursue numerous studies of interest to the fire and emergency services.

Highlights of its recent work address the following issues: computer fire modeling to support investigation of major fires, including firefighter deaths; investigations into fire behavior as it relates to fire and emergency services operations; development of improved standards for personal alert safety system (PASS) devices and personal protective ensembles; cardiovascular and biomechanical responses to firefighting, wind-driven fire in structures; sleep deprivation; and ethanol.

Recently, NIST developed the majority of the test methods for the new NFPA 1801, *Standard on Thermal Imagers for the Fire Service,* used by the fire protection engineering profession. NIST, as an engineering and science organization, does not have fire and emergency services policy issues as its primary focus. (See Figure 7.6.)

RAND Institute

The fire service program of the RAND Institute was the first systematic examination of fire deployment since the development of the insurance industry's grading scale at the turn

FIGURE 7.6 In this NIST fire test, flames from a simulated house with combustible exterior walls ignite a similar "house" six feet away. *Courtesy of NIST*

of the 20th century. The program developed basic rules for describing fire and emergency services response time performance looking at community fire fatalities and injuries per structure fire in New York City. Current requirements for quantifying staffing and response times are linked to the early research. Unfortunately, the work has never been updated.

Private Research: Protective Equipment

One area in which the private sector has led in fire and emergency services safety research is in protective apparel. Highlights include the development and implementation of fire resistive fabrics, adoption of turnout pants, improved ergonomic design of gear to include better wrist coverage, and pants and hoods. Helmets began to include integrated eye protection, further reducing injuries. The research that led to those improvements relied heavily on national firefighter casualty data (NFIRS). In fact, the availability of reliable national data was instrumental to improvements in firefighting protective apparel.

Similar efforts continue in collaboration with academic partners today. One example is the University of North Carolina's Textile Protection and Comfort Center's work (n.d.). With federal sponsorship, the government-industry-academic partnership continues to develop the next generation of protective equipment. The center is testing new firefighter protective equipment for chemical, biological, radiological, nuclear, and explosive resistance.

Barriers to Research

It can be argued that as an institution, the fire and emergency services itself presents a major barrier to fire research. In general, organizations do not have staff with the training and education to design or engage in research. The specialized nature of those skills are usually found only in the largest organizations and predominantly among civilian employees. Even state-level fire marshal offices do not have professional research staffs. The shortage of personnel attuned to research leads to little demand for research and little or no critical analysis of available data.

At the bottom line is this: if the fire and emergency services does not know what questions to ask, it cannot define its research needs.

POOR PARTICIPATION IN NFIRS

Because of the fire department's limited participation in NFIRS, there is poor documentation of local fire and emergency services experience, which delays discovery of national problems and limits the effectiveness of national policy and programs. Participating fire departments in NFIRS help in gathering more accurate fire incident data nationally. NFIRS is the best way for organizations to share their experiences with others. Failure to participate means that emerging issues take longer to be discovered and that demonstrating the need for resources is more difficult locally and nationally.

DEPARTMENTAL BIAS

One of the problems of implementing policies and practices without first conducting research on their efficacy is that no one can determine to any reasonable degree whether or not those new policies and practices are effective or even harmful. Some questions are uncomfortable for departments, individuals, or interested organizations within the fire and emergency services. Fear of finding the "wrong answer" has caused some organizations to oppose research projects, which is not helpful. The fire and emergency services must keep its focus on protecting the firefighter, not on protecting its organizations.

Supporting Fire Policy Research

Tremendous contributions, such as the development of smoke detectors, have emerged from basic fire science research. However, there is a clear and unrecognized need for policy research, which would deal with managerial and departmental questions of action for the fire and emergency services.

DEFINING POLICY RESEARCH

policy ■ A definite course of action meant to guide and determine present and future decisions, usually in the form of an overall plan that sets out general goals and procedures.

The term **policy** refers to a definite course of action meant to guide and determine present and future decisions, usually in the form of an overall plan that sets out general goals and procedures. Research is sorely needed to help guide fire and emergency services decision makers in the development of effective policy. Fire and emergency services research with direct policy implications should be clearly linked to real-world procedures and practices. The skills and expertise required for such research is interdisciplinary, but would be based predominantly in the social sciences, not engineering. Social sciences, particularly public policy and management, offer the ability to make connections between raw scientific information and the behavioral, departmental, and policy context, which is necessary to implement effective change and prioritize competing objectives.

It is especially important for decision makers to design rules and regulations for firefighter safety. Though there are clearly defined engineering research departments serving the fire and emergency services, there is no policy-focused equivalent department that could assist. There are, however, academics and institutions with the capacity to address the questions related to policy development.

Without policy research, there is no way to put findings from the science community into use for the rank-and-file firefighters. For example, knowing that cardiac arrest is a leading cause of firefighter deaths does not help a department understand the effect that staffing, working schedules, or protective equipment have on improving or aggravating the problem.

SCHOLARLY INFRASTRUCTURE FOR RESEARCH

To illustrate the role that research plays in supporting a profession, it is instructive to examine the case of law enforcement, another local service, and its relationship to the academic and research communities. Some in the fire and emergency services wonder at the effectiveness with which law enforcement is able to secure resources and advance its agenda. A comparison of higher education programs with a law enforcement focus versus a fire and emergency services focus can help shed light on the reasons.

To start, there are 29 doctoral programs in criminal justice and only two devoted to fire and emergency services issues. (One is offered by Oklahoma State University [OSU] in fire and emergency management administration and the other by Capella University in public safety with a concentration in the area of fire service administration. OSU sponsors a research symposium and the only refereed journal in the fire and emergency services, the *International Fire Service Journal of Leadership and Management*.) In addition, criminal justice has 152 master's degree programs (O'Connor, 2009), but there are only 12 for the fire and emergency services. Those higher education programs not only indicate the professionalization in both fields but also serve as the pool of talent for which a research agenda is designed and executed. Approximately 610,000 local law enforcement officers serve in the United States, versus over 1,000,000 firefighters, of whom 294,000 are full time (U.S. Census Bureau, 2009). Table 7.2 offers a comparison of academic programs in the two fields.

On the one hand, the shortage of fire and emergency services institutions and the overwhelming reliance on part-time academic faculty make any support of scholarly research a difficult challenge. On the other hand, law enforcement is funded through the

TABLE 7.2	Comparison of Higher Education in Criminal Justice versus Fire Service Fields	
ACCREDITED DEGREE PROGRAMS	**CRIMINAL JUSTICE**	**FIRE SERVICES/FIRE SCIENCE**
Bachelor's and associate's	536[1]	115[3]
Master's	152[2]	12[4]
Doctoral	29[2]	2[4]

Note 1—(American Society of Criminology, 2006)

Note 2—(O'Connor, 2009)

Note 3—Regional Accredited Schools: Bachelor's Degrees at: http://www.usfa.dhs.gov/nfa/higher_ed/resources/resources_schools_bachelor.shtm; Regional Accredited Schools: Associate's Degrees at: http://www.usfa.dhs.gov/nfa/higher_ed/resources/resources_schools_associate.shtm; Degrees at a Distance Program at: http://www.usfa.dhs.gov/nfa/higher_ed/feshe/feshe_ddp.shtm

Note 4—Accredited Schools: Graduate Degrees at: http://www.usfa.dhs.gov/nfa/higher_ed/resources/resources_schools_graduate.shtm

National Institute of Justice, a central coordinating funding and standards-making body that supports extensive research activities and technology through academies and partnerships with law enforcement agencies.

Currently, the fire and emergency services has some of these funded resources in place. However, they are separate and disjointed compared to law enforcement efforts. The National Fallen Firefighters Foundation has done the best job so far in trying to coordinate federally funded activities through its own research agenda. NFPA and NIOSH have helped develop standards in the past, along with building and fire codes. FEMA's Assistance to Firefighters Grant (AFG) program has supported research at the University of Maryland. Department of Homeland Security (DHS) has supported Worcester Polytechnic Institute (WPI). NIST has supported positive pressure ventilation research at University of Texas at Austin, and Harvey Mudd College in California has supported research in structural collapse.

The National Fallen Firefighters Foundation maintains a firefighter life safety initiative research database, which summarizes studies from various sources and classifies them according to the 16 Firefighter Life Safety Initiatives. The database focuses specifically on actual firefighter and fire safety research, scientific or quantitative testing reports published, academic studies and articles, industry reports, and investigations of firefighter fatalities and injuries.

Data Systems and Current Research Activities

Data systems are critical to understanding the firefighter safety problem, performing research to identify causes, and evaluating potential solutions.

NATIONAL FIRE DATA SYSTEMS

Contrary to what many may believe, there is no official census, or complete count, of every fire and emergency services incident responded to by departments within the United States. However, there are two primary U.S. fire data systems, both of which are voluntary.

The first is the National Fire Incident Reporting System (NFIRS). As the primary national fire data system, it is the main source of information on line-of-duty firefighter death and injury. The NFIRS is administered by the National Fire Information Council and the USFA, largely through volunteer efforts on behalf of participating departments (over 21,000 of them) and states. Its database contains reports on about half of all national incidents and includes both the characteristics of incidents and any firefighter casualties. Detailed information on firefighter protective equipment and emergency response activity is included.

The second data system is offered by NFPA, which sends its "Annual Survey of Fire Experience" to a sample of departments nationally. Using statistical adjustments, the results of the NFPA survey are combined with NFIRS data. That combined information is then used to form national estimates of fire deaths, injuries, and dollar loss.

USES OF NATIONAL FIRE DATA

In states that use the NFIRS system, it forms the source for state-level reports on fire loss and department activity. Within a participating department, the reports can be very valuable for analysis and understanding of the local fire problem. Such analyses can be used to formulate public education programs, to identify circumstances and incidents causing deaths and injuries to firefighters and the public, and to recognize local trends when used over a sufficient time period. (See Figure 7.7.)

Because NFIRS is a widely used system, it allows departments to compare their experience with the experience of others. There are many uses for NFIRS data:

- Manufacturers use the data to improve product safety.
- Fire departments and state fire marshals use the data to identify local fire problems, target legislation and improved codes, and design training curriculum.
- Federal agencies, such as the Occupational Safety and Health Administration, use the data for understanding the fire problem and designing regulatory strategies.
- Consumer Product Safety Commission (CPSC) uses the data to identify products for recall or modification.
- National Highway Traffic Safety Administration (NHTSA) applies the data to motor-vehicle safety issues.
- National Institute of Standards and Technology (NIST) uses the data for fire research.
- Insurance companies and manufacturers use the data to advise policyholders, set underwriting criteria, and target public education campaigns.
- Model code groups and state fire marshals change building and fire codes based on fire loss scenarios and protection and detection system performance.
- Legislators use the data to decide on actions concerning fire services, such as federal funding and regulation.

FIGURE 7.7 NFIRS analysis helps to identify local trends over a period of time. *Courtesy of Charles R. Jennings*

Diligent participation in NFIRS should be considered an obligation, not of just a single department but of the greater fire and emergency services. By collecting data on incidents nationally, health and safety issues can be identified more quickly, resulting in earlier awareness and avoidance of deaths and injuries.

NIOSH FIRE FIGHTER FATALITY INVESTIGATIONS

Another source of data on firefighter safety is the investigation reports from the National Institute for Occupational Safety and Health (NIOSH). Through its Fire Fighter Fatality Investigation and Prevention Program, detailed reports and findings are developed to prevent similar types of line-of-duty deaths in the United States. Although NIOSH does not collect data on every line-of-duty death, its collection has grown to become a rich data set for research purposes.

CURRENT USFA RESEARCH EFFORTS

The USFA has greatly increased its emphasis on research in the last few years, which includes sponsorship of research by outside organizations. Its research topics and projects are broken into three areas:

- Detection, suppression, and notification
- Health and safety
- Other initiatives related to technology evaluation and transfer, computer modeling, and behavioral fire mitigation

Much of its funding has been directed to outside organizations, including the Society of Fire Protection Engineers, sprinkler manufacturers, the National Association of State Fire Marshals, and others. NIST received funds for many projects in the USFA research program such as field experiments and some data collection and analysis efforts. Some of the USFA's most promising projects are:

- Detection, suppression, and notification
  - Evaluating Positive Pressure Ventilation in Large Structures
  - Fire Suppression of Hose Streams
  - Residential Fire Sprinkler Activation Project
  - Study of Municipal Water Supply Systems
  - Study of Wind-Driven Fires
- Health and safety
  - Building Performance: Structural Collapse Prediction Technology and Building Performance Awareness of Lightweight Construction During Fires
  - Computer-Based Firefighter Trainer
  - Emergency Incident Rehabilitation
  - Emergency Vehicle and Roadway Operations Safety
  - Firefighter Autopsy Protocol
  - Firefighter Wellness-Fitness
  - Protective Clothing and Equipment
  - Study of Risk Management Program Development for the Fire Service
  - Study of the Impact and Mitigation of Sleep Deprivation in Emergency Services

The partial list of research subjects represents a significant and much-needed commitment to research. Findings from some of these projects may be directly usable by departments, whereas others may require additional work to translate the findings into actionable information to make policies and procedures for real-world operations. Other research is useful primarily as inputs to long-term fire science research that may someday provide benefits. Many of these projects are funded on a small level.

FIRE PREVENTION AND SAFETY GRANT PROGRAM

Research was identified as a component of the federal grants program under the Fire Prevention and Safety Grants (FP&S) sponsored by the Department of Homeland Security's FEMA. The research portion of the grant program emphasizes firefighter safety. Research priorities for the program are based upon the 2005 report of fire research priorities produced by the National Fallen Firefighters Foundation.

A list of initial research projects funded through the grant shows the scientific rigor and level of resources necessary to conduct these studies. As the early studies are completed, it is hoped that findings can be communicated to end users, and that guidance for future research and development of policies and procedures can be identified. It will take several years before some of these projects, by the nature of the problems they are seeking to understand, have an impact on firefighter safety in the field.

Grants funded recently under the program include several cardiac and fitness studies, and others on firefighter safety and deployment, firefighter on-scene accountability technology, safety of engineered lumber assemblies under fire conditions, firefighting under wind-driven fire conditions, and new technologies for fire suppression.

Under the 2008 grant program, over $8 million was awarded to 11 organizations including universities, a hospital, and a fire testing laboratory. This is considered a conservative amount, as projects identified as "fire prevention" under the grant program can include a research component. Such a critical grant program deserves greater funding, and ideally other methods of research can be developed to capture and identify characteristics of high-performing organizations as well as evaluation of policies to improve firefighter safety.

FIRE AND EMERGENCY SERVICES–UNIVERSITY COLLABORATIONS

Collaboration between fire and emergency services departments and academic institutions or faculty can be very productive. The presence of personnel skilled in research methodology and documentation who have a commitment to community service makes universities an ideal partner for undertaking research in the fire and emergency services. Research can be undertaken by graduate students supervised by faculty members, working with members of the fire and emergency services. These partnerships can be as modest as providing access to data and observations, and documentation of operations and activities by a single graduate student, to those involving outside funding and multiple staff and faculty from a university.

Planning for a fire and emergency services–university collaboration should have support from the department's top management and envision a process that may range in time over months to years. The department should form some formal conception of a problem or an intended project before approaching an institution. Because of the complex nature of many fire and emergency services issues, multi-disciplinary teams or research centers may be best positioned to address fire and emergency services needs. Even the most "technical" problem usually has an organizational or human behavior component. Research produced under such arrangements is generally of a high quality, can be subjected to outside review, and may be regarded as more impartial than efforts done completely by the fire and emergency services.

Elaborate research projects may require applications for funding with significant lead times. The effort put into collaboration with an academic partner can be very valuable— it can provide access to expertise not found in the fire and emergency services, and it has the potential for funding sources that may not be accessible to the fire and emergency services otherwise.

Fire and emergency services–university collaborations may be more common than they were in the past. Some current examples include the following:

- ***Chicago Fire Department and the University of California at Berkeley.*** Working through a research center, the partnership is called Fire Information and Rescue Equipment (FIRE): Enhanced Decision-Making and Situational Awareness for Urban/Industrial Firefighting. It is looking at application of cutting-edge technologies to improve the effectiveness and safety of incident scene operations, particularly in high-rise buildings. The project components include an in-building data communications network with firefighter position tracking capability; vital signs monitoring for firefighters; a helmet-mounted data display that can receive building floor plans and other information; and electronic incident command system (ICS) technology integrated into a wireless network (Chicago Fire Department, n.d.).

- ***University of Maryland at College Park and the Center for Firefighter Safety Research and Development.*** The collaboration established a specialized research center in 2007 to work on problems of firefighter on-scene location tracking, vital signs monitoring, and creation of a wireless network capable of providing information to personnel on the scene of emergencies. Working with several organizations, a system is currently under test and may be offered to the public soon.

- ***University of Illinois at Champaign-Urbana.*** The university formed a Homeland Security Research Center in 2004. The new center includes a focus on some fire service policy issues, including incident command. (See Figure 7.8.)

- ***Institution of Fire Engineers (U.S. Branch).*** The group, with the sponsorship of the USFA, organized a series of annual research conferences along changing fire service topics. The 2001 Conference, organized jointly with John Jay College of Criminal Justice, focused on behavioral and organizational aspects of firefighter

FIGURE 7.8 Creating policies and procedures to properly utilize the incident command system on every incident is still a major focus. *Courtesy of Martin Grube*

safety. The conference series has been hosted by the Fire Department Instructor's Conference and by the International Association of Fire Chiefs. Currently, the Institution of Fire Engineers is helping develop and promote a national strategy, Vision 20/20, which has a primary focus on fire prevention, and also continues to hold research conferences.

■ *John Jay College of Criminal Justice.* In 2008, the college (a part of the City University of New York) formed a public safety research center aimed at responder coordination and incident management, which was named the Christian Regenhard Center for Emergency Response Studies (RaCERS) after a probationary firefighter killed in the 9/11 attacks. The center has received pilot funding from the Department of Justice's Bureau of Justice Assistance and intends to partner with first responder organizations to pursue meaningful research with policy application. Its research emphasizes learning from field experience and capturing after-action reports and information for analysis and development of curriculum.

GRANTS AND EDUCATIONAL INSTITUTIONS

There are some sources for funding collaborations between the fire and emergency services and academic institutions. The greatest source of funding is through the Assistance to Firefighters Grant Program (FIRE Act), administered by the Department of Homeland Security's USFA. Although most of the funds are directed at equipment and training for organizations, in 2005 a research component of the program focused on firefighter life safety initiatives. Through a competitive process, grants can be awarded to fund applied research. The research component of the program is crucial, and hopefully it will be expanded to enable more researchers to participate on a wider array of topics.

Numerous other sources of funding for research in various disciplines may be accessible to academic institutions, but those funding sources are not specifically focused on the fire and emergency services.

To repeat: current emergency responder techniques must be studied scientifically, and then the findings from those studies must be translated into policy and procedure guidance for local departments. Such studies should lead the way to when, how, and what benefits arise with use of any proposed new techniques.

Future Research Needs

Numerous areas are in need of research in the fire and emergency services, many bearing on the firefighter safety problem. Some of the major areas for further and future research follow:

■ *Building and fire codes.* Right now little information exists on the relative contribution of codes to firefighter safety. For example, research must consider not only the codes themselves but also the effectiveness of their enforcement, such as examining the data on the dates of inspections to measure the "deterrent" effect of inspections on fire incidence. A recent research report on the topic (Hall, Flynn, and Grant, 2008) was produced with funding from the FIRE Act grant program in the Fire Protection Association Research Foundation. The Insurance Services Office (1996, 2011) is implementing a Building Code Effectiveness Grading Schedule (BCEGS), which may make contributions in the building and fire codes area.

■ *Fire prevention.* Currently little information is published on the impact of public fire education and fire losses. No research has been done on the impact of company inspection programs and their contribution to firefighter safety. Anecdotally, the increased familiarity with buildings should enhance firefighter safety. Of course, fewer fires mean fewer opportunities for firefighters to be injured while firefighting. Similarly, the role of sprinklers in firefighter safety has not been fully studied.

FIGURE 7.9 Fire and emergency services deployment requirements affect firefighter safety and survival. *Courtesy of Martin Grube*

- *Deployment research.* Basic research on deployment analysis was done in the 1970s, but has not yet been systematically updated or validated, and it needs to be. For example, deployment analysis considers staffing, response times, and apparatus. Several published papers have presented alternate approaches to measuring deployment and performance of emergency response systems. There have been several regulatory or standards-related activities in deployment research: publication of NFPA 1710, *Standard for the Organization and Deployment of Fire Suppression Operations, Emergency Medical Operations, and Special Operations to the Public by Career Fire Departments,* along with NFPA 1720, *Standard for the Organization and Deployment of Fire Suppression Operations, Emergency Medical Operations, and Special Operations to the Public by Volunteer Fire Departments;* and the Occupational Safety and Health Administration's enactment of the two-in/two-out regulation in 1997. These deployment requirements are perceived to contribute to firefighter safety. A recent study report, *Report on Residential Fireground Field Experiments* (Averill et al., 2010), investigated the effectiveness of crew size, arrival time and response time, and task completion on residential structure fires to achieve the standards of NFPA 1710. (See Figure 7.9.)
- *Managing the community fire problem.* There is little guidance for officials to manage their fire problem globally or to balance resources applied to code enforcement, public education, response, and other tasks. In reality, organizations operate with constrained resources, meaning that time, money, and personnel are limited. The environment requires proven sound guidance on making choices to balance needs and to decide on what safety-related interventions will have the greatest impact.
- *Professional standards.* The fire and emergency services has made tremendous progress with development and adoption of professional standards for firefighters and officers. The effect of adopting and meeting those standards needs evaluation in order to provide feedback to standards makers. Research is needed to provide guidance to chiefs and state officials on decisions to adopt and require such standards. Little research shows the relationship between professional certifications and

FIGURE 7.10 The relationships between training, education, and performance in the field should be studied. *Courtesy of Travis Ford, Nashville Fire Department*

performance in the field. Studies of professional standards would help to justify the added expense of pursuing higher levels of training and education. (See Figure 7.10.)

■ *National Incident Management System (NIMS).* NIMS—including its component part, the incident command system (ICS)—is generally recognized as a positive step to managing multiple-casualty and mass-casualty incidents. An evaluation of NIMS/ICS adoption, training requirements, and effectiveness would be very useful in making modifications to the system and for identifying areas for improvement. The link between use of NIMS and incident scene safety has not yet been thoroughly studied.

■ *Incident scene management technologies.* The growth in information technology and wireless communication on the fireground holds the promise of improved safety and greater operations effectiveness. Systems being developed include wearable computers, personnel tracking systems, and portable wireless networks. Competing technologies, and the demands they place on users and budgets, will need thorough and systematic evaluation before they can be widely recommended for use. Transfer of military command, control, communications, computers, and intelligence (C4I) systems to public safety use needs to be evaluated.

■ *Emergency response vehicles.* Responding to and returning from alarms remain dangerous despite many improvements to emergency vehicle safety. Fully enclosed cabs have been supplemented with trends toward roll cages and stability-control systems to prevent rollover accidents. However, research is needed to determine whether newer emergency vehicles with enhanced safety features actually reduce the incidence or severity of accidents.

■ *Chronic versus acute illness and mortality.* Although a good understanding of acute injuries to and illnesses in firefighters is gathered through NFIRS and other reports, little solid information is available on occupational illnesses that may result in sickness or premature death of firefighters, and more information is needed.

■ *Organization of services.* In many small fire departments there are problems related to interoperability, adequate staffing levels, response times, cost-effectiveness, and the inability to provide specialized services. Leaders must ask, "Are adequate numbers of qualified/certified personnel available to fill required leadership roles? Is there a minimum or optimal size for a fire department? Should consolidation of small departments become a national policy priority?" Research can lead the way in answering these questions. (See Figure 7.11.)

FIGURE 7.11 Research can lead the way in determining the minimum or optimal size for a fire department. *Courtesy of Travis Ford, Nashville Fire Department*

REPORTING SYSTEMS AND DATA

NFIRS is a good system. It would benefit the fire service even more than it does if there were 100% participation, at least for fires and incidents causing casualties. Proper funding would help as well. Future changes in the reporting system should be carefully considered and impact on research and analysis community assessed. National organizations should seriously consider supporting NFIRS participation as a component of firefighter safety initiatives. The NFFF and USFA especially should consider supporting NFIRS-based research and assist in publicizing findings from the results. Publication of results from NFIRS-related studies will help with showing the importance of participation. (See Figure 7.12.)

Grassroots systems have emerged for collecting data related to safety issues from fire and emergency services organizations. They include Firefighter Close Calls, a Web site and privately run list with national participation; Honor Their Sacrifices, a Web site that gathers, presents, and disseminates firefighter line-of-duty death information; and the National Fire Fighter Near-Miss Reporting System (n.d.). The latter is jointly sponsored by IAFF and IAFC; however, it should be noted that the Near-Miss Reporting System is funded by AFG and is based on an aviation industry model of human factors and accident prevention that promotes a safety culture. All those systems offer case studies and hold potential for further review, classification, and development of material useful for raising awareness and for inclusion in training and education programs. They could become extremely useful for identifying areas for improvement in regard to policies and procedures, equipment, and training and education.

EMERGING ISSUES

Numerous emerging issues will continue to challenge the fire and emergency services. Research will be needed to understand these challenges and to design responses to prepare

FIGURE 7.12 If there were 100% fire and emergency services participation, NFIRS would better benefit the fire service by helping identify issues. *Courtesy of Travis Ford, Nashville Fire Department*

to deal with them. Emerging issues that will require close coordination with law enforcement and a major change in our normal operational approach include:

- Terrorism
- Hazardous materials response
- Large-scale violence
- School violence
- Hostage incidents
- Violence directed at firefighters
- Green technologies (e.g., solar panels, roof gardens)
- Fire development and behavior

Years of research on fire dynamics and fire suppression are beginning to offer useful guidance for firefighters. The ongoing research projects identified in this chapter will grow in prominence as current studies are completed.

There is a need for reliable, comprehensive data on the fire and emergency services, staffing, and equipment across the United States. Information on emergency response vehicles, stations, and staffing could be collected by the U.S. Census Bureau as part of its Census of Governments. Those data are needed by Homeland Security and federal and state officials to get a more accurate picture of the U.S. fire and emergency services and its current capabilities.

There is a need for learning from the experience of other countries. The U.S. fire and emergency services has among the highest death rates for firefighters of western industrialized nations. Field-testing and assessment of equipment and policies used outside the United States are needed to evaluate their applicability to the U.S. fire and emergency services. There is a need for systematic review and dissemination of best practices from other nations.

Summary

Data collection and research touch every area of the fire and emergency services as they relate to firefighter health and safety. Fire and emergency services safety has improved since the 1970s, when it first received focused attention. However, the rate of firefighter fatalities per fire has been roughly steady for the last 15 years. Many policies and procedures have been advocated without well-defined research into how, when, and to what degree they will be effective. A lack of data and inconsistent data collection have hampered understanding why fatalities have reached the 15-year plateau. The lack of progress in reducing fire and emergency services line-of-duty deaths should give everyone pause to consider other ways of doing things.

Research is a systematic, planned inquiry consisting of observation and experimentation. The value of research to the fire and emergency services has been recognized for over 60 years. The greatest progress in firefighter safety has been made when the fire and emergency services has worked with research institutions to understand problems and design solutions. Unfortunately, the fire and emergency services does not have an adequate academic infrastructure to support additional research or to attract researchers. In addition, the fire and emergency services is not unified in expressing its needs and therefore discourages research into some topics.

NFIRS is a primary fire data system. To increase the benefits it can provide, it needs greater participation. USFA has numerous research efforts in progress, some of which are directed at emergency responder safety.

There are many areas of interest in need of research in the fire and emergency services safety area. The bigger issues of prevention versus suppression, professional qualifications, and organizational structure must be considered. Major known problems that need further work relate to physical fitness, driver training, inspection of apparatus, and adherence to established safety practices.

Future technologies such as wireless tracking and electronic accountability systems offer promise, but need to be carefully evaluated before being adopted widely. Homeland Security needs to dictate that policy research be undertaken to ensure fire and emergency services is an effective component of our national response, which also calls for better collection of basic information on the nation's fire and emergency services organizations.

International research is needed to better understand why the U.S. fire and emergency services has such a poor safety record when compared to that of other nations.

Review Questions

1. Describe the two leading actions being taken at the time of death for firefighter fatalities. Tell why knowing about those actions is important to solving the problem.
2. Describe the difference between fire and emergency services policy research and other types of research.
3. List the ways that fire departments might find resources to assist in doing research.
4. Define the need to collect information on protective equipment and other factors involved in a firefighter death or injury.
5. Discuss the safety record of the fire and emergency services over the past 10 to 20 years.

References

American Society of Criminology. (2006). *A Partial List of Undergraduate Programs in Criminology, Criminal Justice and Related Fields*. Retrieved May 22, 2009, from http://www.asc41.com/links/undergradPrograms.html

Angerer, P., Kadlez-Gebhardt, S., Delius, M., Raluca, P., and Nowak, D. (2008, December 1). "Comparison of Cardiocirculatory and Thermal Strain of Male Firefighters During Fire Suppression to Exercise Stress Test and Aerobic Exercise Testing." *American Journal of Cardiology, 102*(11), 1551–1556.

Austin, S. P., Edwards, S. P., Compton, D., Dyer, R. A., Gabriele, R. J., Jones, D. L., et al. (1998). *Blue Ribbon Panel Review of the US Fire Administration*. Retrieved

October 1, 2006, from http://www.lrc.fema.gov/starweb/lrcweb/servlet.starweb

Averill, J. D., Moore-Merrell, L., Barowy, A., Santos, R., Peacock, R., Notarianni, K. A., et al. (2010, April). *Report on Residential Fireground Field Experiments.* NIST Technical Note 1661. Retrieved January 31, 2011, from http://www.nist.gov/customcf/get_pdf.cfm?pub_id=904607

Bacharach, S. B., Bamberger, P. A., and Doveh, E. (2008, January). "Firefighters, Critical Incidents, and Drinking to Cope: The Adequacy of Unit-Level Performance Resources as a Source of Vulnerability and Protection." *Journal of Applied Psychology, 93*(1), 155–169.

Bacharach, S. B., and Zelko, H. (2004). *On the Front Line: The Work of First Responders in a Post 9/11 World: A Study of Work Conditions and Emotional Health Among NYC Firefighters and Fire Officers.* Ithaca, NY: Cornell University/Smithers Institute.

Borsché, C. E. (2004, November). "Kindling an Interest in Saving Lives." *Space Center Roundup 07, 43*(11). Retrieved October 3, 2006, from http://www.jsc.nasa.gov/roundup/online/2004/1104_p4_7.pdf

Boxer, P. A., and Wild, D. (1993). "Psychological Distress and Alcohol Use Among Firefighters." *Scandinavian Journal of Work, Environment, and Health, 19*(2), 121–125.

Chicago Fire Department (CFD) at the University of California at Berkeley. (n.d.). *Fire Information and Rescue Equipment (FIRE): Enhanced Decision-Making and Situational Awareness for Urban/Industrial Firefighting.* Retrieved April 15, 2009, from http://fire.me.berkeley.edu/about_fire.htm

Economist, The. (2009). "Market Failure." Adapted from *Essential Economics,* published by Profile Books. Retrieved April 15, 2009, from http://www.economist.com/research/Economics/alphabetic.cfm?letter=M#marketfailure

Fahy, R. (2005, June). *U.S. Firefighter Fatalities Due to Sudden Cardiac Death, 1995–2004.* Quincy, MA: National Fire Protection Association.

Fahy, R., and LeBlanc, P. (2006, June). *Firefighter Fatalities in the United States—2005.* Quincy, MA: National Fire Protection Association.

Federal Emergency Management Agency (FEMA), U.S. Fire Administration (USFA), and National Fire Data Center. (1997, June). *Uses of NFIRS: The Many Uses of the National Fire Incident Reporting System.* Document FA 171. Retrieved April 15, 2009, from http://www.usfa.dhs.gov/downloads/pdf/publications/nfirsuse.pdf

Hall, J. R., Flynn, J., and Grant, C. (2008, July). *Measuring Code Compliance Effectiveness of Fire-Related Portions of Codes.* Quincy, MA: Fire Protection Research Foundation. Retrieved April 15, 2009, from http://www.nfpa.org/assets/files/PDF/Research/CCEReport.pdf

Hersman, M. F. (1974, June). "Science and the Public Sector: A National Policy Overview." *PNAS:*

Proceedings of the National Academy of Sciences USA, 71(6), 2565–2570. Retrieved April 15, 2009, from http://www.pubmedcentral.nih.gov/articlerender.fcgi?artid=388501

Hometown Heroes Survivors Benefits Act of 2003. (2003, December 15). Public Law 108-182, 117 Stat. 2649, passed by 108th Congress. Retrieved April 15, 2009, from http://bulk.resource.org/gpo.gov/laws/108/publ182.108.pdf

Insurance Services Office (ISO), Inc. (1996, 2011). *ISO's Building Code Effectiveness Grading Schedule (BCEGS®).* Retrieved January 15, 2011, from http://www.isomitigation.com/bcegs/0000/bcegs0001.html

Jardini, D. R. (1998, Fall). "Out of the Blue Yonder: How RAND Diversified into Social Welfare Research." *RAND Review, 2*(1), 3–9. Retrieved April 15, 2009, from http://www.rand.org/content/dam/rand/pubs/corporate_pubs/2007/RAND_CP22-1998-08.pdf

Kales, S. N., Soteriades, E. S., Christophi, C. A., and Christiani, D. C. (2007, March). "Emergency Duties and Deaths from Heart Disease Among Firefighters in the United States." *New England Journal of Medicine, 356*(12), 1207–1215.

Klein, R. A. (2001, November). "Risk Assessment and Firefighter Safety: Patterns of Injury in the UK: The UK and USA Compared." *Fire Engineers Journal (Leicester, England), 61*(215), 13–23.

Levine, R., Gann, R., Steckler, K., Ohlemiller, T., Grosshandler, W., Kashiwagi, T., et al. (2003, December). "Fire Science." In Richard N. Wright, *Building and Fire Research at NBS/NIST 1975–2000* (pp. 221–246). NIST BSS 179. Retrieved April 15, 2009, from http://www.fire.nist.gov/bfrlpubs/build04/PDF/b04009.pdf

Mazuzan, G. T. (1994, July 15). *The National Science Foundation: A Brief History.* NSF Org: OD/LPA, File nsf8816. Retrieved April 15, 2009, from http://www.nsf.gov/about/history/nsf50/nsf8816.jsp#chapter4

Mendez, A. M., Molloy, J., and Magaldi, M. C. (2006). "Health Responses of New York City Firefighter Spouses and Their Families Post–September 11, 2001 Terrorist Attacks." *Issues in Mental Health Nursing, 27,* 905–917.

Mourchid, Y. (2006, April). "The Fire Service and Higher Education: Occupation v. Profession." *Firehouse Magazine,* pp. 140, 142, 144.

National Bureau of Standards/NASA. (1986). "Space Technology for the Fire Department." Originally published in *Spinoff 1986,* 50–53. Currently NTRS Document ID: 20020090888. Retrieved January 5, 2009, from http://ntrs.nasa.gov/archive/nasa/casi.ntrs.nasa.gov/20020090888_2002148571.pdf

National Commission on Fire Prevention and Control. (1973, May 4). *America Burning: The Report of the National Commission on Fire Prevention and Control.* FA-264. Washington, DC: U.S. Government Printing Office #1973-O-495-792. Retrieved April 15, 2009, from http://www.usfa.dhs.gov/downloads/pdf/publications/fa-264.pdf

National Fallen Firefighters Foundation. (2005). *Report of the National Fire Service Research Agenda Symposium, June 1–3, 2005.* NIST GCR 08-918. Retrieved April 15, 2009, from http://www.fire.nist.gov/bfrlpubs/fire08/PDF/f08035.pdf

National Fire Protection Association. (2010). *Firefighter Fireground Injuries by Nature of Injury.* Retrieved November 8, 2006, from http://www.nfpa.org/itemDetail.asp?categoryID=955&itemID=23635&URL=Research%20&%20Reports/Fire%20statistics/Fire%20service&cookie%5Ftest=1

National Firefighter Near-Miss Reporting System. (n.d.). *Saving Lives Through Lessons Learned: Fire Fighter Fatalities: Learning from the Past,* p. 3. PowerPoint presentation. Retrieved May 19, 2009, from http://www.firefighternearmiss.com/Resources/Presentations/

National Institute of Standards and Technology. (2005, March). *The Economic Consequences of Firefighter Injuries and Their Prevention: Final Report.* NIST GCR 05-874. Retrieved April 15, 2009, from http://www.fire.nist.gov/bfrlpubs/NIST_GCR_05_874.pdf

National Research Council. (1961). *A Study of Fire Problems: 1961.* Publication No. 949. Washington, DC: National Academy of Sciences.

National Research Council of the National Academies Committee to Identify Innovative Research Needs to Foster Improved Fire Safety in the United States. (2003). *Making the Nation Safe from Fire: A Path Forward in Research.* Washington, DC: National Academies Press.

National Volunteer Fire Council. (2006). *Heart Healthy Firefighter Program: 2004 and 2005 Summary of Screening Results.* Retrieved May 20, 2009, from http://www.healthy-firefighter.org/files/documents/2006nvfcsummarycharts.pdf

National Volunteer Fire Council. (2007). *Keep It Strong: Heart Healthy Firefighter Resource Guide.* Retrieved April 15, 2009, from http://www.healthy-firefighter.org/files/documents/2005resourceguide.pdf

O'Connor, T. R. (2009, February 24). "MegaLinks in Criminal Justice." In *Graduate Schools in Criminal Justice Mega-List.* Retrieved May 21, 2009, from http://www.drtomoconnor.com/jusgrad.htm

Pearce, D. W. (Ed.). (1992). *The MIT Dictionary of Modern Economics* (4th ed.). Cambridge, MA: MIT Press, p. 267.

Singleton, R. A., and Straits, B. C. (2004). *Approaches to Social Research* (4th ed.). New York: Oxford University Press.

Textile Protection and Comfort Center at North Carolina State University. (n.d.). *Pyroman.* Retrieved April 15, 2009, from http://www.tx.ncsu.edu/tpacc/protection/pyroman.html

U.S. Census Bureau. (2005a, June). "Finances of Special District Governments, 2002: 2002 Census of Governments," Vol. 4, Number 2, *Government Finances.* GC02(4)-2. Washington, DC: U.S. Government Printing Office.

U.S. Census Bureau. (2005b, December 9). *Table 1: State and Local Government Finances by Level of Government and by State 2001–02.* Retrieved November 8, 2006, from http://www.census.gov/govs/estimate/0200ussl_1.html

U.S. Census Bureau. (2009, December). *2007 Public Employment Data.* Retrieved June 18, 2009, from http://ftp2.census.gov/govs/apes/07locus.txt

U.S. Fire Administration. (1987, November 30–December 2). *America Burning Revisited.* U.S. Government Printing Office: 1990–724–156/20430. Retrieved April 15, 2009, from http://www.usfa.dhs.gov/downloads/pdf/publications/5-0133-508.pdf

U.S. Fire Administration. (2002a, June). *America at Risk: America Burning Recommissioned.* FA-223. Retrieved April 15, 2009, from http://www.usfa.dhs.gov/downloads/pdf/publications/fa-223-508.pdf

U.S. Fire Administration. (2002b, July). *1947 Fire Prevention Conference.* Retrieved June 17, 2009, from http://www.usfa.dhs.gov/about/47report.shtm

U.S. Fire Administration. (2006, July). *Firefighter Fatalities in the United States in 2005.* FA-306. Retrieved April 15, 2009, from http://www.usfa.dhs.gov/downloads/pdf/publications/fa-306-508.pdf

U.S. Fire Administration. (2009a, February 17). *Find a Regionally Accredited School or Online Program.* Retrieved April 15, 2009, from http://www.usfa.dhs.gov/nfa/higher_ed/resources/resources_schools.shtm

U.S. Fire Administration. (2009b). *Fire Statistics: Fire Departments.* Retrieved June 17, 2009, from http://www.usfa.dhs.gov/statistics/estimates/nfpa/index.shtm

U.S. Fire Administration. (2009c, September). *Firefighter Fatalities in the United States in 2008.* Emmitsburg, MD: U.S. Fire Administration/FEMA.

U.S. Fire Administration. (2010, May 20). *About NFIRS.* Retrieved June 15, 2010, from http://www.usfa.dhs.gov/fireservice/nfirs/about.shtm

Wingspread Conference Report. (1986, October). *Wingspread III: Statements of National Significance to the Fire Problem in the United States: Conference on Contemporary Fire Service Problems, Issues and Recommendations.* Retrieved November 20, 2009, from http://nationalfireheritagecenter.com/1986wingspread.pdf

Wingspread Conference Report. (1996, October). *Wingspread IV: Statements of Critical Issues to the Fire and Emergency Services in the United States.* Retrieved November 20, 2009, from http://www.nationalfireheritagecenter.com/1996Wingspread.pdf

Wingspread Conference Report. (2006, April). *Wingspread V: Statements of National Significance to the Fire Service and to Those Served.* Retrieved November 20, 2009, from http://www.nationalfireheritagecenter.com/2006Wingspread.pdf

Womack, J. W., Green, S., and Crouse, S. (2000). "Cardiovascular Risk Markers in Firefighters: A Longitudinal Study." *Cardiovascular Reviews and Reports, 21*(10), 544–548.

Martin J. DeLoach

Courtesy of Travis Ford, Nashville Fire Department

KEY TERMS

accountability
   systems, *p. 196*

geographic information
   system (GIS), *p. 192*

global positioning system
   (GPS), *p. 194*

OBJECTIVES

After reading this chapter, the student should be able to:

- Identify the cultural roadblocks that delay or prevent a fire and emergency services department from embracing new technology.
- Identify current emerging technological equipment that can help make the fire and emergency services safer.
- Identify computerized technology that can make the fire and emergency services safer.

PEARSON

myfirekit™

For practice tests and additional resources, visit **www.bradybooks.com** and follow the **MyBradyKit** link to register for book-specific resources.

Register for **MyFireKit** by following directions on the **MyFireKit** student access card provided with this text. If there is no card, go to **www.bradybooks.com** and follow the **MyBradyKit** link to Buy Access from there.

Technology and the Fire and Emergency Services

The utilization of new and emerging technology should be considered when identifying ways to reduce firefighter deaths and injuries. However, concentrating on today's technology would make it obsolete in a very short time. Therefore, this chapter begins with some ideas on how to meet the challenges that new ideas pose in a very traditional profession.

SOME HISTORY OF FIREFIGHTING TECHNOLOGY

History tells us that technology can provide an answer to a recognized problem or expressed need. Simple tools and weapons, each one a technological innovation, go back to the beginning of humanity's first attempts at addressing the challenges of hunting and gathering food. Then came the invention of simple machines, such as levers, wheels, and pulley systems, which likely improved life expectancy as well as quality of life. Over time, improved and new technologies continued to help make life ever more efficient and safe. Little is known about the technology of firefighting in ancient history beyond buckets and axes. But some suggest that firefighters may have used a water pump as early as 200 B.C. In contrast, it is clear that a fire pump was used in Europe during the 1500s.

Early pumpers used cisterns as a source of water, as many rural departments do even today. Eventually, water was made available in wooden pipes under city streets; a "fire plug" was pulled out of the top of a pipe, or a new hole was drilled into it and a suction hose inserted when the water was needed to put out a fire. Advancements in that technology brought fire hydrants with water pressure that could increase when a fire alarm was sounded. However, that method proved to be unreliable, finally causing cities to develop today's valved hydrant systems, which are kept under pressure at all times. (See Figure 8.1.)

Change is *always* a challenge!

FIGURE 8.1 Advancements helped in developing current fire hydrant systems. *Courtesy of Danny R. Yates, Deputy Chief Fire Suppression, Nashville Fire Department*

FIGURE 8.2 Horses remained an important part of the U.S. fire service up until the early 1900s. *Courtesy of George Russell, Nashville Fire Department*

In the earlier years firefighters pushed and pulled their engines to the emergency scene themselves. By the mid-1800s, many had advanced to horse-drawn engines. It is worth noting that although the first steam engine had already been built by the mid-1800s, it took almost 50 more years for firefighters to finally give up their horses for motorized engines. (See Figure 8.2.)

Technological advancements arrived at the fire station relatively quickly after that. Along with the first full-time paid firefighting crews came a telegraph system for dispatching, alarm box assignment cards, and sliding poles in fire stations. The dispatching system over the wires permitted companies to respond collectively to emergency scenes, which benefited the efforts to extinguish fires with quicker responses. The box assignment cards identified who was expected to respond for particular types of emergencies. And the ability to slide down poles to the bay floor, rather than using stairs, allowed firefighters in multistory buildings to respond faster. However, use of the poles caused injuries over the years and many fire stations have removed them. (See Figure 8.3.)

Another historical innovation is the fire ladder. At first, ladders that were extended by hand were used. As these grew in height in tandem with the ever taller city structures, they grew in weight and were eventually built on two large wheels that could be suspended behind a fire truck. Before long, a ladder that could turn on a wheeled chassis was invented. It was mechanically extendable and attached directly to the fire truck. In the mid-20th century, an aerial platform was added. (See Figure 8.4.)

There have been many more improvements and changes in the fire and emergency services since then. Few were accepted immediately. Even today, one can find firefighters who react to technology by ignoring it—even technology as simple as the seat belt. Change is always a challenge.

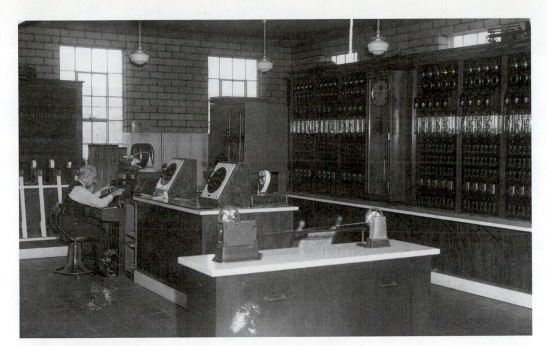

FIGURE 8.3 Early telegraph systems used for dispatching allowed for several fire stations to be notified at the same time. *Courtesy of George Russell, Nashville Fire Department*

FIGURE 8.4 The first long extendable ladders were pulled to the scene by horses. *Courtesy of Larry H. McCown, Nashville Fire Department*

ROADBLOCKS AND SOLUTIONS

Being required to enter areas that are immediately dangerous to life and health (IDLH) creates work traditions and practices that become entrenched in the profession's culture. People who have fought together knowing that if something goes wrong they might die are hard pressed to change the ways that worked out well during previous incidents. In addition, the driver/operator might not want to learn anything new because he or she has survived a lengthy career and now is coasting to retirement. The new company officer may be overwhelmed with responsibilities and not willing or able to add anything

FIGURE 8.5 Firefighters swore by their horses to know the address by hearing the alarm tone.
Courtesy of George Russell, Nashville Fire Department

more to his or her plate. Thus, it may be any one of the informal or formal leaders who prevents a department, shift, or crew from moving forward and adopting new ideas and technology.

Keep in mind that you are entering a profession that is not that far removed from using leather buckets to douse flames. In fact, it may be surprising to learn that fire companies fought verbally—and physically—to keep their horses rather than change to steam- or gasoline-driven vehicles. It was argued that the horses "knew" the address of the fire by the alarm tone and were more reliable than the new horseless carriages. Well, we all know how that argument concluded. It would be wise to reflect on it when becoming involved in one of the fire station discussions on what new technology the department might consider using. (See Figure 8.5.)

That same reliance on past experience as an obstacle to change often is experienced by the new recruit, fresh out of an academy with cutting-edge ideas. Anyone who has been in the fire and emergency services for any length of time has seen rookies sadly learn to embrace the reality of "we don't care what you think; this is the way things are done here." As a leader, one must realize that it is not rookies' ideas that are poor or incorrect, but instead it is their delivery combined with their status that stops them before they are given a chance.

That same dynamic may occur when newly promoted officers immediately set out to implement all of the great ideas they have developed over time on the job. They try to use the new title—and even some coercive power—to facilitate change, but in the end seldom see it accepted in the department.

Unfortunately, what both the new recruit and the newly promoted officer may miss is a key skill of the change agent. First and foremost, it is important to have prepared solutions to utilize technology for a proposed problem. The change agent must demonstrate the need for change to the members of the department. Only after demonstrating how beneficial technology is and creating an awareness of a need, can the agent begin to advocate it. Then the advocate must get the attention of the department and the interest of those who have adequate responsibility to make the changes necessary. Change may take considerably more effort when coming from the bottom of the ladder; however, when done correctly, change can be initiated.

Once you have identified a problem and a possible solution, enlist the assistance of a few officers and firefighters who are advocates for change to try the new product or technology. Provide some clearly written evaluation sheets and data collection tools. Their experience with the new technology could then be presented to the rest of the department as the beginning of a quality discussion about the possibilities of its implementation in the future. If the new technology does not overcome the perceived problem, that must be presented honestly to the group as well. Do not hide efforts that fail. In a fire and emergency services career, you are given only so many chances to sell a new idea. Being honest when you are wrong will help you sell another new idea down the road.

For the best chance at successful change, pay attention to what works in other departments as well as what worked previously in your department. Avoid repeating what did *not* work in previous efforts to change. Observations on how the department's culture accepts changes and new ideas will be a strong indication of how to approach the next innovation.

LEADING THE CHANGE

Setting Goals for New Technology

Goals help identify and set the expectations of the department. Goals benefit a department the most when they are treated as a continuous process in which the department continually seeks to challenge its own practices.

In a lecture (Brandon, 2007), Domino's Pizza CEO David Brandon used a true story as a negative illustration of goal setting. The incident occurred during his first few days as the CEO of Domino's. Mr. Brandon was worried about retention of employees and asked his human resources director what the company's turnover rate was. The numbers astonished Mr. Brandon, and he quickly asked for suggestions on how to improve them. But instead of providing suggestions, the man defended the numbers by telling Brandon they were within industry standards. He was fired.

A goal should not reflect departments that accept mediocrity. Instead, a goal should stand for a clear vision of how to become better, not how to stay the same. It should push toward the next level of excellence and encourage the department to improve at all levels of operation.

The process of goal setting can be a valuable tool to help introduce and sell new technology, because it allows the department to see a finished product before starting down the path to adoption. When doing research for new technology, find a comparable department that has been on the cutting edge of purchasing, evaluating, and implementing the technology. Such a department will give you a goal to shoot for and can help with setting the goals of your entire department.

In the fire and emergency services, comparisons must be made to a similar size department that has similar service demands. For example, there would be very little benefit to compare the Chicago Fire Department to a department of 20 to 30 firefighters protecting about 25,000 people.

The process should not be used without a clear objective and must be done using a transparent method that exposes the results, good or bad. Applied correctly, the goal setting and research process will allow the department and all its participants to grow and improve in the delivery of service.

Technologically Astute Firefighters

Many individuals who are entering the fire department today are attracted to new technology. Most accept technology and its benefits quickly, rather than waiting for others to test or adopt the new product and report their findings. Does that create problems in the fire department? It can sometimes, because the fire department has a semi-military environment and mostly older members at the chief officer rank. So, yes, it can create problems, along with frustration in both the new employees and senior staff.

A leader needs to understand the basic culture of the fire department when making a decision about the adoption of advanced technology. Remember that one's boss may be only peripherally aware of the computer age. Proposals on new technology and how it can improve the fire department must take into account who has to be sold on the idea in order to acquire funding and acceptance.

New and Emerging Technology

Ways must be found to encourage everyone to use affordable existing technology in each fire department. If a certain task is a source of risk for a firefighter and technology can make the task safer, then that technology should be adopted, if at all possible. Firefighters should search for and evaluate technology that fits their department and financial capabilities, using an approach similar to that of risk management for their decisions about what to purchase or request for their department.

In evaluating any new ideas and technology, firefighters must keep the basics in mind. Technology will come along that appears to work and do miraculous things; however, for it to be accepted, it must enhance the fundamental business of emergency scene operations.

Even after ensuring that a new technology is both operationally and financially appropriate, cultural change will still be required. A serious look must be taken to evaluate the difficulties and roadblocks directly corresponding to implementation.

TECHNOLOGY AND TRAFFIC SAFETY

According to the U.S. Fire Administration, over the past 30 years, almost 25% of firefighter fatalities have occurred in driving-related incidents (USFA, 2009a). Safe driving habits are a critical skill needed in emergency vehicle response. But before a department purchases the latest and greatest technological devices available for a new vehicle, it needs to ensure that the individual selected to drive it has personal driving habits that use all the available basic safety devices, such as a seat belt. (See Figure 8.6.)

Seat Belts

Year after year firefighters are killed or injured by being ejected from a vehicle. The National Fire Data Center (USFA, 2002, pp. 158–159) describes one such incident in

FIGURE 8.6 For many years, emergency response has been the cause of many firefighter line-of-duty deaths and injuries. *Courtesy of George Russell, Nashville Fire Department*

TABLE 8.1	States with Primary Enforcement Seat-Belt Laws, June 2008	
Alabama	Alaska	California
Connecticut	Delaware	District of Columbia
Georgia	Hawaii	Illinois
Indiana	Iowa	Kentucky
Louisiana	Maine	Maryland
Michigan	Mississippi	New Jersey
New Mexico	New York	North Carolina
Oklahoma	Oregon	South Carolina
Tennessee	Texas	Washington

Source: National Center for Statistics and Analysis. (2008, September). Traffic Safety Facts: Seat Belt Use in 2008—Overall Results. DOT HS 811 036, p. 3. Washington, DC: National Highway Traffic Safety Administration. Retrieved January 5, 2010, from http://www-nrd.nhtsa.dot.gov/pubs/811036.pdf

which the victim was in the officer's seat of an engine, responding with lights and siren to a medical emergency, when the engine collided with a passenger car. All other injured responders were wearing seat belts and survived the crash. So did the civilian driver. However, the victim was not wearing a seat belt and, upon his engine striking the car, was ejected through the windshield of the engine. He received appropriate medical care on scene, but was pronounced dead upon arrival at the hospital. The incident report would have ended differently, like so many others, had seat belts been utilized. Shockingly, similar incidents continue to occur every year.

According to the National Highway Traffic Safety Administration's (NHTSA) National Center for Statistics and Analysis (NCSA), there are currently 26 states, plus the District of Columbia, that have a primary enforcement law, and 23 states that have secondary laws (NCSA, 2008, p. 3). In 2008 a survey reported that states having a primary seat-belt law had an average compliance of 88%, and states having secondary laws had only 75% compliance of seat-belt use (NCSA, 2008, p. 2). And though in the last 20 years most states have enacted either a primary or secondary seat-belt enforcement law, New Hampshire has yet to do so (NHTSA, n.d.). A primary enforcement law basically means that vehicle occupants can be ticketed simply for not using seat belts; whereas a secondary enforcement law requires occupants to actually be stopped for some other violation such as expired license tag before they can be ticketed for not wearing seat belts. (See Table 8.1.)

Every single firefighter must embrace the idea of using seat belts, a technology that is well over one hundred years old. The laws are on the books. The department policies and procedures are written. It is well past time for all firefighters to buckle up. In fact, it may be time for the fire and emergency services to request new technology that will *not* allow any emergency response vehicle to start unless all seat belts are buckled properly.

Personally Owned Vehicles

Of all the deadly emergency vehicle collisions, over 25% involved personally owned vehicles used by volunteer or paid-on-call firefighters (USFA, 2002, p. 28). According to the Federal Emergency Management Agency (FEMA, 2004, p. 164), it is common practice for many firefighters to travel to an incident scene using their private vehicles.

The problems involved with personally owned vehicles traveling to and from the emergency scene have caused many departments to take a second look at their policies and procedures. However, neither eliminating nor limiting personal vehicle responses to the emergency scene has been accomplished in a consistent way in the United States. Therefore, all vehicles that are personally owned and approved by a fire and emergency services department should be required to go through an annual inspection similar to the Department of Transportation (DOT) inspection for emergency response vehicles. In addition, such an inspection should occur any time the vehicle has been involved in a significant collision. It is a good idea to follow the recommendations provided by the department's insurance company.

In addition, each department should take into account the driver's age, not permitting anyone under 21 to respond in "emergency traffic" mode in a privately owned vehicle. Volunteer departments can obtain information about the proper use of personally owned vehicles from the volunteer fire insurance services.

All driving records should be checked regularly as well. Michigan, for example, provides a technologically based service that will alert the department of any driving infractions of its members once they receive a ticket. Such a service helps keep members who have bad driving habits from being able to operate emergency response vehicles. If a state does not have a system to keep driving record information current, there should be a way to work with local police to run quarterly checks for driving violations on all members.

Emergency Response Vehicles

The newest standards incorporating technology and safety for emergency response vehicles are NFPA 1901, *Standard for Automotive Fire Apparatus* (NFPA, 2009), and, if you are looking to refurbish an older emergency response vehicle, NFPA 1912, *Standard for Fire Apparatus Refurbishing* (NFPA, 2006). These standards establish a minimum set of common items that each new emergency response vehicle purchased will contain, such as the following: reflective striping around the outside of the vehicle, scene lighting, the diagonal striping of bright red and yellow across the back, minimal rotating lights, and directional lights on the back of the vehicle. The standards include air-bag systems in new fire trucks that have been optional and excluded by many departments due to cost factors. (See Figure 8.7.)

The bright red and yellow reflective diagonal striping that is required for 2009 and later emergency response vehicles really helps approaching motorists see the emergency vehicles at night. A recent report, *Emergency Vehicle Visibility and Conspicuity Study* (USFA, 2009b), suggests that increasing emergency vehicle visibility and conspicuity holds

FIGURE 8.7 New emergency response vehicle standards require diagonal striping across the back. *Courtesy of Travis Ford, Nashville Fire Department*

promise for enhancing firefighter safety regardless of being on the emergency scene and outside of the vehicle or responding. The old-style emergency response vehicle with a lot of diamond plate on the back is difficult, if not impossible, to retrofit with reflective stripes. However, reflective dots are available for installation on older vehicles, which can add some visibility until funding is available to purchase new vehicles.

Tenders

Tenders are trucks that have a permanently mounted water tank with the capabilities of dispensing potable or non-potable water. They are vital to rural fire and emergency services. This type of service is predominantly performed by volunteer or paid-on-call departments, although some paid departments that cover a large geographical area benefit from their use. According to a National Fire Data Center report, second only to personally owned vehicles, tender-style vehicles have the most collisions resulting in death or injury of firefighters (USFA, 2009a).

Some tenders have high centers of gravity due to the water tank being mounted above the chassis of the truck. Water weighs approximately 8.4 pounds per gallon, and many of these trucks carry upward of 3,000 gallons of water, which amounts to around 12.6 tons of weight. Untrained responders or responders who overestimate their skills fall victim to crashes and collisions while operating these types of vehicles. Water baffles are crucial to reducing and minimizing these types of incidents.

Newer vehicles have appropriate baffles that minimize the sloshing effect of driving a vehicle with that much liquid. In terms of technology, once again it is back to basics: make sure the department properly trains all drivers/operators to understand the limitations of the tenders, and make sure that the baffles are in place. If a tender does not have baffles, look to retrofitting with baffles to make them safer to drive.

Driving Speeds

Speed is an additional factor that has been a catalyst for firefighter deaths across the United States. Departments must have policies and procedures that limit the speed during an emergency response to no more than 10 miles over the posted speed limit. Officers need to embrace and enforce the policy on maximum speed.

New emergency vehicles can have a second speedometer installed in front of the company officer seat. A mid-priced global positioning system for the emergency vehicle, which shows the vehicle's traveling speed, may be purchased as well. Those devices are available for under $200 in most stores today.

Directional Signaling

Directional signaling devices can be and have been a great tool to direct oncoming traffic away from the workspace along roadways. It is crucially important to ensure that the directional device indicates the proper direction away from the firefighters. These units have been in service for about 20 years for departments able to afford them. They have proven to be very good at assisting in initiating the traffic to shift in the direction of the arrows or cycling lights. Because of its effectiveness, this form of electronic technology is turned on properly or else not at all; if it directs the oncoming traffic the wrong way, it can have devastating results.

Traffic Signaling

Devices that can change the traffic signal from red to green for the approaching emergency response vehicles have been available for some time. Some jurisdictions have written ordinances making it illegal for a person to have the electronic device in a personal vehicle much like some states have outlawed radar detectors.

The units available today are reliable, and most of the transmitter issues have been resolved to ensure only the proper units can change the cycle of the light for the approaching firefighters. These units can identify the emergency response vehicle that asked for the light to change for departments that may have emergency crews approaching from both sides of a traffic light.

A new product on the market is a light installed under the normal traffic light that will flash a symbol that looks like a fire truck as emergency response vehicles approach. The system and the inventor believe that residents can become accustomed to the flashing light displaying a fire truck to recognize emergency response vehicles are approaching the intersection. The system does not change the cycle of the light; it only adds a flashing light to notify people an emergency response vehicle will be attempting to drive through that intersection very soon.

THE COMPUTER AGE

The computer age has yet to be fully embraced by the fire department and vice versa. A lot of vendors at the national trade conference shows offer inspection programs, software for identifying underground utilities, hazard alerts, and so on. Unfortunately, very few of them work together or with any individual department's own information management software. And the customization of the software to correlate information from a drawing program, inspection program, or accountability program to information management software is often prohibited by the vendor, or it is too costly for the custom program writing to stay within the confines of a taxpayer-supported operation.

Ideally any software program should write back to a main source, allowing an inspector to use an inspection program and have it populate the reporting software that all the staff is already trained to use.

geographic information system (GIS) ■ An electronic system that coordinates hardware and software to allow information about critical infrastructure above and below ground— including waterways, sewers, water mains, electric service, gas lines, water hydrants, or any utility installed in an area—to be available and displayed on a computer for analysis.

Geographic Information Systems

Geographic information systems (GIS) are an increasingly important technology for fire departments. These electronic systems have been around for a long time in the private sector and in some government agencies, and in recent years have become more reliable and accessible. The true benefits—and how to provide them to firefighters in the situations that make them most valuable—are just coming to be understood by fire and emergency services leaders. (See Figure 8.8.)

The basic idea behind GIS is that different data are organized based on a geographic area of interest. That area can be an entire community or one smaller, such as a single residential parcel. In most cases the information that firefighters want is in the assessor's office, the planning and zoning office, the water department, the public works department, and the local building department. When GIS has access to that information, it can make it accessible when, where, and even how it is needed. For example, GIS can offer a

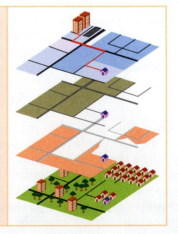

Fire Service Planning and Analysis

- Understand the geographic distribution incidents.

- Compare incidents to demographic characteristics and trends.

- Identify high-risk areas for prevention activities.

- Locate other at-risk areas for locating stations and outreach programs.

FIGURE 8.8 Geographic information systems allow fire departments to better serve the community. *Courtesy of Don Oliver, Fire/Rescue Chief, Wilson Fire/Rescue Services*

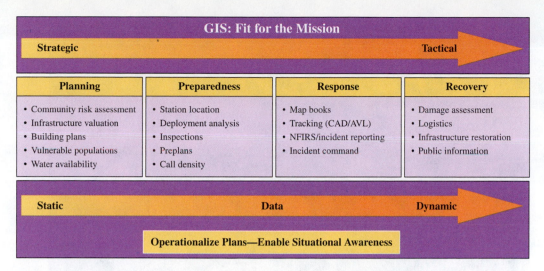

FIGURE 8.9 Geographic information systems fit into the fire department's mission at every level. *Courtesy of Don Oliver, Fire/Rescue Chief, Wilson Fire/Rescue Services*

The figure contains the following text:

GIS: Fit for the Mission

Strategic — Tactical

Planning	Preparedness	Response	Recovery
• Community risk assessment	• Station location	• Map books	• Damage assessment
• Infrastructure valuation	• Deployment analysis	• Tracking (CAD/AVL)	• Logistics
• Building plans	• Inspections	• NFIRS/incident reporting	• Infrastructure restoration
• Vulnerable populations	• Preplans	• Incident command	• Public information
• Water availability	• Call density		

Static — Data — Dynamic

Operationalize Plans—Enable Situational Awareness

geographic area map upon which an overlay shows underground water and electrical systems and underground gas lines, as well as the location of hydrants, fire stations, abandoned buildings, hazardous materials, automatic external defibrillators (AEDs), and a wide variety of other data far too numerous to list. Access to good current information about risk, structures, and systems in the community can lead to better decisions by firefighters. (See Figure 8.9.)

One example of such data available to an on-scene incident commander is the actual up-to-date construction details of a burning building. That information might have prevented the tragic deaths of nine Charleston, South Carolina, firefighters on June 18, 2007 (NIOSH, 2007). The numerous modifications that were made to the structure since it had been built created confusion during the fireground operations. The size and location of the fire were not identified early, and the firefighters became lost and trapped inside the burning structure. If the incident commander had known about the construction changes earlier, based on facts provided by a GIS, the trapped responders might have been located and rescued, or they might even have decided not to enter the structure at all. It would benefit an incident commander's decision-making process on the emergency scene to know whether a structure fire has a truss constructed roof or a traditional stick-built roof. With GIS, the truss structures involved in that 2007 tragedy could have been identified in the incident commander's vehicle or by the first arriving responders. (See Figure 8.10.)

For dispatching, GIS can be used to identify which vehicle is closest to a particular location. When used by the fire department, it can allow the dispatcher to identify the best in-service vehicle to send to the reported emergency. The system can be used to inform incident commanders of when they can expect to have the next units on scene because the dispatchers can monitor their individual progress. The system could incorporate a community's entire fleet of public service vehicles to know where the closest police car is for a domestic situation or a traffic control issue. The system could be used in conjunction with the automatic dispatch of mutual-aid departments if they all had the transponders and a single dispatch point.

Other GIS applications include disaster modeling. It can be used before hazard risk assessment and probability for different types of events through software packages publicly available, such as Hazards U.S. (Hazus) and Consequences Assessment Tool Set (CATS). These two programs can be customized for local use. Hazus allows the fire and emergency services to prepare for the worst by forecasting the most probable physical and economic damage that can occur in a community. Firefighters in the field can use CATS

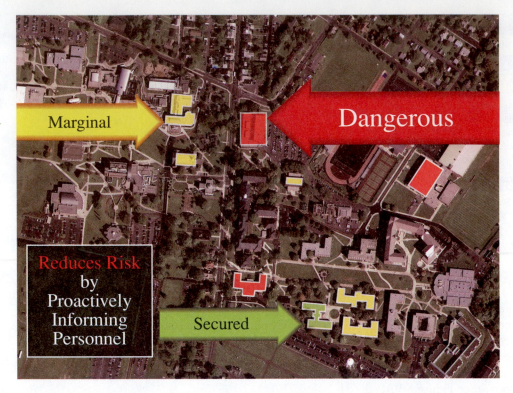

FIGURE 8.10 An example of how geographic information systems allow firefighters to make strategic and tactical decisions quicker. *Courtesy of Don Oliver, Fire/Rescue Chief, Wilson Fire/Rescue Services*

Marginal

Dangerous

Reduces Risk by Proactively Informing Personnel

Secured

to predict the outcome of human-made or natural disasters including earthquakes, tornadoes, hurricanes, hazardous materials release, and so on. (See Figure 8.11.)

In addition, the community vulnerability assessment tool (CVAT) is a program developed to help communication and reduce a community's vulnerability to natural and human-made disasters. The hazard data can be based on historical events and then broken down into frequency, magnitude, and area of impact. It can include events that involve coastal erosion, earthquakes, drought extremes, flash floods, hail, hurricanes, lightning strikes, snow and ice, storm surges, tornadoes, wildfires, wind, and so on.

GIS is great for tabletop discussions with department officers, and its implementation will work toward better relations with other government agencies. The task will require discussions with the leaders of those other agencies to identify the hardware and software needed to access the information that they have already collected and that the fire department and its staff will find invaluable. Any department that has the funding to install computers in its emergency response vehicles should be working on the possibility of adding a GIS into its information system for firefighters.

Global Positioning Systems

global positioning system (GPS) ■ A satellite system that provides reliable positioning and navigation on a continuous basis in all types of weather conditions, 24 hours a day, anywhere.

Global positioning systems (GPS) are satellite systems that provide reliable positioning and navigation on a continuous basis in all types of weather conditions, 24 hours a day, anywhere. The GPS can be used to specifically identify the location of individual firefighters at an emergency scene. The system requires a truck-mounted computer unit and a transponder attached to the individual. Some departments have attached the transponders to their self-contained breathing apparatus. The incident commander can call for additional people who are on scene and redirect their work efforts by knowing exactly where everyone is during the emergency, as long as firefighters are not already inside the structure. Departments that have the funding should take advantage of the information provided by GPS. Informed decisions are always better than those made while uninformed.

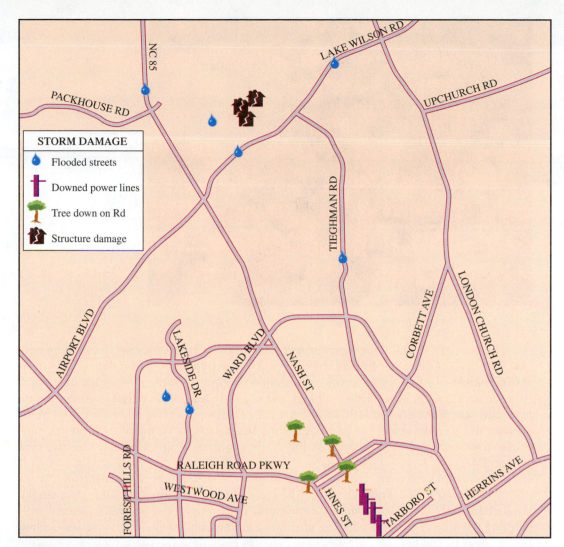

FIGURE 8.11 Geographic information systems can be used to identify areas of the community affected by natural disasters. *Courtesy of Don Oliver, Fire/Rescue Chief, Wilson Fire/Rescue Services*

Thermal Imaging Cameras

The first thermal imaging cameras were developed in the late 1950s and early 1960s. The main benefits of those devices were the ability to see military targets or opposing forces in low light and smoke-filled battlefields. But it was very apparent that the ability to see in those conditions would be extremely valuable to the fire department as well. In fact, the U.S. Fire Administration and the National Institute of Standards and Technology (NIST) researched the performance of thermal imaging to enhance emergency response safety by examining the capabilities and limitations of thermal imaging cameras (Amon, Bryner, Lock, and Hamins, 2008). Even more recently, NFPA 1801, *Standard on Thermal Imagers for the Fire Service,* was developed and published in 2010.

The principle that allows images invisible to the human eye to be seen is based on the wavelengths of light. The electromagnetic spectrum of visible light is a very narrow part of the full spectrum. The thermal imaging cameras detect the wavelengths of light found in the infrared range of the electromagnetic spectrum of light (750 nm to 100 μm). Note that the cameras can give misleading information due to shielding, reflections, water, and so on.

FIGURE 8.12 This dog survived a house fire after being located in the structure with a thermal imaging camera. *Courtesy of Martin Grube*

The uses of thermal imaging cameras by the fire and emergency services are numerous. The typical structure fire will have a lot of by-products of incomplete combustion throughout the building until proper ventilation is applied. In the past an incident commander would walk around the building and use only his past experience to determine how and where the crew should attack the fire. The camera is an additional tool he or she can use to help apply all the basic techniques necessary to make the best decision at the time a decision must be made for the responders to safely attack the fire. (See Figure 8.12.)

The movies have portrayed fire scenes as an area that can be walked through without any breathing protection or mask. The truth is that a building fire will have smoke banked down to the floor in a matter of minutes. The firefighters who are challenged with entering the smoke-charged building often are crawling on the floor looking and feeling for the heat of the fire. Helmet-mounted cameras can allow the fire attack crews to see the area that is on fire earlier in the process. The early detection will assist the first crews to find and extinguish fires in the early stages.

Accountability Systems

accountability systems ■ Methods used at an emergency scene to keep track of all of the firefighters on scene. A system can be large and complex with computer downloading capabilities or as simple as a sign-in sheet on a chalkboard. All emergency operations are required to have an accountability system in place to comply with the National Incident Management System (NIMS).

The fire department has been implementing **accountability systems**, methods used at an emergency scene to keep track of all of the firefighters on scene, since it began to embrace the FireScope model out of California (Chase, 1980) known as incident command system (ICS). The main reason for such systems is to know where responders are in the event of a collapse, Mayday, or major problem or to have such information as how many people are in the "danger area" and where they were working prior to the problem. The importance of such information to the safety of responders cannot be overestimated.

The early systems of keeping track of on-scene responders generally consisted of a tag attached to the back of every firefighter's helmet and given to one specific responder who was responsible for them until the incident was considered complete. A later system consisted of hanging the tags on a board that was used both to track each responder and to identify the task to which he or she was assigned.

The simple tag and board system still works fairly well at a small incident that involves only a small number of crews. However, at large multiple-alarm assignments that utilize ICS, a simple tag system becomes problematic, especially when multiple crews and outside agencies are working over a large area. (See Figure 8.13.)

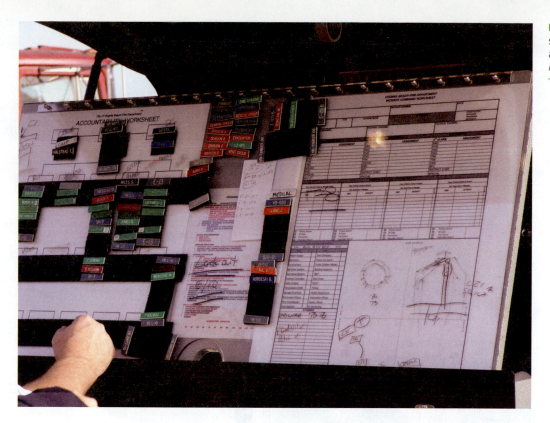

Several new systems connected to computers have been introduced. In one such system, transponders tied into an on-scene GIS can locate each and every responder on scene (but not in a structure), as well as document the times that each responder has been working. In another system, responders wear a scannable bar-coded tag that can include medical and emergency contact information, as well as the training level of each responder, which the incident commander can match with the proper task assignment. The information accumulated in the computer during the incident can be downloaded and printed later for use during the after-action review.

Self-Contained Breathing Apparatus (SCBA) Testing

Maintenance inspection and flow testing by computer are required by federal regulations for respiratory protection, which include 29 CFR Parts 1910.134 and 1910.156. Another document that deals with maintenance and testing of SCBA is NFPA 1852, *Standard on Selection, Care, and Maintenance of Open-Circuit Self-Contained Breathing Apparatus (SCBA)*, which specifies 12 different performance tests and performance levels for passing each test. (See Figure 8.14.)

Computers can help with maintenance of SCBA. Today annual SCBA flow tests to ensure the equipment works as designed are required and include examination of regulators, tanks, and mask assemblies. The annual fit test must be performed on all members who are required to wear breathing equipment. Currently, testing is being conducted at NIST on the thermal performance of SCBA face pieces to increase protection based on the documented line-of-duty deaths due to thermal exposures (Roberts, 2009c). (See Figure 8.15.)

Other important advancements made to the SCBA are in the chemical, biological, radiological, and nuclear protection area along with increased voice communication and heads-up displays.

FIGURE 8.14 Computer testing ensures performance standards are met.
Courtesy of Travis Ford, Nashville Fire Department

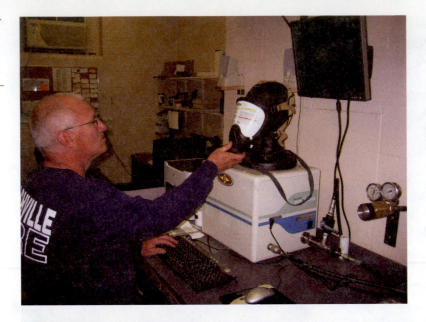

FIGURE 8.15 Annual face piece fit testing is now a requirement.
Courtesy of Travis Ford, Nashville Fire Department

TRAINING TECHNOLOGY

One way to avoid fire department training tragedies is by using training simulation technology. Computerized driver-training simulators are based on the same principle that airlines use to train their pilots and the U.S. military to train its tank drivers: it simply does not make sense to risk the lives of trainees. Nor does it make sense to put a military plane or tank at risk of permanent damage or of the great expense associated with repair.

FIGURE 8.16 The Kentucky Fire Commission driving simulator travels around the state, allowing individuals to train and refresh their driving skills. *Courtesy of Charles W. Shaw, Coordinator, State Fire Rescue Training*

Simulation technology and the fire department would benefit greatly from a close relationship. With the rising costs of emergency response vehicles and the lawsuits that are won against governmental agencies because of death or injury during training, the benefits of having a computer-based driver-training simulator may outweigh its initial cost. Simulators prevent real collisions or other incidents that injure individuals or damage equipment. Simulator technology used to train and refresh driving skills would avoid adding extra use of and maintenance time and expense to the vehicle inventory. It would allow for accurate records of training time for all department members. And tractor-trailer mounted units can be shared easily among several departments. (See Figure 8.16.)

Burn simulators are available to emulate most of the fires that firefighters are exposed to in their careers. For example, in 2005 a public safety training facility was opened at the Wayne County Community College in Taylor, Michigan. The facility includes a pit fire, a vehicle fire simulator, an airplane fire simulator, a ship-fire training simulator, a power substation simulator, a multistory burn building, and several buildings for search-and-rescue evolutions. It built a lake for water rescue simulations and buried a series of plastic tubes for confined space training. Such facilities are not affordable for most fire departments. However, shared resources or community resources such as a flashover simulator can and will provide live fire training to maintain responder situational awareness. (See Figure 8.17.)

An example of a virtual reality simulator may be found at the Houston Fire Department Officer Development Center. Recently opened, it delivers synchronized content to multiple viewers (Roberts, 2009b).

NIST is developing a virtual reality simulation of many different types of fire situations that will allow individuals to make strategic and tactical decisions on computers without risk to themselves. The fire modeling software is known as Fire Dynamics Simulator and Smokeview (NIST, 2009).

Remember, training is meant to improve skills and abilities of firefighters. No one should be injured and certainly no one should die in any training event.

TECHNOLOGY AND EXTINGUISHING FIRE

The trucks in use today cost half a million dollars or more, but still deliver the personnel to the scene in order to extinguish the fire. It can be demonstrated that some of the

FIGURE 8.17 Flashover simulators help firefighters enhance their situational awareness. *Courtesy of Martin Grube*

increased costs involved in producing a modern fire engine are equated to the technological advances in the chassis, braking, visibility, passenger safety systems, improved pumps, frame size, and so on.

Water is still the most cost-effective medium available to extinguish fires. However, there are other options. Compressed air foam, for example, makes water more efficient. It is an additive that reduces or eliminates the surface tension of water so it will soak into the item that is on fire. Compressed air foam systems generally make water more efficient, so it takes less water to complete the job.

When water is not available or when water would react with the burning product and make the fire worse, it may be necessary to use a water alternative. One category of alternatives includes chemical products that react with fuel to either suppress combustible vapors or form a foam that prevents contact with oxygen.

Most people believe that the elements of fire are heat, fuel, and oxygen. But science has discovered a fourth element in fire: the chemical chain reaction within the fire itself. Newer water alternatives address that fourth element, which works on the principle of disruption much like Halon products did in the late 20th century.

New products that can alter the traditional ways of fighting fire usually aim to make firefighting safer and more efficient. Whatever the new products may be, they need to be evaluated by each department to see whether they will work for it. If they do, it will take operational changes as well as special training to establish appropriate guidelines that address the proper utilization of these water alternatives and additives.

FUTURE TRENDS

The demands placed on the fire department will continue to evolve in the areas of planning, preparedness, response, and recovery. Technology now and in the future will continue to be

FIGURE 8.18 Fire behavior will continue to change and become more of a challenge for firefighters. *Courtesy of Dave Brasells*

important in meeting those demands, especially as it becomes more sophisticated, portable, and affordable. Just a bit of what the future holds is described next.

Fire Behavior

Approximately 20% of the firefighter deaths at structure fires over the past 10 years (not including the World Trade Center tragedy) have been the result of structural collapse (Stroup and Bryner, n.d., p. 1). The U.S. Fire Administration and the National Institute of Standards and Technology are researching the technology needed to predict structural collapse on the emergency scene of various building construction types. Their data have lead to the development of a prototype for monitoring the health of buildings. They report that "additional research is underway to continue development of the building monitoring system and examine specific construction types and scenarios of concern to firefighters" (Stroup and Bryner, n.d., p. 1).

Fire behavior will continue to change and become more of a challenge as individuals adopt new "green" energy conservation technology, including weight of green roofs, solar panel installations, photovoltaic systems, solar shingles, and so on. (See Figure 8.18.)

Radio Interoperability

Multiple fire departments and outside agencies at the same emergency scene often have communication problems, mainly due to each department having to operate only on its own radio frequency. Even coordinating the activities of resources within the same department on the small incident scene is sometimes problematic. A new model for communication among departments focuses on interoperability, or the ability among multiple agencies to talk to one another whenever they need to do so. The new model could help firefighters communicate more easily via mobile satellite communications and would involve mutual-aid radio talk groups, or Satellite Mutual Aid Radio Talkgroup Program (SMART™) channels (IAFC, 2009).

SAFECOM, as a program of the Department of Homeland Security, provides information to agencies on interoperability communications. SAFECOM has indicated that the process of achieving interoperability may take up to two decades, but that important interim measures that can help move the United States toward that goal are already in place (SAFECOM, n.d.). SAFECOM is currently involved in a three-part multiband radio research project that includes laboratory testing, short-term demonstrations, and pilot projects. Results on the SAFECOM will be published in 2010. According to an article in *Fire Chief* (Roberts, 2009a), the 14 lead departments involved in the pilot are:

- 2010 Olympic Security Committee (Blaine, Washington, and Vancouver, B.C., Canada)
- Amtrak (northeast corridor)
- Boise Fire Department (Boise, Idaho)
- Canadian Interoperability Technology Interest Group (Ottawa, Ontario, Canada)
- Customs and Border Patrol (Detroit)
- Federal Emergency Management Agency (multiple locations)
- Hawaii State Civil Defense (Honolulu)
- Interagency Communication Interoperability System (Los Angeles County, California)
- Michigan Emergency Medical Services (Lower Peninsula areas)
- Murray State University (southwest Kentucky)
- Phoenix Police Department and Arizona Department of Emergency Management (Greater Phoenix area and Yuma County)
- Texas National Guard (Austin, Texas)
- U.S. Marshals Service (Northeast region)
- Washington Metro Area Transit Authority Transit Police (District of Columbia)

Accountability Systems

The USFA and NIST tests show that the personal alert safety system (PASS) devices under certain conditions cannot fulfill their primary mission (USFA, 2007). However, there are other possibilities for their use in the future. With the proper emergency technologies, PASS devices could provide firefighter location, emergency scene accountability, gas analyzers, and physiological or stress monitors (USFA, 2007). (See Figure 8.19.)

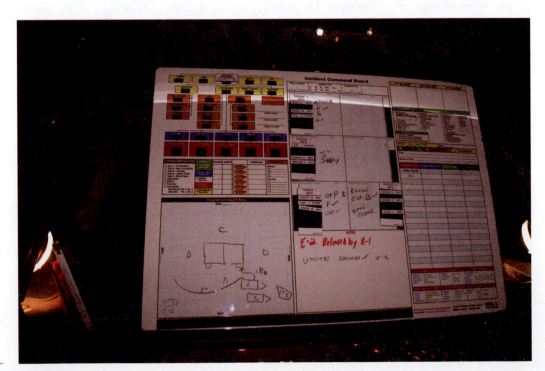

FIGURE 8.19 There can never be enough emergency scene accountability to help ensure firefighter safety. *Courtesy of Martin Grube*

Note that some of the location and tracking capabilities, along with physiological monitoring, began back in 1999 after the Worcester Cold Storage Warehouse fire. According to the Worcester Polytechnic Institute (WPI, 2009), a portable disposable wireless sensor can be carried into a building and placed in selected rooms. The WPI researchers estimate that the warning time to firefighters for flashover can be expanded from eight seconds to well over a minute by using this type of sensor (WPI, 2009).

Another form of accountability is having one system to help identify firefighters regardless of whether they are responding locally, regionally, or on a national basis as part of mutual aid. The U.S. National Grid can be used during any emergency incident for interoperable response capabilities for fire departments.

Future trends in research and development for the fire and emergency services could include innovations in firefighter PASS including personal distress alert signals, vital signs monitoring, tracking devices for multistory buildings, and devices built into SCBA, turnout gear, and other protective ensembles. Look for environmental sensor monitoring and fire robots capable of not only extinguishing fires but also of assisting in hazardous materials incidents and building collapses. And perhaps not too far in the future, mobile computers that produce live video streams of an emergency scene so that responders can view the incident before arrival will be available.

CHAPTER REVIEW

Summary

A fire-related death travels far into a community. Many firefighters are active in their places of worship, coach sports, volunteer regularly, have children, provide mentoring to youth groups, and so on. A loss of even one is immeasurable. So, it is imperative to stay abreast of changing technology throughout one's fire and emergency services career.

Remember, these are rapidly progressing, technological times that require constant adaptation and adjustment. Technological innovation historically has come about due to a recognized problem or expressed need. However, successful introduction of new technology into the fire and emergency services must be approached in the proper way. The assurance of firefighter safety alone will not guarantee success in gaining acceptance of new technology.

Emergency response is a complex career with an end product of dedicated service. The diversity found from one department to the next demands that each item or change in equipment and technology will require a great deal of effort from everyone who is interested in implementing the item into his or her department.

Always remember to identify the problem that technology will solve for the department before introducing the technology for possible adoption. The department always will have limited funds, so choosing the best new tool at the correct time is important. Even today there are still departments that do not use a large-diameter hose for water supply. So, carefully analyze the benefits and the operational changes required to implement new technology that can benefit the department. If the new technology sits on the shelf, it will not do anyone any good.

All of the new technology to keep firefighters safe will not have an impact on reducing the number of line-of-duty deaths and injuries unless an individual has been given the proper knowledge, skills, and abilities to use the technology properly. Firefighters need to understand the limitations of the technology as well as how to operate it properly.

Review Questions

1. Discuss the need for acceptance of new technology into the fire and emergency services.
2. Explain the benefits of setting goals.
3. Describe factors to consider when integrating new technology into the fire and emergency services.
4. Define geographic information systems.
5. List areas that computer technology has benefited the fire and emergency services.

References

American Psychological Association. (2003, October 23). *Fire Trucks Are Supposed to Be Red, Right? Not if You Want to Reduce Accidents*. Retrieved March 4, 2009, from http://www.psychologymatters.org/solomon.html

Amon, F., Bryner, N., Lock, A., and Hamins, A. (2008, July). *Performance Metrics for Fire Fighting Thermal Imaging Cameras: Small- and Full-Scale Experiments*. NIST Technical Note 1499. Gaithersburg, MD: National Institute of Standards and Technology.

Blankinship, D., Richter, K., and Navarro, M. (2008, November 15). "Fire Data Modeling: The Path to an Interoperable Future." *IAFC On Scene*, 22(20), 1–2.

Brandon, D. (2007, September 17). "Leading Change." From the *Success to Significant Speakers Series*. St. Louis, MO: Webster University, School for Business and Technology. Retrieved December 27, 2009, from http://www.webster.edu/depts/business/index_speak.php?page=speakers/video.php&speaker=david_brandon

Buchanan, E., Buckman, J. M., III., Scott, G., Windisch, F., Curl, L., Fulmer, D., et al. (2005, November). *Lighting the Path of Evolution: The Red Ribbon Report, Leading the Transition in Volunteer and*

Combination Fire Departments. Fairfax, VA: International Association of Fire Chiefs. Retrieved November 30, 2008, from http://www.iafc.org/associations/4685/files/vcos_RibbonReportRed.pdf

Buckman, J. M., III., Bettenhausen, R., Fulmer, D., Holman, T., Ray, S., Buchanan, E., et al. (2006, September). *The White Ribbon Report, Managing the Business of the Fire Department: Keeping the Lights on, the Trucks Running, and the Volunteers Responding*. Fairfax, VA: International Association of Fire Chiefs.

Calderone, J. (2003). "Fire Apparatus." In R. A. Yatsuk and J. G. Routley (Eds.), *Firefighters* (pp. 78–111). Emmitsburg, MD: National Fallen Firefighters Foundation.

Chase, R. A. (1980, May). *FIRESCOPE: A New Concept in Multiagency Fire Suppression Coordination*. General Technical Report PSW-40. Berkeley, CA: Pacific Southwest Forest and Range Experiment Station, U.S. Department of Agriculture.

Federal Emergency Management Agency. (2004, August). *Emergency Vehicle Safety Initiative*. FA-272. Washington, DC: U.S. Department of Homeland Security.

Federal Emergency Management Agency. (2009, October 7). *Hazus: FEMA's Methodology for Estimating Potential Losses from Disasters*. Washington, DC: U.S. Department of Homeland Security. Retrieved December 28, 2009, from http://www.fema.gov/plan/prevent/hazus/index.shtm

Granito, J. (2003). "Origins of Firefighting in the United States." In R. A. Yatsuk and J. G. Routley (Eds.), *Firefighters* (pp. 14–35). Emmitsburg, MD: National Fallen Firefighters Foundation.

International Association of Fire Chiefs. (2009, February). *A SMART™ Model for Interoperable Communications: Satellite Mutual Aid Radio Talkgroup Program*. Reston, VA: IAFC Technology Council and Skyterra Communications. Retrieved December 27, 2009, from http://www.iafc.org/associations/4685/files/techCouncil_SmartInfoPaper.pdf

Kehlet, R. (n.d.). *The Use of the Consequences Assessment Tool Set*. Redlands, CA: ESRI. Retrieved January 5, 2010, from http://proceedings.esri.com/library/userconf/proc97/proc97/to400/pap383/p383.htm

Meyer, R. (2008, November 15). "Geographical Information Systems: What Are They and What Can They Do for Me?" *IAFC On Scene, 22*(20), 1, 3.

National Center for Statistics and Analysis. (2008, September). *Traffic Safety Facts: Seat Belt Use in 2008—Overall Results*. DOT HS 811 036. Washington, DC: National Highway Traffic Safety Administration.

National Fallen Firefighters Foundation. (2005–2009). *Everyone Goes Home: 16 Firefighter Life Safety Initiatives*. Emmitsburg, MD: Author. Retrieved November 26, 2009, from http://www.everyonegoeshome.com/initiatives.html

National Highway Traffic Safety Administration. (n.d.). *The Benefits of State Seat Belt Laws*. From *Almanac of Policy Issues* archives. Retrieved November 29, 2008, from http://www.policyalmanac.org/economic/archive/seatbelts.shtml

National Institute for Occupational Safety and Health. (2007). *Nine Career Fire Fighters Die in Rapid Fire Progression at Commercial Furniture Showroom–South Carolina*. Fatality Assessment and Control Evaluation Investigation Report #F2007-18. Morgantown, WV: Author.

National Institute of Standards and Technology. (2009, November). *Fire Dynamics Simulator and Smokeview (FDS-SMV)*. Gaithersburg, MD: Author. Retrieved December 27, 2009, from http://fire.nist.gov/fds/

National Oceanic and Atmospheric Administration. (n.d.). *Risk and Vulnerability Assessment Tool (RVAT)*. Charleston, SC: NOAA Coastal Services Center. Retrieved January 5, 2010, from http://www.csc.noaa.gov/rvat/

Reh, F. J. (n.d.). *How to Use Benchmarking in Business: Who's Best? How Good Are They? How Do We Get That Good?* Retrieved November 29, 2008, from http://management.about.com/cs/benchmarking/a/Benchmarking.htm

Roberts, M. R. (2009a, July 7). "DHS Announces Final Phase of Multiband Radio Testing." In *Fire Chief*. Chicago, IL: Penton Media Inc. Retrieved December 28, 2009, from http://firechief.com/technology/communications/final-multiband-radio-test-20090707/

Roberts, M. R. (2009b, November 1). "Virtual Reality." In *Fire Chief*. Chicago, IL: Penton Media Inc. Retrieved December 28, 2009, from http://firechief.com/training/ar/simulator-warehouse-200911/

Roberts, M. R. (2009c, November 5). "NIST to Determine Face-Mask Weaknesses." In *Fire Chief*. Chicago, IL: Penton Media Inc. Retrieved December 28, 2009, from http://firechief.com/health-safety/ar/face-mask-weakness-investigation-20091105/

SAFECOM. (n.d.). *Interoperability*. Washington, DC: U.S. Department of Homeland Security. Retrieved December 27, 2009, from http://www.safecomprogram.gov/safecom/interoperability/default.htm

Stroup, D. W., and Bryner, N. P. (n.d.). *Structural Collapse Research at NIST*. Gaithersburg, MD: National Institute of Standards and Technology. Retrieved December 27, 2009, from http://www.fire.nist.gov/bfrlpubs/fire07/PDF/f07084.pdf

U.S. Fire Administration. (2002, April). *Firefighter Fatality Retrospective Study*. National Fire Data Center Report FA-220. Emmitsburg, MD: Author.

U.S. Fire Administration. (2007, October). *Personal Alert Safety Systems (PASS) Research*. Emmitsburg, MD: Author. Retrieved December 27, 2009, from http://www.usfa.dhs.gov/fireservice/research/safety/nist2.shtml

U.S. Fire Administration. (2009a, January). "Historical Overview: On-Duty Firefighter Fatalities 1977–2009." Emmitsburg, MD: Author. Retrieved February 14, 2009, from http://www.usfa.dhs.gov/fireservice/fatalities/statistics/history.shtm

U.S. Fire Administration. (2009b, August). *Emergency Vehicle Visibility and Conspicuity Study.* FA-323. Emmitsburg, MD: Author. Retrieved January 7, 2010, from http://www.usfa.dhs.gov/downloads/pdf/publications/fa_323.pdf

U.S. Fire Administration. (2009c, September). *Firefighter Fatalities in the United States in 2008.* Emmitsburg, MD: Author. Retrieved November 19, 2009, from http://www.usfa.dhs.gov/downloads/pdf/publications/ff_fat08.pdf

Worcester Polytechnic Institute. (2009, November 30). *WPI Receives $1 Million to Develop System Aimed at Preventing Firefighter Injuries and Death.* Worcester, ME: Author. Retrieved January 31, 2011, from http://www.wpi.edu/news/20090/fire.html

Wright, T. (2006, August). *Benchmarking for Success: Leading Community Risk Reduction.* Emmitsburg, MD: U.S. Fire Administration (Executive Fire Officer Program). Retrieved January 5, 2010, from http://www.usfa.dhs.gov/pdf/efop/efo39732.pdf

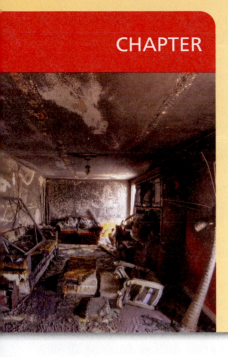

CHAPTER 9

Fatality and Injury Investigations

Dale R. (Rob) Rush

KEY TERMS

documentary evidence, *p. 215*
evidence, *p. 213*

expert witness, *p. 216*
fact witness, *p. 216*

**testimonial
evidence,** *p. 215*

OBJECTIVES

After reading this chapter, the student should be able to:

- Recognize that the lessons learned from a fatality and injury investigation must be communicated to prevent future incidents of a similar nature.
- List the basic functions commonly performed in an effective fatality and injury investigation.
- Identify and understand the makeup of an investigative team.
- Explain the need for effective documentation of a scene.

PEARSON
myfirekit™

For practice tests and additional resources, visit **www.bradybooks.com** and follow the **MyBradyKit** link to register for book-specific resources.

Register for **MyFireKit** by following directions on the **MyFireKit** student access card provided with this text. If there is no card, go to **www.bradybooks.com** and follow the **MyBradyKit** link to Buy Access from there.

Conducting Proper Fatality and Injury Investigations

According to the U.S. Fire Administration (2009), in 2008 alone a total of 118 firefighter deaths occurred in the United States. Fireground operations accounted for 28 of those deaths. However, not one of the structures involved in fatal fires was reported to have had fire sprinkler systems installed. Responding to and returning from alarms accounted for 24 deaths. Stress and overexertion, usually resulting in heart attacks or other sudden cardiac events, continued to be the leading cause of fatal injuries with 52 deaths.

The vast majority of firefighter fatalities and injuries are predictable and preventable. Unfortunately, most firefighter deaths and injuries are not thoroughly investigated. Starting in 2004 the U.S. Fire Administration (USFA) set the goal of reducing firefighter fatalities by 50% within 10 years. If the fire and emergency services is to meet that goal, it is critical for department administrators to understand the need both for thorough fatality and injury investigations and for the proper training of personnel who are to investigate the incidents for event failure.

The need to conduct a comprehensive fatality and injury investigation extends well beyond the normal idea of investigation for fire origin and cause determination. In fact, procedures of a line-of-duty death investigation can and should be applied to other situations as well, such as incidents that result in serious injuries and incidents that could have resulted in death or injury under different circumstances.

Does the department have health and wellness fitness protocols in place? Those protocols can allow for proper testing to determine whether items such as drugs or alcohol may have been a factor in a stress- or overexertion-related death. Annual physical testing for heavy metals as well as other toxins in the blood can help to prevent future problems. Does the department test for those issues regularly through annual wellness checks? Does the department investigate the findings by the medical examiner to determine whether there were issues that could necessitate steps to discontinue unsafe practices to prevent like incidents from occurring in the future? A close call should be identified as a warning to prevent the same situation from reoccurring.

Proper investigation and the reporting of findings of all firefighter deaths, injuries, and near misses can contribute significantly to the future reduction of tragic incidents. (See Figure 9.1.)

FIGURE 9.1 Proper investigation and reporting can help reduce firefighter line-of-duty death and injuries.

NFPA Investigative Standards

There are several NFPA standards and guides that can assist in the development of protocols and procedures to help prevent line-of-duty deaths and injuries from occurring and to assist in developing methods to ensure that such an event, should it occur, is properly investigated. They include:

- NFPA 921, *Guide for Fire and Explosion Investigations*
- NFPA 1002, *Standard for Fire Apparatus Driver/Operator Professional Qualifications*
- NFPA 1033, *Standard for Professional Qualifications for Fire Investigator*
- NFPA 1451, *Standard for a Fire Service Vehicle Operations Training Program*
- NFPA 1500, *Standard on Fire Department Occupational Safety and Health Program*
- NFPA 1582, *Standard on Comprehensive Occupational Medical Program for Fire Departments*
- NFPA 1583, *Standard on Health-Related Fitness Programs for Fire Department Members*
- NFPA 1915, *Standard for Fire Apparatus Preventive Maintenance Program*

In addition, a variety of American Society for Testing and Materials (ASTM) standards can assist in the development of training and protocols. Occupational Safety and Health Administration (OSHA) has regulations that may apply to fire department training and investigation efforts.

In addition to the preceding standards listed, departmental policies and procedures must be developed and followed in order to ensure the completion of each required task. Those policies and procedures should include:

- Agencies that must be notified
- Insurance benefits that are available and guidelines for filing claims
- Guidelines for the preparation of press releases and dealing with the media
- Contacts for community religious representatives
- Guidelines for securing permission from family members for an autopsy

OUTSIDE INVESTIGATORS

There are many parties interested in the facts related to a firefighter death or serious injury. Among them are the following:

- Local fire department and its firefighter union
- Municipality's risk manager
- Municipality's worker's compensation insurer
- Equipment manufacturer (if product involvement or failure may have been involved)
- Federal, state, or provincial occupational safety and health agency

With so many having a potential interest in the final determination of the death or serious injury, accuracy is paramount. Intense pressure may be applied to those responsible for the investigation by family members, politicians, media, community groups, police and fire administrative personnel, among others. As such, it is necessary to conduct a complete and precise fatality and injury investigation.

It may be wise to utilize outside assistance in the composition of an investigative team, which can help eliminate any hint of disparity or claims of cover-up. Outside agencies that can offer suggestions and directives throughout the completion of the investigation include the following:

- ***Bureau of Alcohol, Tobacco, Firearms and Explosives (ATF).*** Under certain criteria, the ATF can provide investigative assistance to local agencies on scenes involving firefighter death and serious injury.

- *Federal Bureau of Investigation (FBI).* The FBI will generally become involved in cases that may have circumstances related to civil rights violations (church fires, for example).
- *International Association of Arson Investigators (IAAI).* The IAAI can provide information relative to performing a proper, relevant investigation. It has an accredited, certified fire investigation program.
- *International Association of Fire Fighters (IAFF).* The IAFF has an occupational health and safety department that develops knowledge within the fire department so that firefighters themselves can recognize and control safety and health hazards associated with the profession.
- *National Institute for Occupational Safety and Health (NIOSH).* NIOSH has been named as the testing, approving, and certifying agency for a number of types of personal protective ensembles, including self-contained breathing apparatus.
- *National Institute of Standards and Technology (NIST).* Under the Building and Fire Research Laboratory, NIST can offer technical assistance in understanding the dynamics of fire behavior and provide analysis of an event utilizing fire modeling.
- *National Transportation Safety Board (NTSB).* The NTSB investigates and determines the facts, conditions, circumstances, and cause or probable cause of incidents that occur in connection with the transportation of people or property.
- *U.S. Department of Transportation (DOT).* The DOT has interest in two types of incidents:
  - Emergency vehicle incidents where vehicle design or maintenance defects may be a factor
  - Hazardous materials transportation incidents
- *U.S. Fire Administration (USFA).* Although having no investigative authority, USFA does have interest in collecting and distributing information relevant to firefighter safety and health that may be of interest to the fire and emergency services at large. (See Table 9.1.)
- *State fire marshal's office.* The office can provide investigative assistance to the local authority and possibly provide resources unavailable to local agencies.
- *Local and state law enforcement.* If the scene is determined to be criminal in nature, the involvement of law enforcement may be required in any fatality and injury investigation.

STANDARD OPERATING PROCEDURES AND GUIDELINES

An investigative team must look for and determine whether and how certain operations such as the following were conducted:

- Emergency scene safety
- Operational mode
- Risk management and task prioritization
- Technical correctness
- Staffing and support
- Command and control

TABLE 9.1	USFA/NFPA Report Categories for Nature of Death	
Cardiac arrest	Fracture	Hemorrhage
Drowning	Asphyxiation	Pneumonia
Aneurysm	Electrocution	Burns
Internal trauma	Heatstroke	Gunshot
Stroke	Crushing	Other

The department must have defined operational modes based on the incident commander's size-up of an incident scene. The incident commander must always clearly communicate the operational mode to firefighters. The offensive mode is an interior attack on a fire. The defensive mode is used when a fire is so fierce that interior operations are precluded, the structures are unsafe, or there is insufficient staffing. The department must explicitly state in standard operating procedures/standard operating guidelines (SOPs/SOGs) the actions expected of firefighters for each offensive or defensive operational mode. Marginal operations occur when conditions on the emergency scene are changing from offensive to defensive mode. During those times it is easy to lose track of firefighter locations and inappropriately mix tactics. The incident commander must continually size up the incident to ensure that the tactical approach matches the operational mode.

The incident commander should relay the information about the conditions of the emergency scene and operational mode to the fatality and injury investigation team because such information may have an impact on the incident investigation. The size-up process conducted by the incident commander is similar to the tactical evaluation by the investigative team.

Investigators must determine whether the affected agency followed proper protocols during the incident. To do so, investigators need to obtain copies of all department SOPs/SOGS that govern emergency scene activities before evaluating scene tactics, which is a vital portion of the investigation.

Investigators should consider the following three questions:

1. Did the firefighters follow the departmental SOPs/SOGS?
2. Did the department's SOPs/SOGS direct firefighters to take proper courses of action?
3. Did the incident commander follow proper SOPs/SOGs and make appropriate assignments, given the information available at the time?

The purpose of an investigation is to seek the truth, not to place blame. The goal of the investigation must always be to detect and correct operational deficiencies.

The Investigative Team

Investigators of line-of-duty deaths and injuries must have a firm understanding of their mission. The investigators must have the support and independence necessary to perform a thorough and unbiased investigation. The investigative team needs to separate emotions from facts to prevent a biased analysis of the incident. Any instruction that attempts to alter that mission is inappropriate. Any suggestion that a bias or cover-up is involved is a serious accusation. All necessary steps must be taken to avoid any such accusations. Using a task force concept and outside resources will help to legitimize findings. (See Figure 9.2.)

TEAM RESPONSIBILITIES

Some of the basic functions commonly performed in each investigation are:

- Leadership/coordinating
- Safety assessment
- Searching the incident scene
- Evidence collection and preservation
- Interviewing witnesses
- Photography
- Note taking
- Mapping and diagramming
- Final documentation

FIGURE 9.2 The investigative team must conduct an unbiased investigation into the cause of any firefighter line-of-duty death or injury. *Courtesy of Martin Grube*

TEAM COMPILATION

For building fatality and injury investigation teams, refer to NFPA 1033, *Standard for Professional Qualifications for Fire Investigator* (2009b), which details the knowledge and skills necessary to establish the proper professional qualifications for fire investigators. The selection of team members should be a cooperative effort among all interested parties. For a concise investigation, an investigative team usually requires a minimum of three to five individuals, each chosen for his or her professional skills and abilities. (See Table 9.2.)

There may be a need for team members with specialized skills as well. For example, if the health of a firefighter is suspected to be a factor in the incident's outcome, it may be advantageous to include on the team a physician who specializes in occupational medicine and inhalation toxicology, and who is familiar with the functions of protective clothing and equipment.

A matrix for a basic team could be set up as shown in Figure 9.3. It is meant to establish a chain of command for information flow. However, there may not be a need for an individual for each of the depicted functions, because some team members may possess knowledge in two or more of the category functions.

Once the team is chosen and the investigation team leader has been selected, the team leader should hold a pre-investigation team meeting prior to any on-site investigation functions. The team leader should address any issues dealing with jurisdictional boundaries

TABLE 9.2	Qualities of a Good Investigative Team Member	
■ Integrity	■ Experience	■ Meticulousness
■ Communication skills	■ Motivation	■ Empathy
■ Sound judgment	■ Analytical ability and writing skills	■ Curiosity
■ Tact and diplomacy	■ Perseverance	

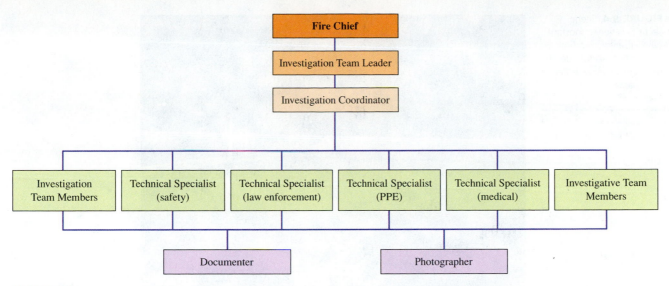

FIGURE 9.3 A basic organizational chart for the flow of investigative information. *Used with permission of the National Fire Protection Association*

and then assign specific responsibilities to the team members. The team leader should have made previous arrangements for any necessary housing and feeding of investigative team members. Some investigations may be lengthy based on the size and complexity of the incident.

The investigative team should be advised of the condition of the scene and the safety precautions required. All investigative personnel should be equipped with appropriate safety equipment before commencing on-site work. The team leader should ensure that there is a complete set of basic tools available for use. The team will be split up to perform different functions, such as cause determination, witness interviewing, mapping and diagramming, and so on. As such, the team leader should assign times for team members to meet and share information that may be pertinent to each function as it has been gathered since the last meeting.

If a fire is the cause of firefighter death, injury, or near miss, it will be necessary to perform an origin-and-cause determination as to the fire's causation in conjunction with the determination of factors that led to the death, injury, or near miss. Guidelines found in NFPA 921, *Guide for Fire and Explosion Investigations* (2008a), identify the proper steps to perform in an investigation of a fire death or injury incident.

If any criminal responsibility is suspected for the cause of the incident, it is vital that personnel of appropriate law enforcement agencies be included on the team. Local law enforcement will possibly get involved by way of its homicide or crime scene divisions.

Scene Security

Scene security is one of the most important functions for an incident that requires investigation. Scene security should begin with the arriving crews. Initially, the incident commander and, later, the lead investigator should ensure security of the scene from unauthorized intrusions, trace evidence, contamination, or loss or avoidable movement of physical evidence. (See Figure 9.4.)

NFPA 921 defines **evidence** as "any physical or tangible item that tends to prove or disprove a particular fact or issue." Failure to protect evidence can bring claims of spoliation, which can lead to civil and criminal sanctions. Access should be limited strictly

evidence ■ Any physical or tangible item that tends to prove or disprove a particular fact or issue.

to those at the scene who need to perform active operational functions. Once the scene hazard has been eliminated, the scene should be locked down and preserved for the investigators. Every attempt should be made to protect and preserve the scene as intact and undisturbed as possible. The scene should be cordoned off by the use of flags, tape, or personnel posted at the scene.

Posted personnel should require name, rank, and contact information for all those entering the scene because they will need to be interviewed about their purposes for accessing the scene at a later time. Those personnel will be added to the witness list, which could subject them to testifying in court about their actions inside the scene.

RIGHT OF ENTRY

The ability to access a scene will be dictated by different circumstances. The four methods generally recognized to establish right of entry are:

- Exigency
- Consent
- Administrative search warrant
- Criminal search warrant

Those methods are enumerated by constitutional rights, federal/state statutes, rules, and regulations.

Supreme Court rulings, such as *Michigan v. Tyler,* 436 U.S. 499 (1978) and *Michigan v. Clifford,* 464 U.S. 287 (1984), set the minimum parameters for entering any suspected crime scene. Those rulings establish when exigency ends and the need for a warrant begins. Investigation team leaders should be schooled in those rulings and ensure that their team is on solid legal ground before entering a scene.

Should the team leader utilize consent to enter the property, it should be clearly understood that the consent is voluntary and can be rescinded at any time the consenter chooses. Consent, therefore, should be obtained in writing on a form that clearly outlines the parameters for which the consent is being granted. If consent is rescinded, it may be necessary to obtain the appropriate search warrant. Team leaders should be versed in the laws of consent to ensure that the person giving consent is the legal party capable of doing so.

SCOPE OF SCENE SECURITY

Whether the investigation involves a vehicle accident scene or a fire training burn, explosion, or hostile fire scene, the scope and size of the protected area for preservation must take into account the scene circumstances. For example, a typical single-family-dwelling fire may include the structure and its contents, the immediate surrounding property, and possibly appurtenances such as dumpsters in the area. In comparison, a scene the size and complexity of the terrorist attack on the Alfred P. Murrah Federal Building in Oklahoma City or the World Trade Center in New York City will present challenges that could require greatly expanded areas of security control.

The Investigation

The scene should be investigated utilizing the scientific method as described in NFPA 921. A basic methodology should be used that sets parameters for each investigation. The systematic approach of the scientific method will ensure that the investigators are utilizing a standard that will hold up to the scrutiny of a court case should one present itself. After all the empirical data have been collected and the cognitive and experimental testing is complete to the satisfaction of the lead investigator and the investigative team, a final conclusion can be established. The conclusion will be based on the compilation of information gathered from a thorough scene investigation, gathering of facts and information, and analysis of all evidence.

> PEARSON
> **myfirekit**
>
> Please visit MyFireKit Chapter 9 for more information on the NFPA 921 Scientific Method.

EVIDENCE

Chain of Custody

The chain of custody for evidence is paramount to a particular case when court testimony may be required. The evidence can be the single most important aspect that can make or break the case in court. If the evidence has not been properly preserved, collected, and documented as to security control, it can be dismissed from the court case. Evidence for the defense is most vulnerable to attack. Custody forms or property receipts and continuous control documentation can mean the difference between the court accepting or throwing out the evidence. For that reason an investigation team leader should assign one specific person to properly collect, package, and control the evidence at the investigation scene.

The team member assigned the responsibility for identifying, collecting, packaging, examining, and arranging for testing of the physical evidence should ensure through proper documentation that the chain of custody remains unbroken.

Types of Evidence

There are three types of evidence: demonstrative, documentary, and testimonial. Demonstrative evidence is evidence that incorporates a form or representation of an object. Examples of demonstrative evidence are maps, diagrams, timelines, models, sketches, samples, photographs, and so on. This type of evidence is generally admissible in a court of law on the basis of investigator testimony.

Documentary evidence is any evidence in written form related to the fire, explosion, or other incident. Examples of documentary evidence include business reports, inventory lists, bank records, medical records, insurance policies, witness statements reduced to writings, fire/law enforcement reports, and maintenance records. These records are generally admissible if they are maintained in the normal course of business.

Testimonial evidence is that which is given by a competent live witness speaking under oath or affirmation. The investigator assigned to interview witnesses should make sure that the written statements are signed by the person giving them and are witnessed by a third party.

documentary evidence ■ Any evidence in written form related to the fire, explosion, or other incident.

testimonial evidence ■ That which is given by a competent live witness speaking under oath or affirmation.

There are two types of testimonial witnesses in a legal proceeding: fact witnesses and expert witnesses. A **fact witness** is one who has firsthand knowledge of a crime or dramatic event. His or her testimony is not based on scientific, technical, or other specialized knowledge. The fact witness may be a person who has witnessed the event or has intimate information about the incident or how it occurred. It may be someone who has information about documentary evidence.

An **expert witness**, according to NFPA 921, is "someone with sufficient skill, knowledge, or experience in a given field so as to be capable of drawing inferences or reaching conclusions or opinions that an average person would not be competent to reach. The expert's opinion testimony should aid the judge or jury in their understanding of the fact at issue and thereby aid in the search for truth." Origin and cause investigators on the team and those with specialized skills may fall in the expert-witness category and must have adequate credentials to be accepted as such by the state and federal court systems.

INTERVIEWING

Interviews are conducted for a variety of reasons. Often the success of the investigations will rely on how effectively the witnesses are interviewed. Interviewers must be skilled enough to ask questions in a manner that stimulates the gathering of facts, enhances witness recollection, and is used as a method of determining truthfulness or deception through verbal responses or the study of body motion and unusual behavior. Generally, interviewing is a learned skill. Investigation team leaders should select team members to conduct interviews based on their background and experience to obtain usable, reliable information.

Primary objectives for the interview are:

■ To gain the subject's version of events
■ To obtain the details of the situation
■ To establish independent sources to corroborate the story/facts

Establishing trust with the subject being interviewed is a must. The person being interviewed will have information useful in determining where the situation took place, who was involved, what happened, when all of the events occurred (timeline), and how it happened.

Two types of information can be gathered from an interview: direct and indirect. Direct information deals with events that took place immediately prior to or during the incident. People with direct information have either observed or participated in the incident. (See Table 9.3.)

Indirect information reveals the general circumstances surrounding and contributing to the incident. (See Table 9.4.) It can reveal specific technical knowledge that may explain aspects of the incident. Types of indirect information include maintenance records, inspection logs, medical information, training documents, and technical specifications of equipment.

TABLE 9.3	Witnesses: Sources of Direct Information	
■ Firefighters	■ Safety officer	■ Owners/occupants
■ Fire officers	■ Dispatchers	■ Civilian witnesses
■ Incident commander	■ Police officers	■ News media

TABLE 9.4	Witnesses: Sources of Indirect Information	
■ Coworkers	■ Family and friends	■ Training instructors
■ Maintenance personnel	■ Family or department physicians	■ Technical specialists
■ Rescue personnel	■ Medical examiner	

A thorough investigation of all sources of information is vital in determining the facts surrounding an incident. The accuracy of an investigation often relies on interviewing people with direct information as soon after the incident as possible. If the interview may be delayed or is incapable of being held soon after the incident, get a written statement.

Firefighter reports from first due crews may be particularly important. Those reports should be filled out individually, not as a group. Encourage each witness to include any drawings, diagrams, or sketches that may be helpful in explaining what happened.

One of the most important aspects of the investigation is for the interviewer to determine what is and is not credible. It is not unusual to have contradictions or inaccuracies in accounts, especially when interviewing multiple persons who have different job functions.

Make sure each subject being interviewed understands that he or she should respond only to questions with facts that he or she knows to be true. Those being interviewed should be made aware that the purpose of the interview is to gain facts. Opinions should not be given unless solicited by the interviewer. However, interviewers should take into consideration the psychological stresses the subject being interviewed may be under.

Some of the factors that can have a dramatic impact on testimony evaluation are:

- Length of time between the incident and the interview
- Amount of contact the witness has had with other principal witnesses
- Misinformation or omissions
- Witnesses questioning their own testimony
- Signs of shock, amnesia, or critical incident stress
- Health issues or use of any prescription/nonprescription drugs or alcohol
- Personal interest in ending the interview, evading responsibility, or diversion

Because of these factors, there can be discrepancies in answers. It is useful for interviewers to establish a chart for witness answer comparison to help in further defining an accurate response. An example of such a chart is shown in Figure 9.5.

Follow-up interviews may be necessary to clarify questions of conflicting testimonies. They can be useful to fill in gaps in information and explore factors in the incident that had previously been covered. All follow-up interviews need to be closely documented.

Obtaining accurate information is the result of proper interviewing techniques. Comfort for the witness and how questions are asked will directly affect the amount of information received. Remember: the goal of the interview is to find out what happened and how to prevent it from happening again.

EXAMINATION OF THE SCENE

As soon as a fatality is determined and the victim is found to be beyond medical aid, every effort should be made to minimize operations in close proximity to the victim, including

Witness Answer Comparison Chart

Witnesses	Fact 1	Fact 2	Fact 3
Emergency Responder 1	✓		✓
Emergency Responder 2	✓		✓
Officer 1	✓	✓	✓
Incident Commander	✓	✓	
Witness		✓	✓

FIGURE 9.5 Sample chart used for witness answer comparisons.

foot traffic, hose lines, equipment, and so on. Obviously, if there is any chance of resuscitation, the survival of the victim must take priority. It is beneficial to the entire fire and emergency services death investigation for the body to be left in place until it is properly documented and examined by the investigative team. Only severe emergency conditions should force premature removal of the body, such as imminent collapse of the building or other uncontrollable circumstances.

Once the scene has been documented photographically, and through notes, diagrams, and sketches, the body should be placed in a new, unused sealed body bag. All debris associated with or adhering to the body should be transported in the body bag and preserved for trace evidence, volatiles, projectiles, and so on. Search the area under where the body had been located for evidence that may have fallen loose while the body was being moved. Critical evidence often is found within arm's reach of the body. A careful search must be made of the entire room or area.

After the body has been removed from the emergency scene, a standard investigation for origin and cause should be performed according to the policies and procedures for any scene investigation. Search of the emergency scene for specific evidence related to the death should be conducted, encompassing the entire scene.

Any notification of personnel outside the department should not take place until the immediate family members of affected personnel have been notified.

A search system may be developed to conduct the investigation by dividing the scene into specific areas. The common search methods include spiral, parallel, grid, and zone.

- *Spiral.*   An outward spiral search begins in the center of the scene area and continues in an outward circular pattern. In an inward spiral search, the pattern begins at the outer portions of the scene and progresses in a circular pattern toward the center.
- *Parallel.*   In a parallel search, members form a line parallel to each other and walk in a straight line, at the same speed, from one end of the scene to the other.
- *Grid.*   A grid search is conducted using two parallel searches offset by 90 degrees, performed one after the other.
- *Zone.*   A zone search is divided into sectors, and each sector is examined individually. Be sure to overlap search areas to ensure complete coverage.

FIRE MODELING/FIRE DYNAMICS ANALYSIS

A very valuable tool that is available to analyze the complex fire dynamics issues involved in a firefighter death or serious injury is that of fire modeling or fire dynamics analysis as described in NFPA 921.

Two fire dynamics analysis tools, commonly referred to as *fire models,* that are readily available are the Fire Dynamics Simulator (FDS) and the Consolidated Model of Fire Growth and Smoke Transport (CFAST). These peer-reviewed and validated software programs are recognized as authoritative in the field of fire science. Verification is a check of the math; validation is a check of the physics. That is, when the model predictions closely match the results of experiments, using whatever metric is appropriate, it is assumed by most that the model suitably describes, via its mathematical equations, what is happening. It is assumed that the solution of these equations must be correct. (See Figure 9.6.)

The outputs of both programs then can be visualized in a third program, known as Smokeview (SMV). Smokeview visualizes or converts the data from the predictions generated by the FDS and CFAST programs to color graphic visual representations. One feature of Smokeview is that it can display slices (horizontal or vertical) cut through the room or compartment. These slices can show data such as temperatures or gas concentrations.

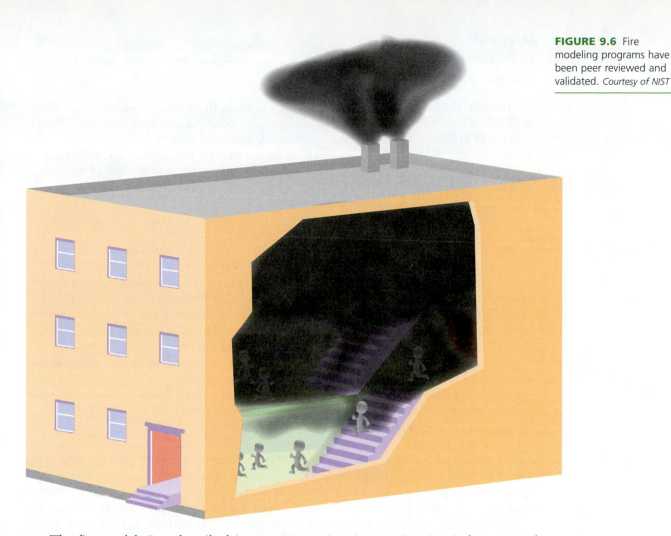

FIGURE 9.6 Fire modeling programs have been peer reviewed and validated. *Courtesy of NIST*

The fire models just described incorporate engineering, engineering judgment, and scientific principles in the analysis of fire and fire effects. Models can be used to simulate or, some would say, predict the characteristics and conditions of a specific fire under specified conditions.

Fire characteristics simulated in the fire effects models include:

- Gas/surface temperature
- Flow rates of gas
- Heat flux
- Smoke obscuration
- Toxic gas movement
- Building elements
- Activation time for sprinklers/detectors
- Various fire parameters

The Fire Dynamics Simulator has been used to assist in the analysis of a number of firefighter fatality incidents, including a townhouse in Washington, DC (Madrzykowski and Vettori, 2000), and a two-story dwelling in Iowa (Madrzykowski, Forney, and Walton, 2002).

Fire modeling software is intended to be used only by those competent in the fields of fire safety, fluid dynamics, thermodynamics, combustion, and heat transfer. The use of the software is to supplement the informed judgment of its qualified users. All results should be evaluated by an informed user.

AUTOPSY

A complete autopsy should be performed for every firefighter line-of-duty death. Autopsies serve four purposes:

1. To assist the investigation in determining the cause of death
2. To help determine eligibility for death benefits under the U.S. Public Safety Officers' Benefit Program, as well as state, provincial, and local programs
3. To advance the analysis of causes of firefighter deaths, thus aiding in the development of improved firefighter health and safety equipment, policies and procedures, and standards
4. To address an increasing interest in the study of deaths that could be related to occupational illness among both active and retired firefighters

Every department should have on hand the U.S. Fire Administration Firefighter Autopsy Protocol, which should be given to the local medical examiner or coroner as soon as possible, if he or she does not currently have it on file. The medical examiner performing the autopsy should follow the protocol for each firefighter line-of-duty death. A number of examinations can be conducted on the victim that may yield valuable information to the investigator, including X-rays and tests for carbon monoxide levels, presence of other toxic products, smoke and soot exposure, and burns.

A complete understanding of the cause of a firefighter's death must include some consideration of toxicological agents that may have been involved and how they may have interacted with biological processes and systems to cause death. Previous conditions often are accompanied by other injuries, which may not themselves have caused death, such as crushing forces or prolonged exposure to heat.

The most common products of combustion are carbon monoxide and either soot or ash. However, acrolein, cyanide, formaldehyde, hydrogen chloride phenol, phosgene, polycyclic aromatic hydrocarbons (PAHs), nitrogen oxides, water vapor, and carbon dioxide can be present. Blood tests should be conducted to detect the presence of ethyl alcohol to determine whether or not the deceased was under the influence of intoxicating beverages. Tests for common narcotics, barbiturates, amphetamines, hallucinogens, or cannabinoids should be done. Tests for other prescription and nonprescription drugs are occasionally performed to detect such compounds as common steroids, analgesics, and other indicators of coexisting illnesses and conditions, as well as any drugs used in emergency resuscitation attempts. The location, distribution, and degree of burns or other injuries should be photographed and shown on a body diagram.

Burns should be thoroughly documented. It is vital to be aware that conducted heat can be more dangerous than radiant or convected heat. Skin can be damaged when it reaches a temperature of 130°F (54°C). Clothing, especially heavier cellulosic fabrics, can transmit enough heat by conduction to cause skin burns even though the fabric does not exhibit any burning or charring.

Upon hospital entry of a victim with fire-related injuries, a blood sample should be taken. Its analysis should include the percent saturation of carboxyhemoglobin (COHb), HCN concentrations, blood alcohol, drugs, and blood PH. The percent of COHb begins to be reduced as soon as the individual is removed from the fire environment, so it is important to take the blood sample as soon as possible. The presence of soot or thermal damage in the upper airways provides information about the fire environment to which the individual was exposed. Lung edema and inflammation can be indications of exposure to irritant gases.

It is important to remember that any clothing or associated items related to the injured firefighter must be preserved for examination. Clothing or other items may be cut off at the scene or at the hospital. The hospital needs to be notified immediately not to destroy any items associated with the victim.

Many fire-injured victims suffer from respiratory burns as well as external surface burns. An investigator should be dispatched immediately to the hospital to obtain a statement from the victim before he or she is intubated. Crews that treated and transported the victim should be interviewed to determine whether the victim made any spontaneous statements.

All injuries should be documented through photographs. All associated clothing removed should be photographically documented. All pertinent medical information compiled by the medical staff should be obtained and examined. Any statements made by the victim to others prior to intubation should be acquired and recorded.

DOCUMENTATION

Proper documentation by the investigator is critical; it provides a method of recall when the investigative team sits down to examine all of the facts and issues to come up with its final conclusion. Methods often used to document the incident are photographs, videotapes, audio recordings, written notes, maps, diagrams and drawings, outside documents that support the case, and security tapes from cameras located at the incident scene.

Items to consider for documentation include the following: the location and position of dead or injured persons; the incident management structure at the time of the incident; the location of the incident command post; the location of windows, doors, and ventilation openings; the location and position of apparatus; the position of hose lines; the location of tools and equipment; the location of pieces broken off from equipment or tools; and areas of debris. The origin and cause of the incident should be documented by the investigator or investigative team.

Photography

Photography provides the investigator with pictures of the scene that he or she can use for recall when reconstructing the scene, use as points of reference when writing the report, or use as a source of evidence location and collection when describing the scene to the court. Visual media, such as digital still and video, is the primary method of recording the scene, which can provide an incontrovertible record of the incident.

A series of photographs and video should be taken to make a visual record of the incident. It needs to include the following: areas of fire suppression, the exterior and perimeter of the scene, structural elements, the interior of the scene, victims or bodies, observers and bystanders of the scene, aerial shots of the incident scene, areas of suppression overhaul, witness viewpoints, collection of evidence, and the evidence.

Photographic documentation of the incident scene should be taken as soon as practical, because the scene may be altered, disturbed, or destroyed. Such attention to the scene may require investigators to don SCBA and full protective gear to enter the scene to record it on film. Photos should be taken as soon as practical when the building is in imminent danger of collapse, the structure must be demolished, or contents of the building are creating an environmental hazard that requires immediate attention.

There should be a photographic documentation during the entire incident to show all processing. Evidence must be documented when discovered as layers of debris are removed. Still photos should be documented on photo logs, indicating the picture number, when the picture was taken, who took the picture, condition the picture was taken in (film type, lighting, lenses used), position from which the photo was taken, and what the picture actually shows.

Organization of the photos is important. There may be a great deal of time between when the photographs are taken and when they are analyzed. Someone other than the lead investigator may have taken the photos, so the photographer should be identified in case he or she is needed for court testimony. As many pictures should be taken from as

FIGURE 9.7 Photos should be taken from all four angles of the room and both sides of the door. This is an example of a picture taken from two angles. *Used with permission of the National Fire Protection Association*

many angles as necessary to recall the incident scene and document all items involved in the investigative process. Photos should be taken from all four angles of a room and from both sides of the door. When the finished product of a floor plan is complete, it can be copied, and directional arrows can be drawn to indicate the direction from which each of the photographs was taken. Corresponding numbers are then placed on the photographs. (See Figure 9.7.)

Note Taking
Note taking is a method of documentation in addition to photographs and drawings. The following items may need to be documented: dates and times; names and addresses; make, model, and serial numbers; statements and interviews; photo log; identification of items; types of materials; and investigator's observations. Field notes are to be used to enhance investigator recall. Again, such notes are not designed to constitute the incident report, but may be subpoenaed for future court proceedings.

Diagrams and Drawings
Sketches, diagrams, and plans for structures or property can be drawn up by the investigator or obtained through other means such as the town building department. These diagrams and drawings can be useful in conducting witness interviews. At times it may be advantageous for the witness to do his or her own rough sketch of what he or she is giving testimony to. The sketch will further solidify witnesses' statements should they attempt to alter their statements at a later date. There are two categories for drawings: two dimensional, which includes floor plans, elevation drawings, site plans, and the like, and three dimensional, which is more realistic. Small-scale models of the incident scene can be built to enhance testimony for court and give a better perspective as to the case issues. The disadvantages of a three-dimensional model is it can be time consuming and costly.

Generally, freehand diagrams or drawings done with minimal tools are completed at the scene. Field sketches include features found at the scene. Diagrams are more formal drawings that are usually drafted after the scene is completed using the incident scene sketches and investigator's photographs for recall. Diagrams can be drawn using traditional methodologies or computer-based drawing programs.

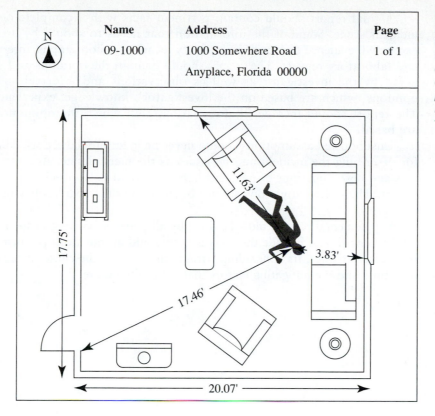

| N | Name 09-1000 | Address 1000 Somewhere Road Anyplace, Florida 00000 | Page 1 of 1 |

FIGURE 9.8 At the minimum, a simple diagram must be completed on all investigative incidents, especially those involving death or injury. *Used with permission of the National Fire Protection Association*

Sketches and diagrams should include the following:

- Compass orientation
- Scale
- Symbols
- Legend
- Name of person performing the sketch
- Date and time
- Case number
- Address

At the minimum, a simple diagram must be completed on all investigative incidents, especially those involving death or injury. A diagram should be produced showing the fire-damaged areas of the contents after reconstruction was done. The body should be documented on a diagram using measurements to show the exact location where it was found. (See Figure 9.8.)

Investigation Reports

The purpose of an investigation report is to effectively communicate the observations, analyses, and conclusions made during the investigation. Reports should include the following:

- Date, time, and location of the incident
- Date and location of the examination
- Date the report was prepared
- Scope of the investigation
- Tasks completed
- Nature of the report (preliminary, interim, final, summary, supplementary)

A final report should contain pertinent facts with a complete description of the incident scene. Some of the information that needs to should be provided are items examined, evidence collected, injury/body locations, photographs, interviews, diagrams, and laboratory reports. The report should contain the opinions and conclusions rendered by the investigator as well as observations and information relevant to the opinions, which are based on the investigator's knowledge, experience, and training. The report should include the foundation(s) on which the opinion and conclusions are based.

Unsupported assumptions should never be in an official report. The report is solely for describing the incident and its issues as the facts dictate. Accusations of negligent acts and determinations of personal liability should not be included in the body of the report. These determinations should be made by other parties beyond the investigating entity.

Supplemental reports must be done by all persons assisting in the investigation. The final report should include the facts gathered and all functions performed and observed by the investigator. Its concluding paragraph must list the cause of the death, injury, or near miss and the mitigating factors that led to the failure.

Summary

The information addressed in this chapter can serve as the foundation for establishing effective investigative teams capable of providing accurate and detailed information about the events that led to the failure causing a firefighter death, injury, or near miss. The fire and emergency services has suffered far too many deaths. Countless injuries continue to occur as a result of factors such as inefficient operations, ineffective equipment and maintenance, fatigue, lack of training, and complacency.

Many of these factors go unidentified, which results in the continued use of methods, policies, and procedures that allow for repeated failure. The importance of fatality and injury investigations and funding to properly train personnel in the techniques needed to be effective, could help in reaching the USFA's goal to reduce the number of firefighter deaths and injuries in the United States. Fatality and injury investigations will result in the ability to recognize areas in which concentration can be given in training or updating practices and establishing new SOPs/SOGs for incident operations to avoid future incidents.

Review Questions

1. Discuss what the minimum recommended number of members is for investigating a fatality or injury.
2. List the two essential needs of an effective investigative team.
3. Discuss the basic functions commonly performed during an effective fatality or injury investigation.
4. Describe three types of evidence.
5. Define the different search pattern systems that may be used to conduct a fatality or injury investigation.

References

De Haan, J. D. (2006). *Kirk's Fire Investigation*. Upper Saddle River, NJ: Pearson Brady.

Icove, D. J., and De Haan, J. D. (2003). *Forensic Fire Scene Reconstruction*. Upper Saddle River, NJ: Pearson Brady.

Madrzykowski, D., Forney, G. P., and Walton, W. D. (2002, January). *Simulation of the Dynamics of a Fire in a Two-Story Duplex—Iowa, December 22, 1999*. NISTIR 6854. Gaithersburg, MD: National Institute of Standards and Technology. Retrieved December 16, 2009, from http://fire.nist.gov/CDPUBS/NISTIR_6854/duplex.htm

Madrzykowski, D., and Vettori, R. L. (2000, April). *Simulation of the Dynamics of the Fire at 3146 Cherry Road NE, Washington, DC, May 30, 1999*. NISTIR 6510. Gaithersburg, MD: National Institute of Standards and Technology. Retrieved December 16, 2009, from http://fire.nist.gov/CDPUBS/NISTIR_6510/index.html

Michigan v. Clifford, 464 U.S. 287 (1984). Retrieved December 14, 2009, from http://supreme.justia.com/us/464/287/

Michigan v. Tyler, 436 U.S. 499 (1978). Retrieved December 14, 2009, from http://supreme.justia.com/us/436/499/

National Fire Protection Association. (2007a). *NFPA 1451, Standard for a Fire Service Vehicle Operations Training Program*. Retrieved December 14, 2009, from http://www.nfpa.org/AboutTheCodes/AboutTheCodes.asp?DocNum=1451

National Fire Protection Association. (2007b). *NFPA 1500, Standard on Fire Department Occupational Safety and Health Program*. Retrieved December 14, 2009, from http://www.nfpa.org/aboutthecodes/AboutTheCodes.asp?DocNum=1500

National Fire Protection Association. (2007c). *NFPA 1582, Standard on Comprehensive Occupational Medical Program for Fire Departments*. Retrieved December 14, 2009, from http://www.nfpa.org/AboutTheCodes/AboutTheCodes.asp?DocNum=1582

National Fire Protection Association. (2008a). *NFPA 921, Guide for Fire and Explosion Investigations*. Retrieved December 14, 2009, from http://www.nfpa.org/AboutTheCodes/AboutTheCodes.asp?DocNum=921

National Fire Protection Association. (2008b). *NFPA 1583, Standard on Health-Related Fitness Programs for Fire Department Members.* Retrieved December 14, 2009, from http://www.nfpa.org/AboutTheCodes/AboutTheCodes.asp?DocNum=1583

National Fire Protection Association. (2008c). *NFPA 1915, Standard for Fire Apparatus Preventive Maintenance Program.* Retrieved December 14, 2009, from http://www.nfpa.org/AboutTheCodes/AboutTheCodes.asp?DocNum=1915

National Fire Protection Association. (2009a). *NFPA 1002, Standard for Fire Apparatus Driver/Operator Professional Qualifications.* Retrieved December 14, 2009, from http://www.nfpa.org/AboutTheCodes/AboutTheCodes.asp?DocNum=1002

National Fire Protection Association. (2009b). *NFPA 1033, Standard for Professional Qualifications for Fire Investigator.* Retrieved December 14, 2009, from http://www.nfpa.org/aboutthecodes/AboutTheCodes.asp?DocNum=1033

U.S. Fire Administration. (2009, September). *Firefighter Fatalities in the United States in 2008.* Emmitsburg, MD: Author. Retrieved November 19, 2009, from http://www.usfa.dhs.gov/downloads/pdf/publications/ff_fat08.pdf

Brian P. Vickers

Courtesy of Brian P. Vickers

OBJECTIVES

After reading this chapter, the student should be able to:

- Describe the brief history of fire and emergency services grants.
- Identify grant team needs and common success traits and downfalls in grant applications.
- Perform an internal needs, external needs, and financial assessment.
- Identify and research support information; create proper structure of narrative for grant applications.
- Recognize communications downfalls related to grant programs.
- Create objective communications related to grant narratives.

Proper Grant Program Funding

Organizations typically fall into one of three categories: for-profit companies, nonprofit tax-exempt organizations (as recognized by the U.S. Internal Revenue Service), and public sector tax-supported organizations (generally government agencies).

Very few for-profit companies, nonprofit organizations, and public sector organizations have unlimited funding, but there are vast differences between the three types of organizations. For example, the for-profit sector has the ability to change products and services to increase the profitability of the company. Some nonprofits, such as membership or fraternal organizations, sell products and services that generate income. Other nonprofits, such as volunteer fire departments, offer their services at no cost to the user. In both nonprofit and public sector organizations, the organization's financial well-being is based on the generosity of donors and the ability of the organization to secure grant funding.

Public safety organizations, whether volunteer or tax-supported fire and emergency services organizations, are more at risk than other organizations because of the diversity of needs for personal safety equipment and vehicles, in addition to operating expenses such as utilities and, of course, payroll in many cases. As municipal budgets and volunteer donations shrink, the risk to the safety of emergency responders increases. The ability to replace safety equipment, maintain training, or even have enough personnel available to safely handle incidents depends on a solid funding base. Particularly when tax support or donations decrease, volunteer and government-supported fire and emergency services organizations turn to grants for revenue. In good times or bad, the ability to secure grants is critical to the safety of the community, the fire and emergency organization, and its members.

A **grant** is basically a monetary award to an organization for a particular purpose to allow the organization to continue or expand its operations. A **purpose** is the grantmaker's stated goal or reason, including the identification of specific deficiencies to be reduced or eliminated through grant programs. Under normal circumstances the only organizations that can receive grants must be certified nonprofit or not-for-profit because their purpose is not to enrich stockholders but to benefit the community at large. Grant programs are found in two categories, government and private. Government grants come from all levels including local, state, and federal sources. Private grants come from foundations that were created by either for-profit companies or individuals. Both have the same premise of using charitable giving in the form of grants to other nonprofit organizations in an attempt to correct certain deficiencies found in the world.

Prior to the widespread existence of grant programs and knowledge about them becoming more mainstream, thanks to the Internet, many fire and emergency services organizations unfortunately just did without. The only options for increasing funds were locally based, such as tax-based funding and fund-raising activities.

Even now, many volunteer fire and emergency services departments are solely funded through activities such as pancake breakfasts, bingo, raffles, and mailing and subscription campaigns—all based on local generosity. Those budgets are highly volatile because the funding depends on the discretionary money of the community. As economies go up and down, so do funding levels. Major capital expenditures such as stations and vehicles must be financed, but a fire department may have to save for years to create a down payment large enough to make an ongoing loan payment but small enough to afford without risking a default. Other areas of preparedness (such as personal protective equipment, maintenance of existing buildings and vehicles, and insurance) would also suffer. (See Figure 10.1.)

Grant programs were created solely for public safety agencies to attempt to ensure a minimal operating level across the country. Many private grant foundations have been in

grant ▪ A monetary award to an organization for a particular purpose to allow the organization to continue or expand its operations.

purpose ▪ The grantmaker's stated goal or reason, including the identification of specific deficiencies to be reduced or eliminated through grant programs.

FIGURE 10.1 Emergency response vehicles are major capital expenditures. *Courtesy of Brian P. Vickers*

existence for decades, but traditionally have not been pursued by public safety agencies because of the perceived difficulty in writing a proposal and completing the application. More grant programs are now available through the Internet and other resources. Information also exists on how to complete successful applications.

Still there is no golden arrow, no single element to guarantee grant funding. Securing grant funding is more involved than simply putting words on an application. The path to grant funding success is a long one. However, for those willing to undertake the tasks necessary to create highly competitive grant applications, the reward is an increase in the fire department's ability to reduce risks to life safety for the public and emergency responders.

Barriers to Grant Funding

There are many reasons why grant applications do not receive funding, but there are several easily avoidable barriers that will help focus the applicant's efforts. The top barriers to receiving funding include the following categories: ineligible applicant, ineligible project, low project priority, multiple projects in one application, missing information, and lack of clarity.

■ *Ineligible applicant.* Every **grantmaker** (an organization or agency that funds and manages a grant program) has a list of organizations that may request funding and possibly a list of those that it will not fund. If an applicant is in the latter category, then by definition it will not receive funding based on the program's rules. For example, some corporate foundations limit projects to organizations within a particular city, state, or province.

grantmaker ■ A for-profit or nonprofit organization or government agency that funds and manages a grant program.

- *Ineligible project.* Just as grantmakers have a list of eligible applicants, they have a list of projects that they will and will not fund. If an application asks for something that the grantmakers state they will not fund, the application will of course not be funded. For example, many grantmakers will not give awards to help an organization with operating expenses, such as rent, utilities, or insurance. Because those types of costs are relatively constant and recurring, the grantmaker does not want to give funding that handles only one month of operating costs. Grantmakers would rather enhance long-term capital goals to help ensure the organization continues to operate for a long time by allowing it to concentrate its existing funds on operating costs.
- *Low project priority.* Within their list of eligible projects, grantmakers place greater emphasis on activities that they would like to fund over some others. The lower the priority of the project the applicant is seeking funds for, the less likely the grant application will score well. Clearly, any application that asks for a higher priority project will be more competitive.
- *Multiple projects in a single application.* Grantmakers have limited funds to distribute, and most like to spread it around as much as possible to organizations that need assistance. When applicants ask for funding to correct multiple deficiencies simultaneously instead of focusing on one at a time, it is less likely those applications will be funded.
- *Missing information.* Each grantmaker has its own requirements for information on the application. For example, some grantmakers require specific financial information, such as audited financial statements. Grantmakers will provide information about their requirements on their Web site or elsewhere. Any information missing from the application negatively affects its overall scoring and eligibility. The required information that is most commonly missing is answers to questions related to the project and its needs.
- *Lack of clarity.* When an application is done too quickly or without planning, the applicant's ability to communicate the overriding need for the project can be lost. The reviewers need to understand the deficiency to be corrected and how the proposed project is a solution to that deficiency. The applicant also may want to explain why the organization has not taken on the project without grant funding.

deficiency ■ A condition in which a gap exists between the fire department's expected/required level of preparedness and its actual operating level, indicating the need for improvement.

Generally, a **deficiency** is a condition in which a gap exists between the fire department's expected/required level of preparedness and its current operating level, indicating the need for improvement. It is a gap between the actual and needed level of safety provided to members of a fire and emergency services organization. A successful grant application will be clear about both the deficiency and the proposed solution.

Assembling a Grant-Writing Team

Just like any other team, a grant team or committee needs members with unique skills such as critical thinking; organizational, computer/Internet, and communication skills; plus subject-matter knowledge. One person is not able to complete all tasks needed to create a competitive grant application because the process requires multiple types of people to complete as many tasks as possible in the shortest time frame possible. Most grant application periods are open only once a year, so if the deadline is missed, the next chance to enter an application is at least 11 months away (assuming the program is even available the next year). Sometimes there are so few opportunities to go after grant funding for certain types of applicants or projects that wasting any of them is a critical mistake. Due to once-a-year applications, it is essential that the grant team prepare the application before the grant's opening period.

Depending on what needs arise, there will be dozens of opportunities for departments to apply for each year. Having the right team will help the department apply for all grants for which the organization is eligible.

TEAM LEADER AND MEMBERS

All teams need a leader, someone responsible for the planning and coordination of grant application efforts and research supporting those applications. The leader may do some of the other tasks involved, but his or her main goals are to ensure no opportunities are missed and that all efforts taken will focus on the goal of improving the organization.

The rest of the team will need to include people with solid research skills to identify funding sources and requirements either on the Internet or at the library. Computer skills are useful for managing a tracking system so that all leads can be stored centrally and updated quickly and easily.

Of course, having some people who write well is another plus because many applications require answers to dozens of questions on all sorts of topics, which will be the culmination of the work of other team members.

THE "ONE-PERSON TEAM"

If the team ends up being one person, then being highly organized will be more critical than if there are others who can be depended on to contribute.

Using technology can help the lone grant writer stay organized and track potential programs that may help meet organizational needs. For example, most e-mail programs have a scheduling tool that can set reminders of program deadlines. The Internet is a powerful tool that can be used to research information about available grants and their application requirements. Saving bookmarks of Web sites can help locate new information as quickly as it is posted. Keeping electronic documents is the most compact way of storing information that can help increase the level of applications, because paper copies take up a lot of space and are not easily retrievable from anywhere other than the filing cabinet they are in. But, as computer-savvy people will insist, all files need to be backed up. Keeping oneself centered with readily available information by using common technologies will help the team of one stay focused and ready to move on any opportunity that comes along, and also reduce the likelihood of missing one.

If the bulk of the effort to create the application is left to one person, objectivity is missed because communications will all be done from one viewpoint. An outside viewpoint is always needed to help create an objective argument. So, all grant teams, regardless of the number of members, always need others to read and critique the proposal.

Necessary Pre-Application Effort

The grant application process can vary, depending upon the grantmaker's requirements and the requirements of the applying fire department, which may include legal or accounting review. Some common fundamental steps in the grant application process include the following: determining the long-term goal of grant funding, conducting an internal needs assessment, conducting an external needs assessment, determining financial needs, and preparing the application. (See Figure 10.2.)

LONG-TERM GOAL OF GRANT FUNDING

The short-term goal of applying for grants is to bring in needed funding for equipment purchases or training for the purpose of increasing firefighter safety and survivability. However, too many times an application falls short because the applicant's focus is entirely on the short-term (equipment) rather than the long-term goal of increasing firefighter safety and survivability. Successful applicants, on the other hand, focus on the big

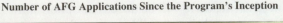

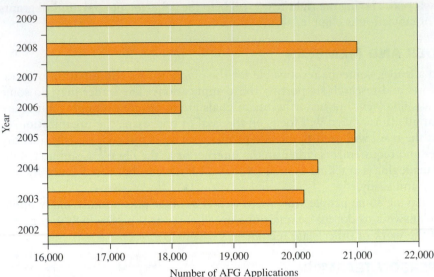

picture of improving the fire department and the safety of its members. They are concerned with creating solutions to long-term deficiencies. Many grantmakers do not give funding for repetitive operating costs for that very reason.

The difference between success and failure often comes down to fractions of a point when the applications are reviewed and scored. The person taking on the grant-writing role for the fire department does not want to take shortcuts because most applications are accepted only on an annual basis. Any misstep costs the most expensive commodity: time. Even when following all of the steps to create competitive applications, there is still no guarantee of success, but the grant writer does not want to be second-guessing for another year when an obvious mistake has been made. To capture all possible points and be highly competitive, the applicant must know about all aspects of his or her fire department and its operations before attempting to put together an application. Often, grantmakers require applicants to detail the sustainability of the proposed grant. They want to make sure the award will positively impact the community and fire department long-term.

INTERNAL NEEDS ASSESSMENT

The first step on the roadway of successful grant applications is to determine the deficiency, or the need or problem to be solved. In the case of firefighter safety and survival, the needs assessment involves determining the health and safety need within the fire and emergency services organization and how the requested funds help the fire department meet that need.

That may sound overly simplistic. In some cases it is. But thousands of applicants omit the important first step of identifying the internal need, resulting in disappointment when grants are awarded. All grantmakers want to know where the applicant needs to be in terms of operational safety and how the grantmaker's funding will help achieve that higher level of operational safety. What applicants tend to skip are the sometimes unasked questions such as:

■ Where is the fire department in terms of personnel safety and survivability right now?

■ Where does the fire department need to be in terms of personnel safety and survivability?

- What specific steps has the fire department already taken to reach its safety and survivability goals?
- Why does the fire department need the grantmaker's money to accomplish the task?

To help the applicant answer those and similar questions, the grantmaker's application sometimes includes an assessment process so that the applicant can provide all the pertinent information for a competitive application.

Just as in personal improvement, the first step in making improvements in a fire department is to look inward first. Then take a look outward. Too often grant applicants try to either gloss over issues or blow them out of proportion. Both are obvious evidence to a reviewer that no assessments were performed and the applicant is just after some free or discounted items. Claiming emergency responders are going to die or be severely injured on a daily basis if a situation is not corrected may come across as overly sensational, especially if the deficiency has existed for some time and no injuries or fatalities have been attributed to it. Applications such as those tend to demonstrate that they have no specific basis. Just like an attorney calling an expert witness, applicants have to bring in objective studies and knowledge from outside sources to show that the deficiencies not only exist but also are risks to the fire department and the community.

Purpose of the Internal Needs Assessment

The purpose of the **internal needs assessment** is to check the department's actual safety and survival situation against the ideal situation. It is the process of analyzing the inside workings and preparedness of the organization, based on national, state, or local standards and requirements for equipment, training, staffing, and overall readiness. With the internal needs assessment in hand, the grant team will ultimately be able to communicate any deficiencies it has found to a grantmaker. The internal needs assessment provides information to convince the grantmaker that the application presents a solution for an objectively defined real situation, rather than just the applicant's possibly exaggerated opinion. At the internal assessment point in the application process, the applicant is not yet creating solutions for deficiencies, but taking an objective look at the department for those deficiencies.

Personal Protective Ensemble (PPE) Example

The personal protective ensemble (PPE) provides a useful example of an internal needs assessment for a grant to support safety and survivability. The applicant's first step is to list all personal protective equipment and manufacturing dates in the department's inventory. Although PPE for structural firefighting is the main component of a department's PPE inventory, the internal needs assessment also should consider equipment needed for other types of emergency services such as wildland firefighting, technical rescue, emergency medical services, and hazardous materials services. (See Figure 10.3.)

The internal needs assessment would then determine the condition of each item of PPE and whether it meets current standards, such as NFPA 1851, *Standard on Selection, Care, and Maintenance of Protective Ensembles for Structural Fire Fighting and Proximity Fire Fighting*, and NFPA 1971, *Standard on Protective Ensembles for Structural Fire Fighting and Proximity Fire Fighting*.

Other Factors to Consider

The internal assessment process should be repeated for every piece of the department's safety-related equipment, facilities, or personnel. Those organizational resources include self-contained breathing apparatus (SCBA), rescue tools, hose, vehicles, stations, and of course the most important piece of equipment, the people (FEMA and NFPA, 2002).

internal needs assessment ■ A process of analyzing the inside workings and preparedness of the organization, based on national, state, or local standards and requirements for equipment, training, staffing, and overall readiness. Its purpose is to check the department's actual safety and survival situation against the ideal situation.

(a)

(b)

(c)

(d)

FIGURE 10.3 Fire departments must consider the various types of emergency responses when conducting an internal assessment. *(a) Photo by Miranda Simone, (b) Courtesy of Michael Russell, Nashville Fire Department, (c) Courtesy of Captain Terry L. Secrest, (d) Courtesy of Travis Ford, Nashville Fire Department*

An organization is only as strong as its people, and in the fire and emergency services the best tool any fire department has is a well-trained and disciplined emergency responder. No amount of equipment will make up for the firefighter who is ready to handle the calls with the knowledge, skills, and physical capabilities needed to perform safely and properly on an incident scene. People handle incidents using equipment, and if the people are not ready, the equipment purchased from a grant award is not a real solution to safety and survival.

In making a people assessment, the following questions need to be answered:

- What incident types does the fire department handle regularly?
- Is the training for those incidents conducted by certified instructors?
- Do emergency responders train often enough to keep skills ready and practices safe?
- What are the certifications each member has?
- Have medical exams been done as required to ensure no hidden health risks?
- Are immunizations current for all personnel?

EXTERNAL NEEDS ASSESSMENT

An **external needs assessment** is a process of analyzing the preparedness of the organization relative to its risks, based on the emergency incidents expected in the community, such as transportation accidents, building types and sizes, hazardous materials, and life safety risks. Properly conducted external needs assessments will give the applicant a 360-degree view of its fire department and all deficiencies that exist by comparing the organization's deficiencies to the types of incidents it is more likely to face based on the composition of the community served. Once the external assessment is finished, a grant-making plan can be created to chip away at those deficiencies.

The fire and emergency services has many measuring sticks to show whether safety deficiencies truly exist. The organizations in the field that guide every aspect of operations include the National Fire Protection Association (NFPA); the Occupational Safety and Health Administration (OSHA); OSHA's parent, the United States Department of Labor (DOL); the U.S. Department of Transportation (DOT); and even the Environmental Protection Agency (EPA). It could take hundreds of hours to find the right codes and recommendations to support the grant writer's project depending on what is felt to be the fire department's greatest need. Depending on what type of project is being pursued, there may be dozens of other documents available to assist in showing that the powers that be agree the fire department really needs what it is asking for, or that it is a viable solution to the defined problem. Many of these documents are freely available for download, while others can be purchased.

Determining the likelihood of a major incident's happening by pure statistical analysis is just as important as knowing the national standard. It requires knowing what is in one's own district and how those local hazards apply to the project. The applicant wants to answer the risk-based questions related to what could happen so that in the solution creation process the fire department can compare the internal to the external assessment and ensure that any plans made will actually accomplish the goal of making improvements to operations. For instance, the external assessment process might include the answers to these questions:

- How many nursing homes or hospitals do you protect and how many residents are there?
- How many schools and students are in the first due area?
- What is the number of warehouses? Chemical plants?
- What is the traffic volume on the interstate, railroads, waterways, and so on?
- What is the current median household income of the response area?
- What is the current poverty level indicator from the government?
- What is the racial or ethnic diversity level in the district, requiring languages other than English to be spoken in outreach programs?

The answers to these and many other similar questions will build a case of risk to the fire department and its members while responding to incidents at these locations as well as the likelihood of having certain types of incidents. For instance, knowing how many

external needs assessment ■ A process of analyzing the preparedness of the organization relative to its risks, based on the emergency incidents expected in the community, such as transportation accidents, building types and sizes, hazardous materials, and life safety risks.

FIGURE 10.4 Building use determines equipment and emergency response vehicle needs when writing grants.
Courtesy of Brian P. Vickers

buildings with volatile chemicals in them would help determine whether or not an invest-ment in hazardous materials technician training and the related expense for equipment would be a worthwhile return on investment. With only one location in the district and no incidents that anyone can remember in the past 20 years, the physical and perceived likelihood of one now is not very high. In contrast, having several high-rise housing com-plexes occupied by senior citizens means that evacuation time may be high, so additional thermal imaging cameras, radios, high-rise hotel packs, and other tools are more pressing needs. (See Figure 10.4.)

Some of the pertinent statistical values that will be great supportive material in a grant application will change on a monthly basis. Often *current* means one to two months old, such as unemployment-level statistics from the state or federal government. County or city statistics may be available in a more recent time frame and, if so, they should be used while citing the source of that information. Writers should take care and remember to update statistics in each subsequent grant application they write. Even though prior applications are a starting point for the next grant, the information must be as current as possible.

With a completed assessment, an applicant can use the answers to the preceding questions and current statistics to show the need for improvements based on what is likely to happen. Proper preplanning will show problem areas and what difficulties the organization will face when incidents occur at those facilities, so do not forget to use them if they are part of the process. Certainly, citing specific incidents in which having what is being asked for would have possibly made a difference is a good thing to do; it shows a demonstrated need for the project at hand. Avoid having too many of these listed because it will make the narrative sound like a series of war stories. The external needs assessment, if properly constructed, can be used to show that tomorrow's war stories have not happened yet and, if they do, the applicant already knows that the fire department will not have the training or equipment to properly mitigate that risk safely and effectively.

To meet the goal of increasing safety and survivability for fire and emergency services, the grant team needs other types of information to help create the needs portion of the application. Just like the straightforward assessments on equipment age and condition inherently show life safety risks for the responders, other information can be used to show risk. Many times people claim to have a gut instinct when a safety issue exists, and they may be right.

Grantmakers, however, do not fund on gut instinct but on demonstrable deficiencies. For instance, if the department thinks there is an issue with juvenile fire setting, statistics will be needed from the fire marshal stating how many arrests have been made, or even how many investigations are underway in which the suspect or suspects might be juveniles. Contacting hospitals to find out how many burn victims they have treated that are under the age of 18 will also possibly point to an issue in the community.

Because the fire and emergency services has become an all-hazard response agency, the fire department can get involved in other injury and safety prevention programs that might help reduce the response load. The organization could undertake helmet awareness campaigns for motorcycle and bike riding, child safety seat education classes, citizen CPR classes, fire extinguisher training, or other topics that the organization feels are not being covered and create a life safety situation for the community. But in order to identify the risk, statistics are needed to support the claim that the fatalities or injuries are happening. The doctors and hospitals in the area know what types of injuries they treat for each age-group, as do the EMS transport agencies. Computerized reporting has made data mining much easier in recent years, so if those groups use such reports, they can normally pull them within a matter of minutes or hours instead of days or weeks of sifting through paperwork. In order to show true need, the frequency of certain events has to be calculated. A high number of smoke inhalation injuries after fires shows that possibly there were no smoke detectors or not enough of them to alert occupants to the fire more quickly and allow escape prior to major smoke conditions occurring in their homes.

If these reports do not show as large of a risk as the organization believes, then the focus needs to move into surveys. Risk assessments are all about what can happen, which is why the metrics just discussed at the beginning of this section are so important. But when it comes to public life safety, one cannot know something is really an issue until it is brought to the forefront by asking the tough questions. For instance, take the following set of scenarios when it comes to prevention and education efforts:

- *Scenario #1.* The applicant knows it has 400 low-income residences in the district who do not have hard-wired smoke detector systems. So, the possibility exists that if there are any detectors, the occupants may not have the funding to replace the batteries every six months. The applicant is requesting 400 detectors with 10-year lithium batteries.
- *Scenario #2.* The applicant sent out surveys through the local day care centers and elementary schools, and based on those surveys it has 300 multilevel homes with no detectors at all, and 100 multilevel homes with only one detector in the

house. The applicant is requesting 700 smoke detectors with 10-year lithium batteries so that every home will have two detectors, one on the main level and one upstairs to increase the likelihood that the occupants hear the alarms as early into the fire as possible. The applicant also is going to be handing out copies of home escape planning paperwork to the residents.

While Scenario #2 is asking for more funding, it has a basis for the request that is supported by documentation. It also has a distribution plan that will include more information to the occupants to ensure that they know what to do if an alarm goes off. While more information could certainly be added given the small portions of the application that were shown, Scenario #2 is the more competitive application because it shows a more in-depth assessment was performed, which grantmakers like to see.

Partner Organizations

Also included in external assessments are any partner organizations in the community that the applicant may work with toward improvements in safety and survival. Certainly the other public safety agencies (such as mutual-aid organizations, private EMS transport organizations, law enforcement agencies, and other similar entities) in the area will be included.

In addition, within the community there are other organizations that the grant team could possibly work with to ensure that enough people are available to assist in creating solutions. Local businesses, schools, and civic organizations will be important groups to identify as partners not only for the narrative portion of an application but also during the application research process discussed later in the chapter.

Partnerships cannot be emphasized enough. When grant applications articulate multiple partnerships contributing to the success of grant implementations, it shows greater organizational effort, commitment, and ownership of program delivery.

FINANCIAL ASSESSMENT

The most probing questions that grantmakers might ask include:

- If there is such a great need, why has the applicant not funded it already with the fire department's money?
- What is the applicant doing with the funding it does get that has hampered its ability to handle this particular need?

The answers are not as simple as stating that the project costs more than the annual budget. The grantmaker wants to know details about the internal and external financial conditions of the fire department. The applicant's assessment must be specific to the fire department and community served. When stock markets fall or the United States goes into a recession, it is not a unique argument to say "the economy is down in our area," because nearly everyone can make the same claim. Just like internal assessments show that the applicant has a need, the **financial assessment** has to show why the applicant cannot afford the project or even any parts of it. It is the process of analyzing all things financial within the fire department, including the current sources of funding, risks to the stability of those sources, ability or inability to increase funding levels, and current expenditures.

In the financial assessment process, the first step is to detail the sources of existing funding:

- Is the fire department tax-funded, donation/fund-raiser based, or a combination?
  - How much is received from each source?
  - How has each source changed in value over recent years?
- What factors are affecting the revenue sources for the fire department and driving their changes?
  - What are unemployment rates for the community?
  - How is the real estate market in the community?

financial assessment ■ The process of analyzing all things financial within the organization, including the current sources of funding, risks to the stability of those sources, ability or inability to increase funding levels, and current expenditures.

- Have there been any natural disasters in the community recently?
- What is the taxing authority spending money on since the department is not getting it? (Describe revenue and funding for schools, roads, water, sewer, and so on.)
- What factors are affecting the short- and long-term financial outlook of the department?
  - Are new employers coming into the area? Existing ones leaving?
  - Can new sources for funding be found (creating/increasing tax income)?
- What is the department spending money on now that are not needs? (Payroll, insurance, utility bills, mortgages/rent, fuel, equipment and vehicle maintenance are all needs. Leather helmets are nice but not needs, neither are upgraded apparatus, paint jobs, nor apparatus that do not support a direct fire or EMS function.)

The answers to these questions can then be communicated to the grantmaker to show that all options for local funding have been pursued. The goal is to attribute as much of the applicant's existing budget as possible to other expenses. When a grant application is made for only one thing that is determined to be the single greatest need for the fire department, it is important to make sure that the financial argument is not weakened by the reviewer seeing an abundant amount of excess money in the budget that could be used for any part of the project.

CREATING SOLUTIONS

Now that both internal and external risks for the fire department are known, each set of deficiencies has to be broken down one by one and a list of solutions created that can correct each. At this point, the applicant does not delve into comparisons or even costs. The goal is to be objective and list every possible solution for a given deficiency to safety and survivability. (See Figure 10.5.)

FIGURE 10.5 Identifying potential solutions to deficiencies helps the fire department document the facts objectively.
Courtesy of Brian P. Vickers

Going back to the PPE example, Department A shows that all its gear is more than several years old, and that many sets have rips or tears, or are contaminated from repeated use. There are three basic solutions for the specific deficiency: replace the gear, repair the gear, and properly decontaminate the gear.

Once those potential solutions are identified, writers can assign specific costs, benefits, and disadvantages to each potential solution.

The fire department should recognize that a grantmaker wants to create a long-term solution for the least investment part of grant funds. Such a thought process will enable the fire department to document objectively the facts needed in a highly competitive grant application.

While having the least expensive solution to the stated deficiency is a goal for applicants that may help create a more competitive application, that does not necessarily mean applying for less money just to keep the costs down is the most effective strategy. None of the potential solutions—replacing, repairing, or decontaminating—alone will create a cost-effective and comprehensive long-term solution to the need for personal protective ensemble. Without proper care, replacing the gear will put the organization back in the same situation eight years from now. Replacing only 10 sets instead of 20 is an incomplete solution because 10 individuals will have safe and reliable gear, and 10 will not. Repairs might not be able to be performed cost-effectively or at all on some sets, so those individuals will need new gear anyway. Cleaning the gear can help extend its useful life, but does not repair ripped or torn gear, although it may prevent some damage due to contaminants breaking down the fibers.

In the hypothetical case, the most effective solution would be to apply for replacement of all gear along with a washer or dryer that is built to decontaminate gear in adherence to NFPA standards. Such a solution will save the fire department ongoing costs for cleaning by an outside vendor while still providing the proper maintenance after each incident, minimizing out-of-service time to no more than a few hours during the washing and drying process.

Repeating the process for each deficiency will give a complete picture of every avenue that a fire department could take to improve responder safety and survivability. To continue along the line of PPE is the other types of clothing (such as EMS, wildland, and technical rescue, or hazardous materials response) that the fire department might need for safety. If the damage to the structural gear occurs mainly at wildland fires, and Department A has a larger number of wildland fires every year, then maybe wildland gear should be requested so that the structural gear is not damaged on calls unrelated to its primary purpose. At an average of less than half the cost of structural firefighting gear per set, a fire department can put its personnel in wildland-compliant PPE, which not only will reduce the damage to the structural PPE but also is lighter and easier to move in, which creates less physical stress when responding to wildland calls.

THE NEED TO LIMIT FUNDING REQUESTS

Applicants are cautioned against seeking funding for multiple deficiencies at once. The downfall of multiple requests is twofold:

- The computer scoring metrics will be lower with different item weightings. The lower scoring items will bring the overall score down no matter how many high-scoring items are included.
- Answering the narrative questions becomes more difficult because each question has to be answered once for each item being requested.

Computer Scoring of Grant Applications

Some grantmaking organizations use computer scoring while reviewing grant applications. With computer scoring, each item requested on the application has an assigned priority, with no two items having the same weighting. The weighting is a factor between 1 and 0 that assigns the higher scores for the items that the grantmaker considers higher priority for funding. The top-priority item is always a factor of 1, and the other items start down from there at .99, .98, and so on. The lower the priority of the item, the lower the likelihood of funding, even if there is a statistically supported need in the application.

Asking for too low of a priority reduces the likelihood of funding, and asking for too many projects at one time compounds that effect when compared to asking for a single item even when in the same project area. In keeping with the PPE example, structural PPE is a higher priority than wildland PPE. Asking for both at the same time will result in a lower score than just requesting one type at a time.

Multiple Requests Within One Application

Attempting to get multiple items from different project areas in one application is generally called a shopping or laundry list. Although every now and again a few of these applications get funded, the overwhelming majority do not. Because the goal is to create the highest probability of funding as possible, and adding too many requests into a single application drops the score, multiple requests need to be avoided as much as possible.

Contrary to common thinking, structural PPE and SCBA are from different project areas. PPE is protective clothing, and SCBA is related to breathing air. Although both are important, they have different uses during incidents and are not necessarily always used together. Protective clothing should be used on every call, and SCBA is used only during incidents involving compromised air quality or immediately dangerous to life and health (IDLH) atmospheres. When grant programs assign only one priority weighting to one item, structural PPE is the higher weighting and SCBA is slightly lower as is the case with the Assistance to Firefighters Grant (AFG) program from the Department of Homeland Security. An application for funding for both PPE and SCBA will score lower than an application for only the structural PPE.

Implications for Matching Grant Applications

In **matching grant** applications, the applicant is required to contribute a certain percentage toward the project budget, in either direct monetary support or in-kind support such as volunteer labor. Matching grant applications with "shopping lists" that are computer-scored compare the matching amount to the total requested. For example, an applicant hopes to purchase PPE, SCBA, hose, training, radios, and some hand tools, a total expenditure of $100,000. The applicant requests $95,000 from the grantmaker and will match $5,000. The application is essentially an admission that the applicant has $5,000 readily available. When the computer compares each item being requested to that matching amount, it may find that all of the radios and the hand tools cost less than $5,000. The computer scoring concludes that the applicant has funds available to purchase some of the projects being applied for and assigns a lower value of need to the entire application. Essentially, the scoring reflects that the applicant could handle some deficiencies, but chooses not to. In effect, it looks as though the applicant must have true needs if it is unwilling to spend its own money to make improvements.

matching grant ■ A type of grant in which the applicant is required to contribute a certain percentage toward the project budget, in either direct monetary support or in-kind support such as donated volunteer labor.

The Final Argument Against Multiple Funding Requests

For highly competitive applications, the strategy is to ask for funding for the single project that cannot be handled by the applicant at all. For example, in the same $100,000 request with $5,000 in local funding, the applicant could buy one SCBA for $5,000, but needs a total of 20 SCBAs. Such an approach is not a true solution in the grantmaker's eyes. In most cases, proposing such a project is not an effective strategy.

Funding Sources

As discussed at the beginning of this chapter, the three types of organizations are for-profit companies, non-profit tax exempt organizations such as foundations, and public sector tax-supported organizations (generally government agencies). Although grants are often thought to come from only government agencies, in reality for-profit companies and nonprofits also provide grants.

Regardless of the type of funding organization, all grant programs award funds for specific purposes. Competitive applicants know that the key to success is understanding the grantmaker's specific purpose. Grantmakers exist for all types of deficiencies from childhood literacy, cancer research, and helping the homeless, to historical preservation and environmental preservation. The grantmaker's focus is to award funding that improves its area of interest, knowing that there will be more applications than available funds. Naturally, grantmakers rank applications to determine who has the greater need and cost-benefit combination so that the funds do the most good.

The next section outlines various sources of grant funding, with an emphasis on U.S. federal funding most directly applicable for fire and emergency services safety and survival.

CORPORATE, FOUNDATION, AND CIVIC ORGANIZATION GRANT FUNDING

Large and small companies and large and small foundations (and other not-for-profit organizations such as civic organizations like the Rotary) fund thousands of projects of different sizes each year. Some companies fund projects directly, whereas others fund their own foundations, which award the grants. For example, the conglomerate General Mills established the General Mills Foundation; and since 1954, that foundation has distributed more than $400 million to not-for-profits in communities in which the company operates, including $20 million in 2008 (General Mills, n.d.). Similarly, many wealthy individuals and families establish foundations to manage their charitable giving. Each of those organizations has its own priorities, such as the deficiencies it wants to solve and its own eligibility and application requirements. In addition, each organization has its own deadline and reporting requirements.

Fire and emergency services organizations are aware of large corporations and not-for-profits in their jurisdictions. In many cases, being local is a distinct advantage to securing funding, as in the case of the General Mills Foundation and many others. Identifying other corporate, foundation, and civic funding sources may require the use of a fundraising or development expert or consulting one of the many references that identify funding sources and their requirements. These references include publications such as:

- *The Foundation Directory 2009,* David G. Jacobs (Editor)
- *Annual Register of Grant Support 2009: A Directory of Funding Sources,* Beverley McDonough
- *Grantseeker's Toolkit: A Comprehensive Guide to Finding Funding,* Cheryl Carter New and James Aaron Quick

Particularly in the case of funding by local civic organizations, securing a grant may be as simple as contacting a neighbor who is active in the Kiwanis Club or making a presentation to the Jaycees or Lions Club.

GOVERNMENT GRANT FUNDING

Federal and state grant programs are more straightforward in their purposes and intended applicants, but finding them is no easy task either. Many programs are briefly summarized on a Web page buried in the rest of the state or federal agency's site. The mentality for a long time has been that as long as the information is available, it is the potential

PEARSON
my**fire**kit

Please visit MyFireKit Chapter 10 for links to selected grant programs and support information.

applicant's responsibility to find the opportunities. In some areas, the philosophy is still true. However, information about the majority of available federal programs has been consolidated in one place on the Internet (see www.grants.gov). Information about every existing federal grant program is published there, including program information, funding levels, and other guideline materials. The repository is searchable by either federal agency or keywords, so it is a very user-friendly system. The site and many other opportunity sites also have announcement e-mail lists that can be sent to anyone who wants to receive them, so applicants should not hesitate to subscribe to every one they can find. It will only make the grant application process easier in the long run.

Generally, state programs also are posted on the state Web sites, but sometimes the applicant has to be proactive and contact various state agencies to find out about available funding. In certain cases, federal funding may be allocated to the state for a general purpose and what to do with it is left to the states. A good portion of the funding from the American Recovery and Revitalization Act of 2009 (ARRA) was released in such a manner. Several other grant programs (such as the Rural Fire Assistance and Volunteer Fire Assistance programs offered by the U.S. Department of Agriculture) are handled through the state-level offices.

Preparing the Grant Application

Once a grant program is found, the next thing to do is fully read the instructions for the program. The two terms used interchangeably to describe the complete instructions for a grant program are *program guidance* and **request for proposal (RFP)**. In the private foundation arena, RFP is used more often than not; the federal and state programs vary in their usage depending on the program and agency directing it. (See Figure 10.6.)

Within the RFP will be nearly everything an applicant needs to know. The first step, of course, is to verify that the grant-seeking organization is an eligible applicant and that the project in mind for the application is also eligible, which will usually be apparent in the program's scope section. The **scope** is the section in which the purpose of the program is detailed, as well as what sort of improvements the program is intending to achieve through the awards process. For instance, generally the grant programs administered by the Department of Homeland Security involve preparing emergency responders to mitigate major events whether natural or human-made. The Assistance to Firefighters Grant (AFG) Program's purpose is to fund awards to reduce the frequency of firefighter injuries and fatalities. Knowing the scope will help drive project choices because applications that are more in line with the scope of a program will score higher and may be more likely to be funded.

request for proposal (RFP) ■ The complete instructions for a grant program detailing eligible applicants, eligible projects, prerequisites for applying, matching fund requirements, reporting requirements, and other possible limitations or permissible actions that can be taken under the grant. Also called *program guidance document*.

scope ■ Section of the request for proposal (RFP) in which the purpose of the program is detailed, as well as what sort of improvements the program is intending to achieve through the awards process. The list of organizations or organization types eligible to apply for funding from a grantmaker along with the list of eligible projects that can be funded.

Distribution of 2009 AFG Applications by Service Area

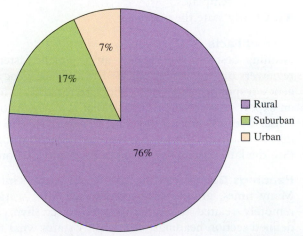

FIGURE 10.6 Pie chart showing the 2009 AFG submittals by service area.

THE GRANT NARRATIVE AS A FORM OF COMMUNICATION

Grant applications can be seen as a specialized form of communication. To summarize parts of college psychology and communications classes, there are six parts of a communication thread: the sender, the idea the sender wants to communicate, the encoding of the idea into verbal or written communication, the receiver, the decoding of said message by the receiver, and sometimes feedback from the receiver to the sender. In the case of writing grant applications, writers do not always receive proper (or any) feedback as a result of the application. All the applicant knows is whether or not the proposal is being awarded: there are no indications whether the readers did not understand parts of the application or if it just was not deemed as competitive as others. Although no message sender can ever truly control what the receiver does receive and comprehend, what can be controlled is how the writer encodes the message (application) so that the receiver (review panel) decodes it properly.

In the case of grant applications, the important points to communicate are why the proposed project is important and why it meets the highest priorities established by the grantmaker. Most grant applications go astray in the so-called encoding of that message, for example, the **narrative** of the application. The narrative is the section of the grant application in which applicants answer specific questions or discuss other parts of the application to help the reviewers gauge their need for the funding. So, while many applications use statistics to initially determine which applications are worthy of funding, the narrative is where the applicants tell their stories in their own words using all of the supporting information that has been gathered using the process detailed earlier in this chapter.

narrative ■ The section of the grant application in which applicants answer specific questions or discuss other parts of the application to help the reviewers gauge their need for the funding.

It is not for a lack of trying that narratives just do not read well. Many times the reason the narrative falls short is emotional. Very few people have to write their feelings or justifications at any point in their lives. Most communications of that type are verbal, and there is constant feedback during verbal communications. The message sender can read the body positioning, eye contact, and, of course, listen to verbal feedback. Written communications are different because there is no chance for the message's intended receiver to read it with the emotion it was written in or to understand completely the context in which the message was sent.

There are straightforward methods for ensuring that a grant application's narrative effectively communicates the grant seeker's message. Approaches for effective grant communications include the following:

■ Use correct facts.
■ Use the headings designated by the application guidance.
■ Use candid arguments.
■ Avoid templates.
■ Clearly state the purpose.

Correct Facts

Nothing destroys the logic of an argument faster than flawed data. Although not all reviewers of grant proposals for fire and emergency services health and safety have emergency responder experience, they are reasonable human beings. A reviewer need not be an expert on a subject to realize that a statement or supporting argument does not make sense. With the limited time that reviewers have to review a single application, applicants do not want to confuse them or cite something they know to be untrue or exaggerated. (See the later discussion on supporting information.)

Headings Designated by the Application Guidance

Many times, grant writers create their own sections to a narrative that are not even remotely related to the proposed project. Every grant application either has clearly defined section headings, or explicitly states what tasks must be accomplished within the narrative. Do not stray from these. The scoring methods employed are metered by how well the writer completes those headings or meets those tasks. Applications that fail to meet those requirements will not score well. Adding sections such as Department History and Community Involvement is not needed unless specified. While mentioning both of those topics in the course of the narrative is beneficial, devoting entire sections to them is not. The flow of the narrative is very important.

Candid/Objective Arguments

Candor is honesty and objectivity, without adding rhetorical flourishes. Being candid does not mean that the writer should be brief to the point of leaving out information. But it does not mean that the writer should appear to be letting his or her emotions rule the writing style either. Applying for a grant is a long and emotional process, but when it comes time to put pen to paper, the applicants are better served by leaving some of the emotion out. Hopefully in the design of the project, the writer encountered a series of facts to use as the basis for developing the solution. That is the time to elaborate on these facts, not bury them in rhetoric.

Templates

Templates are prewritten boilerplate documents about no particular project. Because the primary purpose of an application narrative is to state the applicant's reasoning for needing a project funded by grants, using templates defeats the purpose of a narrative. The narrative is the one chance to shine above the rest. After reading a pile of applications that sound exactly the same, it is impossible to judge applications on their merits because they are identical in form, verbiage, and sometimes length. While the computer scores may vary, the lack of diversity in the narrative compared to other applications will bring the score down significantly.

Avoiding a template is not to be confused with not using the headings defined by the grantmaker. There is a distinct difference. Several organizations have put out true templates complete with blank lines for applicants to simply enter their names and proper quantities. What is also missing from a template is the personality of the organization. After reading thousands of applications, reviewers can easily distinguish between templates and original material. When it comes to points for originality, being your own person will get you farther.

Stating the Purpose

The most important point to be included in a narrative is the applicant's purpose in applying. It is not enough to state that the applicant would like some funding. Reviewers and programs are looking for improvements in fire departments, so there has to be a correlation between the requested project and the end result in the short- and long-term operational improvements in the safety of the department and its personnel. But many times the purpose becomes lost in other statements. The more that is written about anything other than what is needed, how much it costs, and why it is needed will distract the reviewer from the clear purpose of the application: to handle one particular deficiency in fire and emergency services safety and survival. Straying from justifying why the fire department needs something will lower the score as much as applying for the wrong project. Discussing issues that are not part of the proposed solution might convince reviewers that the application is for the wrong project, resulting in a denial.

COMPONENTS OF A GRANT NARRATIVE

A narrative portion of a grant proposal typically has two components: an introduction to the grant-seeking fire department and a project description. For the reviewers, the project description is more important than the introduction. As a result, the introduction should be shorter than the description.

Introduction

Even if not part of the official structure, the narrative should start with an introduction to the grant-seeking organization, usually the fire and emergency services. The introduction establishes a bond with the reviewers through the process of describing the applicant, the applicant's physical location, the community being served, and what the organization does in terms of responsibility to the community. A natural human tendency is to create mental pictures when reading text, because most humans are visually oriented. Descriptive text helps the reader correlate his or her own experiences with the applicant's, which helps create a higher level of understanding of the applicant's challenges and needs on the part of the reviewer. The goal is to create empathy on the part of the reviewer so that the reviewer begins to understand why the organization took the time to put the application together.

The introduction should be brief. Avoid too much organizational history, such as who founded the organization, where, and why. As proud as the organization may be of its heritage, applicants do not want to distract the reviewers from the true message they are trying to communicate, which is explaining the current need of the organization and why the solution in the application is more deserving than the rest of the applications in front of them. Applicants do not want to lose the interest of the reviewers before they even get to the important part of the narrative.

Project Description

The project description section of the narrative offers the deficiency or problem to be solved and a straightforward description of the proposed solution. The description of the proposed solution is sometimes combined with the benefits of the project. As long as the blending can be done with each of the two parts still being distinguishable to show the reviewer that all of the requirements of the narrative are being met in one section versus two, mixing the description and the benefits can be done without a loss of clarity. As applicants write and review the narrative, if it is felt that coming back to the project benefits in another section would cause the narrative to be overly redundant, then include the benefits as part of the description of the project.

SUPPORTING THE NARRATIVE

PEARSON
myfirekit

Please visit MyFireKit Chapter 10 for a supporting documents list.

Grant seekers can increase the likelihood that their application will be chosen by persuading the grantmaker that its funding is going for a sound project. In other words, applicants need to demonstrate that the proposed projects are real solutions to real deficiencies. A way to show the soundness of a proposal is to remove the opinionated argument from the application and cite published reports from unbiased sources.

USFA Reports

One of the largest and most successful fire and emergency services grant programs is the Assistance to Firefighters Grant (AFG) Program. The AFG Program was signed into law and authorized to begin taking applications in March 2001 and began awarding throughout the rest of that year. The basis for the program was developed and refined long before the application period ever opened. The U.S. Fire Administration (USFA) and its parent, the Federal Emergency Management Agency (FEMA), developed the bulk of the rules and priorities of the AFG based on their own surveys.

USFA sent surveys to nearly every fire and emergency services organization in the country in the late 1990s and compiled the results into a report called *A Needs Assessment of the U.S. Fire Service* (FEMA and NFPA, 2002). The report includes not only the survey itself but also the collective answers that definitively explain the priority matrix (which balances community size and age and number of engines/pumpers in the fleet) as well as the emphasis on basic firefighting needs (such as PPE, SCBA, other basic tools and equipment) and basic firefighter training versus technical rescue equipment and training. The report forms a basis for different priorities for grant funding, which vary depending on the classification of the applicant's community as rural, suburban, or urban.

USFA, FEMA, and now the Department of Homeland Security (DHS) constantly conduct surveys and studies showing various widespread needs. The findings are available for free on their Web sites. Reports on nearly every subject possible in the fire and emergency services are available and will stand more steadily in the face of objection than an applicant's opinion. For example, most emergency response personnel "know" that foam is more effective than plain water, and Compressed Air Foam Systems (CAFS) are more effective than the other two. Rather than rely on the reviewer's knowledge, a grantmaker can cite the USFA's report on the subject (Stern and Routley, 1996). For health and wellness projects, the USFA report *Health and Wellness Guide for the Volunteer Fire Service* (USFA, 2009) is a good reference for career, combination, or volunteer fire departments.

NFPA Publications, Codes, and Standards

The National Fire Protection Association (NFPA) publishes reference works such as the *Fire Protection Handbook* (NFPA, 2008), as well as some 300 fire protection codes and standards.

The *Fire Protection Handbook* includes a discussion (or "translations") by the leading fire protection experts of NFPA codes into plain English. The *Handbook* includes every major NFPA standard clearly explained in a simple, easy-to-read format. In addition, the *Handbook* includes in-depth non-code information on a wide variety of fire protection subjects, including safety and survivability. The book is one of the few references that will have to be purchased, but it is well worth the expense. If applicants cannot afford to purchase the book themselves, a collective purchase by a county association with everyone chipping in could be an answer.

It is important for applicants to know what the NFPA codes and standards mean, not just what their numbers and titles are. Be sure to check the NFPA's Web site to ensure that the code cited to support a grant narrative is the current edition. For example, NFPA 1981, *Standard on Open-Circuit Self-Contained Breathing Apparatus (SCBA) for Emergency Services,* was revised in 2002 and 2007 with many significant changes in each edition. Citing the incorrect code or edition will cost the applicant points during the scoring process.

In addition, the individual codes and standards are available for purchase or download from NFPA's Web site. Many related publications have articles detailing the codes and searching their Web sites can be helpful. Because manufacturers and vendors have to comply with codes, many times they have documentation detailing the requirements or at least can inform applicants about where to find them.

In past years, the Occupational Safety and Health Administration (OSHA) within the U.S. Department of Labor was the primary entity that applicants would cite for the statutory requirements section of the application. Technically NFPA standards do not have the force of law, but in recent years DHS has adopted NFPA standards as law relating to incidents of all types and most importantly for its grant programs. In other words, grant award funding has to be spent on solutions that are fully NFPA compliant, and applications that do not meet NFPA standards cannot be funded at all.

Occupational Safety and Health Administration Regulations

The regulations of the Occupational Safety and Health Administration (OSHA) are normally available on its Web site for review. Because OSHA regulations are federal laws, knowing those relating to fire organization operations will help increase scoring in any application. For example, OSHA 29 CFR 1910.120 governs the two-in/two-out policy around which many organizations have developed their rapid intervention team programs. The requirement to have two firefighters outside and ready to go into a structure for every two firefighters that are inside already could be applied to all sorts of projects from thermal imaging cameras to personal protective equipment to SCBA. Section 120 also contains the regulations for respiratory protection during hazardous materials incidents. Section 134 covers EMS agencies in the same situations, so if the applicant were asking for chemical, biological, radiological, nuclear, and explosives (CBRNE) compliant SCBA, such information would help its case tremendously. Section 146 defines hazardous atmospheres when dealing with confined spaces.

Department of Transportation Resources

The U.S. Department of Transportation (DOT) and its National Highway Traffic Safety Administration (NHTSA) are both involved in increasing safety on roadways, on railways, and in skies in the United States. Those agencies have studies available on topics ranging from the hazards associated with specific intersection types, to accidents by type of road or vehicle involved, and even how many accidents on average happen on each type of road or at specific speeds. If an applicant seeks funding for hydraulic rescue tools, rescue vehicles, or even hazardous materials equipment, the narrative should show the potential for a major incident in the response area and how the requested funds will assist in the safe mitigation of that incident. In addition, DOT also recently passed 23 CFR 634, requiring traffic safety vests on all highway workers, including all public safety personnel during incidents on roadways.

FINANCIAL NEED

The last section as required by the most applications is the explanation of financial need. It is the tricky part and where most applicants completely falter despite having done solid work in all other portions of the narrative. In many cases, the person writing the application does not know how the fire department's funds are spent, and the discussion of financial need just falls apart. It becomes very apparent that the writer has no knowledge of why the project cannot be funded locally. And if the writer does not know why, he or she certainly will not be able to convince a panel of strangers.

It is highly recommended that the application's writer see, at minimum, a yearly line-item budget for the fire department to have a good understanding of how funds are spent at the local level.

PRE-SUBMITTAL REVIEW

The fire and emergency services organization can require a review of the grant application by legal and accounting staff or by the chief officer before submittal to the grant-maker. The application's narrative and financials should always be reviewed by someone other than the preparer before submittal. Those reviews should be scheduled to allow sufficient time for needed corrections.

Although it is not necessary to have a Ph.D. to successfully form a narrative, having the written draft work reviewed for logic and proper spelling and grammar is very important. When a writer uses broken arguments, difficult sentence structures, jargon, and localized language, the reviewers have a hard time deciphering the intended message. For example, not everyone knows what a task force is, and for those that do, a task force means something completely different depending on where the applicant's organization is located. The reviewers will not think less of the applicants if any terms that might be misunderstood are explained. Remember, if the reviewers are confused, they will score the application lower.

In addition, someone other than the person who prepared the application budget should review the financials to make sure they "add up," make sense, and are linked to the proposed solution. Electronic spreadsheets make it easy for the numbers to add up; however, a small error in a single cell can lead to major miscalculations.

Reviewers do not go out of their way to think of reasons to deny applicants the money. On the other hand, writers still have to give the reviewers reasons to give their application high scores. For example, mentioning that the organization does not have money for safety training because it spent it all on PPE and SCBA helps the reviewers understand the fire department's situation and its preexisting commitment to fire and emergency responder safety. It shows that the fire department is putting its people and their safety first, which is truly the primary goal of many grant programs such as the AFG Program. If the opposite is the case, and the fire department spent all its money on replacing multiple trucks in the past few years, grant writers will have a very hard time convincing anyone that the organization cannot afford to replace old PPE or SCBA. By purchasing the vehicles, the statement made in the application is basically that the fire department felt that having two or more new vehicles was more important than a safe and compliant personal protective ensemble, so applicants are not leaving much leeway for the reviewers to come to a different conclusion. No matter how writers try to state it, the fire department's actions as described in their narratives will speak even louder.

Use people from outside the fire department as well because they will not get lost in details of the routine organizational operations. Local business owners and high school or college educators are among the potential resources for pre-submittal review of grant applications. The teachers and professors are useful resources because they deal with essays on a regular basis. Although they may not be able to give pointers about the specific merits of applying for a tanker (water supply) versus a pumper (offensive operations), educators can verify that the argument is well documented and flows properly.

The reason for a pre-submittal review by a businessperson is simple: a grant application is basically a business proposal. The grant seeker is trying to convince the grantmaker that the need or deficiency is valid, and the proposed solution is highly cost-effective and beneficial. In the business world, it is called evaluating return on investment (ROI). If the proposed usage of the funding creates the greatest return for the amount invested, then something is deemed to have a very high ROI. In federal grant programs, the ROI is taken in terms of the amount spent versus the increase in safety of the citizens and responders. Applications with the highest ROI have the best chance of receiving funding.

Finally, use the local government's elected representatives in pre-submittal application review. The reason is twofold:

- Convincing local representatives who know the grant-seeking organization and control its funding is a test of how well the application will convince reviewers who are unfamiliar with the grant seeker.
- Allowing local representatives the opportunity to review the application makes them more familiar with the fire department's situation. Perhaps they believe the need is too great to wait for the chance of federal funding and will try to fund the project immediately. Second, if they agree that the need exists, but they cannot fund it, perhaps they know of other funding avenues to be pursued should the proposal not be successful. The key is to communicate with them so that nothing comes as a surprise.

PUBLIC INFORMATION ABOUT THE GRANT APPLICATION

Always keep the general public—the people who will benefit from the application—apprised of the status of a grant application. There is nothing wrong with not being funded; not everyone will be. Yet there is great triumph in letting the community know that the fire department cares about providing it the best possible service, and there is just not enough money in the current grant program. Many areas have come together to establish tax increases to bring up funding levels after learning of a fire department's dire situation. No one can guarantee an award, but fire departments should always put their best foot forward. Grants will not correct every health and safety deficiency in the fire service. Yet a guaranteed way that they will not fix *any* problem will be not to apply or to continue to rely on funding that may be unavailable.

Post-Award Procedures

Post-award procedures include award announcements and fulfilling grantmaker requirements. Do not let this part of the process go unnoticed.

AWARD ANNOUNCEMENTS

One of the best things awardees can do is announce their successes, which serves multiple purposes. The first purpose is the common courtesy of thanking the grantmakers. Corporate and foundation grantmakers welcome positive publicity. Announcing an award from federal and state programs shows that tax dollars are returning to the local area to help protect it. At the same time, an announcement lets the various government representatives know that the programs they vote to fund each year are being used and the deficiencies that were found are being addressed. Such information helps to convince the representatives to continue the funding in subsequent years.

GRANTMAKER REQUIREMENTS

All grant programs have a list of requirements that applicants must comply with when funded. In most programs, compliance is not negotiable, especially with federal and state grants. With the many diverse reporting requirements, creating a compliance checklist for a program if it does not have one in the guidelines might help avoid later issues with current or future awards.

Regardless of the source of the funding, the money must be used for the stated intentions in the grant application. Nearly every program requires extensive financial tracking of all monies related to the awards and purchasing. Most areas of the country require compliance with bid laws: prior to any purchases, awardees must take steps to ensure that they do not violate the bidding regulations. These can vary from single source purchases to advertising open bids in the newspaper for a number of days. In any event, whatever is needed to comply with the grantmaker and local bidding requirements must be done. Most grantmakers defer to local purchasing laws that will be uniform regardless of the source of money.

A certified public accountant may be required to perform an audit, which can cost up to several thousand dollars. Many grant programs allow for these costs to be included as post-award administration expenses that the grant will help fund. When such conditions are required, applicants should include these costs.

Nearly every fire and emergency services federal grant program requires fire departments to report to the National Fire Incident Reporting System (NFIRS), and they must adopt the National Incident Management System for every active year of the grant award's period of performance (POP). The Assistance to Firefighters Grant (AFG) and Fire Prevention and Safety Grant (FP&S) programs are one-year programs, so awardees must report for at least the year following the award. The Staffing for Adequate Fire and Emergency Response Grant (SAFER) program from the DHS has four- and five-year periods of performance, so reports must be filed every single year.

Other required reports might include quarterly updates of purchasing, status of spending, or implementation of initiatives such as training classes or physicals. Most grantmakers operate on an annual basis and do not prevent prior awardees from receiving more awards in subsequent years. As a result, awardees should remember that if they meet all reporting requirements, the grantmaker is more likely to award to them again. No fire department has only one deficiency at any given point in time. Applicants should not want to exclude an opportunity because of their own reporting shortfalls.

TAKING ACTION

The best-laid plans in the world mean nothing if they are never acted upon. No organization ever has enough funding to continue to operate and meet tomorrow's challenges. The fire and emergency services is dynamic. Those incidents faced in the past will not be exactly the same as those coming tomorrow, next year, or several years from now. Fire departments must continue to make improvements because every piece of equipment has a limited lifetime and will need to be replaced at some point in the future. Budgets will not always have the funding to make replacement happen, so grants will have to be used to fill in the gaps between local risks and organization preparedness. Continual improvement is the goal, and constant identification of deficiencies in emergency operations and readiness will keep the fire department ready to apply for any and all grant opportunities it discovers.

Ethical and Legal Considerations

The writers of grant applications must adhere to all applicable federal, state, and local laws during the process of writing the grant as well as when administering any resulting award. Any inaccuracies in the application that are deemed to have been done for the sole purpose of increasing the chances of being funded are considered fraud or, in some states, theft by deception. In federal grant programs, committing fraud by lying in the creation of an application is a felony and could result in prison time. State laws are not much different when dealing with grant programs regardless of who the grantmaker is.

Over the past several years, grant writers, salespeople, and members of fire and emergency services organizations have been convicted of fraud in relation to their grants with violations ranging from falsified information to improper use of the funding. Vehicles and equipment have been confiscated, without regard to the need of the applicant. Because the programs are competitive in nature, any alteration in figures or statements that better an applicant's chances also means another applicant that did not alter its situation might not receive what it deserves in the form of a higher score than the cheating applicant. So it is logical to assume that, had things been done properly by all applicants, the organization that committed the infraction probably would have scored much lower and would not have received the funding.

Grant seekers are advised to take no chances with ethical or legal matters. Double-check all numbers, and include only those that are fully documented and can be proven accurate. Permanent resident population is one of the figures most altered from what the federal government might have recorded; in many areas the census is performed only every 10 years, but in others the census is performed during the fifth or sixth year also. In an expanding rural or suburban area, the population could easily double from the last census. Because population does tie into cost-benefit calculations, applicants want the highest figure that they can claim according to an accepted methodology. The census is one method; but using tax rolls, having the post office run a listing of residential addresses in the service area, or even performing a hands-on survey of the community are ways that applicants can ensure their scoring is as high as possible and still within the realm of legality.

Summary

Grantmakers have decided to use their financial resources to help fire departments reduce the deficiency in firefighter safety and survivability in the local community. Some of the common downfalls that applicants make in grant applications include not understanding the purpose or scope of the grant programs, resulting in a less competitive application. As used in grants, scope is the list of organizations or organization types eligible to apply for funding from a grantmaker along with the list of eligible projects that can be funded.

To create highly competitive applications, applicants must first understand themselves and what measuring tools they are being judged against. Shortfalls are the identified deficiencies to firefighter health and safety that should be corrected within the fire department. Solution lists include items or actions that can correct the deficiency, the cost associated with each possible solution, and a listing of pros and cons for each solution. Such an assessment will help the applicant better understand the improvements that the grantmaker is seeking to make and also help the applicant create a better argument on why it needs to make the improvements. Applicants can expect to perform significant research to uncover studies showing what risks will remain if the improvements are not made, and possibly the short- and long-term effects of those risks to the members of the fire department and their health and safety.

Applicants must know their financial situation, including their existing funding sources and uses. Competitive applications demonstrate that the existing funding of the fire department is exhausted, letting the grantmaker know that there are no local means to pay for the project. Asking for too much at once might demonstrate that the current finances could have handled at least a part of the project.

All grantmakers require some sort of reporting. The documentation ranges from proper bid construction to invoicing, incident reporting, and financial audits. Applicants must understand these requirements prior to accepting to ensure all legal and ethical considerations have been addressed.

Review Questions

1. Go to the National Fire Protection Association (NFPA) Web site and review NFPA 1500, *Standard on Fire Department Occupational Safety and Health Program*. List three main requirements that fire and emergency services organizations must comply with in terms of responder health and safety.
2. Discuss internal and external needs assessment.
3. Identify at least one major support document to support an area of deficiency.
4. Discuss where to find information on the Internet about obtaining a grant for a fire and emergency services organization.
5. Discuss tips for writing an effective grant proposal.

References

Federal Emergency Management Agency and National Fire Protection Association. (2002, December). *A Needs Assessment of the U.S. Fire Service: A Cooperative Study Authorized by U.S. Public Law 106-398.* FA-240. Retrieved March 3, 2009, from http://www.nfpa.org/assets/files/pdf/needsassessment.pdf

Fire Protection Handbook (20th ed.). (2008). Quincy, MA: National Fire Protection Association.

General Mills. (n.d.) *General Mills Foundation.* Retrieved January 31, 2011, from http://www.generalmills.com/en/Responsibility/Community_Engagement/general_mills_foundation_2010.aspx

National Institute for Occupational Safety and Health. (n.d.). *Fire Fighter Fatality Investigation and Prevention Program.* Retrieved March 3, 2009, from http://www.cdc.gov/niosh/fire

Occupational Safety and Health Administration. (n.d.). *Standard 1910*. Retrieved March 3, 2009, from http://www.osha.gov/pls/oshaweb/owastand. display_standard_group?p_toc_level=1&p_part_ number=1910

Stern, J., and Routley, J. G. (1996, December). *Class A Foam for Structural Firefighting*. TR-083. Retrieved March 3, 2009, from http://www.usfa.dhs.gov/ downloads/pdf/publications/tr-083.pdf

U.S. Fire Administration. (1994, January). *Compressed Air Foam for Structural Fire Fighting: A Field Test*. TR-074. Retrieved March 3, 2009, from http://www. usfa.dhs.gov/downloads/pdf/publications/tr-074.pdf

U.S. Fire Administration. (2009, February). *Health and Wellness Guide for the Volunteer Fire and Emergency Services*. FA-321. Retrieved March 3, 2009, from http://www.usfa.dhs.gov/downloads/pdf/publications/ fa_321.pdf

Wil Dane

Courtesy of Martin Grube

KEY TERMS

accident chain, *p. 268*

mutual-aid response, *p. 265*

negative right-of-way intersection, *p. 270*

nonemergency response, *p. 267*

paid-on-call organizations, *p. 264*

violent incident, *p. 265*

OBJECTIVES

After reading this chapter, the student should be able to:

- Identify the percentage of line-of-duty deaths attributed to emergency vehicle crashes.
- Name best practices relative to emergency vehicle safety.
- Explain the value of policy statements, SOPs, and SOGs.
- List the reasons why one should always be seated and seat belted whenever an emergency vehicle is in motion.
- Recognize the criticality of responding to a violent incident or one that could turn violent.
- Identify innovations that can make driving an emergency vehicle safer.
- Describe how a basic understanding of the *Manual on Uniform Traffic Control Devices* (MUTCD) helps driver safety.
- Explain the importance of adequate staffing at an incident, making reference to NFPA standards 1710 and 1720.

PEARSON

myfirekit™

Studying Emergency Vehicle Response

Hardly a week goes by without news of a crash involving an emergency vehicle or a fire-fighter being struck by a vehicle while working at an emergency incident. Even as far back as 1973, when *America Burning* (National Commission on Fire Prevention and Control, 1973) was released, the significance of the risk to firefighters was noted, because they push themselves to the limits of their endurance and beyond.

Approximately 15 years later, when *America Burning Revisited* was released, Dr. John Granito responded to the question of when communities would allocate adequate resources for fire prevention by stating: "When we can diminish the sense of excitement that one gets from red trucks, sirens, and red lights" (USFA, 1987, p. 45). Since Dr. Granito's remarks, it seems that fire departments have used the same language to explain why firefighters are still killed and injured during an emergency vehicle response. Response excitement is just one reason for these deaths and injuries. Numerous other reasons will be addressed in this chapter.

Emergency Vehicle Response Crashes

In the past, line-of-duty deaths (LODDs) during response to emergencies were quite common. That is one reason the dalmatian is part of fire service lore. In order to keep the horses from crashing the steamers when responding to fires, the dogs would run alongside to keep other dogs from chasing the horses and making them go "wild." When horse-drawn steamers were replaced by motorized apparatus, emergency vehicle incidents continued to be a problem. Even today, over 20% (USFA, 2009, pp. 13–14, 17–18) of all firefighter line-of-duty deaths result from responding to, returning from, or operating at emergency incidents. (See Figure 11.1.)

FIGURE 11.1 Dalmatians continue to be associated with the U.S. fire service.
Courtesy of Martin Grube

In 1998 the University of Michigan Transportation Research Institute studied emergency response vehicle collisions in the United States over a three-year period. It noted the following statistics in an average year:

- 2,472 emergency response vehicle collisions
- 6 occupants of emergency response vehicles killed in collisions
- 413 occupants of emergency response vehicles injured in collisions
- 21 civilians killed in collisions with emergency response vehicles
- 642 civilians injured in collisions with emergency response vehicles (Campbell, 1999, p. 5)

Death and injury from emergency vehicle crashes have become a focus of numerous organizations, such as the U.S. Fire Administration (USFA), National Fallen Firefighters Foundation (NFFF), International Association of Fire Chiefs (IAFC), International Association of Fire Fighters (IAFF), and National Volunteer Fire Council (NVFC). As a result, numerous training programs have been developed to address the issue.

The fire and emergency services were involved in an estimated 14,950 emergency vehicle crashes while responding to an estimated 25.3 million incidents in 2008 (Karter and Molis, 2009). The year before there were more than 5,000 accidents involving ambulances (Ludwig, 2008). (See Figure 11.2.)

In addition, the National Volunteer Fire Council has developed a "Best Practices in Emergency Vehicle Safety Self Assessment" (NVFC, 2009, pp. 8–10). The best practices were compiled from input gathered throughout the fire and emergency services community, and any interested fire department can implement them as needed. Though the practices are from a variety of emergency services, they highlight safety in the fire and

FIGURE 11.2 During an emergency response, take precautions to ensure you will arrive on the scene safely. *Courtesy of Martin Grube*

emergency services with a four-step safety engineering approach meant to limit incidents and losses:

Step 1: ***Engineer out the problem.*** For example, since *America Burning* promoted improvements to emergency vehicle design, one such improvement is apparatus rollover prevention.

Step 2: ***Implement loss reduction techniques.*** The proper use of seat belts is a good example of a loss reduction technique.

Step 3: ***Implement administrative controls.*** The consistent enforcement of standard operating procedures/standard operating guidelines (SOPs/SOGs) is a good example of a best practice.

Step 4: ***Train personnel to use the proper safety devices and to do the job correctly.*** For example, training that includes actual on-road driving can help personnel understand the physical and dynamic forces affecting emergency response vehicles.

The goal of any fire department is for members to respond to an incident, render aid, and return safely. "Everyone Goes Home" should be everyone's motto. Interagency training, utilization of a common incident management system, and improving communication among all responders go a long way in helping to meet the goal.

Policies and Procedures

The inclination to operate emergency response vehicles in a reckless manner, under the guise of the urgency of the response, must be stopped. People call firefighters to make a situation better. A firefighter cannot do anybody any good if he or she becomes part of the problem, making the situation worse. Additionally, actions such as not coming to a complete stop at a stop sign or traffic light or not wearing personal protective ensemble (PPE) while the vehicle is responding can actually place firefighters in an unsafe situation that is not worth the additional response time saved.

The attitudes and behaviors of responders who accept dangerous and unsafe acts as "part of the job" and "heroic" must change if fire departments are to reduce the number of firefighter deaths and injuries. One acceptable method of making that change is through policies and procedures. However, the culture of the department has to be receptive to enforcing those policies and procedures.

Policies and procedures provide guidance that can be used when responding to, returning from, or operating on the emergency scene. They also ensure that incident and individual safety is a top priority by minimizing confusion at the emergency scene, maximizing utilization of resources through a coordinated effort of rescuing endangered victims, minimizing property damage, and bringing stability to the scene by reducing overall impact to the community.

At the minimum, policies and procedures should be written to cover emergency response vehicle safety and response to a violent incident, such as the following:

- Using seat belts while the vehicle is in motion
- Approaching and crossing an unguarded railroad crossing
- Backing operations involving vehicles
- Bringing vehicles to a complete stop at stop signs and signals
- Encountering school buses with warning lights activated
- Operating a privately owned vehicle (POV) in a safe and prudent manner
- Responding to and operating during a violent incident
- Operating at any incident that has a high probability of escalating into violence

The National Fire Protection Association (NFPA) offers helpful standards in the area of emergency response. Departments should implement NFPA 1002 and NFPA 1451.

NFPA 1002, *Standard for Apparatus Driver/Operator Professional Qualifications* (2009a), requires driver/operators to be certified as a Firefighter I in accordance with NFPA 1001, *Standard for Fire Fighter Professional Qualifications* (2008), before being certified to drive or operate an emergency vehicle. NFPA 1451, *Standard for a Fire Service Vehicle Operations Training Program* (2007a), provides direction to fire departments on establishing and maintaining emergency vehicle driver/operator training programs.

Other programs that have been developed by fire and emergency services organizations in order to reduce emergency vehicle crashes include the following:

- The International Association of Fire Chiefs created a Web-based program called *Guide to Model Policies and Procedures for Emergency Vehicle Safety* (IAFC, n.d.).
- The International Association of Fire Fighters developed *Emergency Vehicle Safety Program: Improving Apparatus Response and Roadway Operations Safety in the Career Fire Service* (IAFF, 2005).
- The National Volunteer Fire Council created the Web-based program *Emergency Vehicle Safe Operations for Volunteer and Small Combination Emergency Service Organizations* (NVFC, 2009).

SEAT BELTS

All fire and emergency services organizations are strong advocates of seat-belt use by all responders at all times an emergency vehicle is in motion. Additionally, state and federal laws and regulations require seat-belt use. Yet, the prevention of deaths and injuries from vehicle-related incidents is entirely under the authority of the fire department, which means it is firefighters and emergency responders who must aggressively and immediately seek action to help enforce safe driving practices that can reduce responder deaths and injuries.

Nobody joins the fire and emergency services to have an accident or to be involved in a crash that could cause death or devastating injury. Firefighters have a personal responsibility to safely operate vehicles—first and foremost for their own safety, second to their crew, and finally to the community. Just think about the human cost factor alone, not including medical costs, that may be incurred as well as the cost of vehicle repair or replacement in the event of a catastrophic incident.

Numerous firefighters are killed every year in traffic-related incidents during which the responder did not use seat belts. The cultural changes needed in that area are some of the most difficult for some firefighters to accept. Departments must foster a culture of staying seated and "buckling up" seat belts any time an emergency vehicle is in motion. (See Figure 11.3.)

FIGURE 11.3 Use your seat belt every time the emergency vehicle is in motion. *Courtesy of Michael Conder*

However, opponents to seat-belt use contend that certain firefighters who are wearing personal protective ensemble (PPE) cannot find the seat belt in order to properly put it on. Or, they continue, the seat belt does not fit over all of the responder's PPE. Those opponents call on the various manufacturers of emergency response vehicles to change the design of the emergency vehicle seat belts and restraint systems. If seat belts in emergency vehicles prevent responders from being properly belted while the vehicle is in motion, fire departments may need to alter any SOP/SOG that requires full PPE to be worn when the vehicle is in motion. There can be no excuse for not wearing a seat belt any time the vehicle is in motion. Members who can safely wear PPE and be belted in, should do so. Members who cannot wear PPE and be belted in, should not wear the PPE and instead should always wear their seat belt.

Departments must insist that emergency response vehicle manufacturers design seat belts and restraint systems that work easily for firefighters riding in any emergency response vehicle. There needs to be a specific date set whereby emergency vehicles must be retrofitted with seat belts and restraint systems that work for everybody, regardless of individual body type. Until that happens, the fire and emergency services must do whatever is in its power to make sure that all responders are seat belted at all times in any emergency vehicle that is in motion.

Railroad Crossings

Emergency vehicles that approach or cross an unguarded or activated railroad crossing are at serious risk. Over the past few years, several firefighters have lost their lives in crashes with trains; others have had near misses involving trains at railroad crossings. At any unguarded railroad crossing, it is essential that all firefighters in emergency vehicles, including privately owned vehicles (POVs), should consider such a crossing as if it were an activated railroad crossing, learn to bring their vehicles to a complete stop before entering, and verify that it is safe to proceed. Additionally, each driver should perform the following upon approaching an unguarded railroad crossing:

- Silence any siren or air horn.
- Bring the vehicle to a complete stop.
- Make sure any other device that produces noise is silenced or turned off.
- Listen for a train's horn by opening the windows and temporarily removing the hearing protection device on the ear nearer the closest window, while looking both ways to ensure no train is near.

At an activated railroad crossing, the emergency vehicle must be brought to a complete stop even if there is no train in sight. Once stopped, the firefighter should look up and down the track in both directions. If there is no train in sight, the emergency vehicle may proceed cautiously past the signal and across the tracks. If the view of the tracks is blocked, it may be necessary for one of the vehicle's occupants to exit the vehicle, proceed to the tracks on foot, and determine whether it is safe for the emergency vehicle to proceed farther past the activated railroad crossing.

Backing Operations

Backing operations that involve driving a vehicle in reverse is another type of common incident involving emergency response vehicles. Though that type of crash rarely involves any deaths or injuries, some have been reported. A related issue is, of course, the significant damage to the emergency vehicle as well as to the firehouse or station and the costs of repair or replacement.

Whenever possible, responders should avoid backing emergency vehicles. However, on certain occasions there is no other option. When backing, at least one person—and

preferably two—with a portable radio should be assigned to clear the way and warn the driver of any obstructions hidden by a blind spot. The vehicle may be equipped with backup safety devices, but these are no substitute for a ground guide or spotter. When two spotters are used, it is preferred that only one should communicate with the driver, to avoid confusion. The second spotter should stay in view of the first spotter and assist. If the driver/operator does not have or cannot see the ground guide or spotters behind him or her, the driver/operator should not back the vehicle! The department must develop policies and procedures for proper use of hand and radio signals to communicate with the emergency vehicle driver. (See Figure 11.4.)

New buildings meant to house emergency response vehicles today should include drive-through bays, which eliminate the need to back the vehicle. In addition, every department needs to adopt and enforce policies and procedures regarding safely backing emergency vehicles into or out of the fire station and in other situations. NFPA 1500 (2007b) includes language about the safe backing of the apparatus and should be consulted when developing a backing policy or procedure.

STOPPING AN EMERGENCY VEHICLE

NFPA 1500 (2007b) lists eight situations in which an emergency vehicle must come to a complete stop before proceeding safely and for which the firefighter may be held accountable should there be a mishap. The firefighter must come to a complete stop:

1. When directed by a law enforcement officer
2. At red traffic lights
3. At stop signs
4. At negative right-of-way intersections
5. At blind intersections
6. When the driver cannot account for all lanes of traffic in an intersection
7. When other intersection hazards are present
8. When encountering a stopped school bus with flashing warning lights

During any emergency response in which an emergency vehicle does not come to a complete stop in any of those eight situations, the driver/operator and officer should be held accountable for their actions or lack thereof.

Stopping at Intersections

Intersections are another area that accounts for emergency vehicle collisions. One reason for those collisions is the failure of emergency vehicle drivers to bring the vehicle to a complete stop at stop signs or signals. An emergency vehicle driver/operator should never continue into an intersection until he or she knows that all other drivers see the emergency vehicle and are allowing it to proceed.

Not all intersection crashes are the fault of the emergency vehicle driver. Reckless driving by the public, including failure to obey traffic signs/signals, excessive speed, and failure to yield to emergency vehicles, makes the job of emergency response more hazardous. It is estimated that a vast majority of the driving public exhibits poor driving habits (Drive for Life, 2005). Therefore, at every intersection, including those with a green light or no signal at all, drivers should slow their vehicles to a speed that will allow them to stop if necessary.

Driving in congested traffic conditions adds hazards of which the driver must be aware. In those situations the driver must safely maintain control and be prepared to stop the vehicle at all times. In addition, drivers can never assume that civilian drivers will yield them the right of way. Even when a civilian is willing to yield, there simply may be no place into which the civilian can pull over. Drivers should never try to force their way through such congestion to gain the right of way.

Remember, driver/operators are simply asking for the right of way. The emergency vehicle does not have the right of way until it is yielded to. Adding a few seconds to a response time is better than not arriving at all due to becoming involved in a crash.

Traffic Control Devices

A green traffic light that has been in that position for a long time could mean that the light is soon going to change to yellow and then to red. Drivers should be aware of such and plan for the light to change. When the light turns yellow, the emergency vehicle should be prepared to stop. When it turns red, the emergency vehicle must stop.

Another traffic control device that emergency vehicle drivers should be aware of is a flashing "Do Not Walk" sign at an intersection crosswalk, another indicator that the green light is about to change. Those lights usually flash 15 to 20 seconds before the green light turns yellow and are a signal to the driver to be prepared to stop.

Traffic control devices are sometimes used to make an emergency response safer. The simplest traffic control device is a sign informing drivers that emergency response vehicles may enter at a certain location, usually in front of a fire station. The sign also will assist emergency vehicles as they exit the station. Another type of traffic control device is on both the responding emergency vehicle (transmitter) and the traffic light (receiver). It is used to gain control of the intersection by changing the traffic signal light to green for the responding emergency vehicle's direction of travel. There are several models that sense special lights being transmitted by the emergency vehicle, and other models that sense an emergency vehicle's siren.

It should be noted that traffic control devices may be used to get the right of way for responders, but they do not guarantee it. (See Figure 11.5.) In fact, such traffic control devices should never be thought of as a substitute for using proper safe driving techniques. That is, when approaching an intersection with a green signal, the driver/operator should maintain a speed that will allow him or her to stop should another vehicle enter the intersection. And remember, if two emergency response vehicles, each with a traffic control transmitter, approach the same intersection, the vehicle whose signal is first to the traffic light receiver will get the green light. The other response vehicle should receive a red light.

FIGURE 11.5 Traffic control devices do not guarantee the right of way. *Courtesy of Martin Grube*

If the traffic control device is used but the emergency response vehicle does not receive a green light, the driver should consider that another emergency response vehicle is approaching from another direction. Never assume that because the light does not change to green that the system is *not* operating properly and proceed through the intersection. The situation should be handled as any other occasion of a red light: proceed through the red signal only after coming to a complete stop and ensuring that it is safe to proceed.

A newer technology in traffic control devices is one that uses a GPS system to activate green lights for emergency responding vehicles. The vehicle sends a signal that is tracked by satellite, which in turn relays the information back to a monitoring system. The system determines the speed and direction of the emergency vehicle and changes the appropriate traffic light at the intersection the emergency vehicle is approaching.

Stopping for School Buses

Whenever an emergency vehicle encounters a school bus with its warning lights activated, the responder should stop just as he or she would at an intersection with a stop sign or red light. Generally, emergency vehicles must obey the various state or local laws that require vehicles to stop for school buses with flashing signal lights indicating that they are loading or unloading occupants. The emergency vehicle driver should proceed past a stopped bus with its warning lights flashing *only* after being signaled to do so by either the bus driver or a police officer. When motioned to proceed, the driver should do so slowly, continually watching for any bus occupant who might dart out in front of the emergency vehicle.

FIGURE 11.6 Drivers of emergency response vehicles should always be able to come to a complete stop. *Courtesy of Adam Williams, Anderson County Fire Department*

Speed Affects Stopping

One of the most important lessons that drivers of emergency vehicles must learn and remember, especially those vehicles built on a truck chassis, is that the emergency vehicle they are driving does not handle or stop the same as their POVs. Generally, it takes a much greater distance for the emergency vehicle to stop due to the fact it weighs much more than a POV.

Drivers should always be prepared to stop. If a driver cannot come to a complete stop, he or she is traveling entirely too fast and does not have full control of the vehicle. Remember that three factors affect total stopping distance of a vehicle: perception distance, reaction distance, and braking distance. (See Figure 11.6.)

EXCESSIVE SPEED

Among the most frequent causes of an emergency vehicle crash is excessive speed. It usually involves losing control of the vehicle and the driver being unable to bring the vehicle to a stop soon enough to avoid a crash with another vehicle or object.

Excessive speed is exhibited during an apparatus response when:

- The vehicle is unable to negotiate a curve in the road.
- The vehicle is unable to stop before hitting another vehicle or object.
- The vehicle is unable to stop before entering an intersection or railroad crossing.
- A weight shift occurs when the vehicle is slowed, causing it to skid or overturn.
- Control of the vehicle is lost after hitting a pothole, speed bump, or similar defect in the driving surface.
- Control of the vehicle is lost as a result of swaying outside the lane of travel and striking a median or curb.
- Tires on one side of the vehicle (usually the right side) leave the road surface; or tire traction is lost on wet, icy, snowy, or unpaved road surfaces.

Excessive speed should be addressed in safe driving policies and procedures. Then any time an emergency response vehicle is driven, those policies and procedures should be consistently and strictly enforced by the department. (See Figure 11.7.)

PRIVATELY OWNED VEHICLES

The safe response procedures discussed thus far are all applicable to any privately owned vehicle (POV) being operated by a volunteer or paid-on-call firefighter dispatched to an incident. According to line-of-duty death statistics, approximately 25% of firefighter deaths from vehicle crashes involve a POV (USFA, 2002, p. 21).

Most volunteer and **paid-on-call organizations** allow responders to make calls and assist at any event when they can and are needed, while their primary employment is in another field. They do not require firefighters to be on duty at a station. Many of those departments allow members to respond to an emergency from their homes or workplaces in their POVs. Because of the use of those vehicles, volunteer or paid-on-call firefighters are more likely to die in a vehicle crash involving a POV. Many of those vehicle crashes occur in rural areas, which are predominantly protected by volunteer or paid-on-call organizations.

Training in POV safety and properly established policies and procedures should be mandatory in all fire departments that allow members to respond directly to the incident or to the fire station in POVs.

paid-on-call organizations ■ Emergency service organizations that allow responders to make calls and assist at any event when they can and are needed, while their primary employment is in another field.

Responding to Violent Incidents

Because firefighters are killed or injured at emergency scenes that are not initially reported as violent, it is important always to expect the worst whenever responding to any incident. Such an expectation should occur whether responders are called to a local incident (such as a world championship sporting event) or an event of wider proportions (such as Hurricane Katrina). However, because it is impossible to think of all the dangers that may be encountered, it essential for departments to have policies and procedures in place when firefighters are approaching or become involved in any incident involving violence.

On most responses, firefighters arrive on scene, render aid, and then return to service. But occasionally there is an event from which every member on scene does not return, because of either a spontaneous violent act or one that is premeditated and meant for firefighters. The increased number of line-of-duty deaths and serious injuries involving firefighters is cause for great concern. Safety summits have been held with the outcome being a number of recommendations regarding emergency responses to a **violent incident**, which is any emergency wherein a threat or act of violence is the expression of the intent to inflict pain, injury, or other harm to firefighters. Among those recommendations are the following:

violent incident ■ Any emergency wherein a threat or act of violence is the expression of the intent to inflict pain, injury, or other harm to firefighters.

- *Pre-incident planning and training.* There must be communication and cooperation at an incident among all emergency response agencies so that there may be sharing of intelligence about suspected targets when violence is encountered. Pre-incident planning and training by and among agencies—police, fire, and emergency services—must occur in order to make that possible.

- *Dispatch information.* The role of the emergency call taker/dispatcher prior to and during the response must be emphasized. The uncertainties involved during incidents in which violence is a factor are innumerable. The ability of the call taker/dispatcher to reduce those uncertainties is central to the proper gathering of critical information and the safety of responders. Through pre-incident planning and training, responders can learn to recognize the potential for violence through the information they are given.

  In addition, dispatch staffs should be trained to carefully monitor the radio traffic of operations at a violent incident. They also should request a personnel accountability report (PAR) periodically to monitor the safety of responders at the incident.

- *Communicating with other responding organizations.* The use of plain language for communication is essential to the safety of responders. Fire departments that still use a form of communication involving codes or signals should at a minimum verify that all of their **mutual-aid response** departments are familiar with them. A mutual-aid response is an agreement between emergency response organizations to provide assistance across jurisdictional boundaries, which may occur due to an emergency response that exceeds local resources. Additionally, a common radio frequency should be utilized to eliminate the possibility of the message being wrongly understood or missed when passed through a call taker/dispatcher.

mutual-aid response ■ An agreement between emergency response organizations to provide assistance across jurisdictional boundaries, which may occur due to an emergency response that exceeds local resources; requires a firefighter to make a specific request for assistance.

  In fact, every firefighter who has the responsibility of responding to a violent incident must be trained in and use an incident management system (IMS) common to all those responding. Because communication is an integral part of an incident management system, familiarity with the system among all responders helps to minimize communication problems. In addition, face-to-face communication among the various emergency response agencies at an incident prior to entrance should be encouraged.

- *Single-member responses.* Many jurisdictions allow the response of individual responders in a POV or single members in a response vehicle to proceed directly to the emergency scene. Such a practice should be discontinued. A policy prohibiting or at the very least limiting single-resource or single-member response should be implemented instead. The ability of an individual to recognize and retreat from danger is too limited. The old saying "there is safety in numbers" is paramount here because members responding in a group increase individual safety, awareness, and visibility.

- *Incident action plans (IAPs), especially for emergency incidents spread over a large area.* A common incident action plan, which defines the roles of the responders from various organizations, is necessary for a successful and safe outcome to a violent incident. A contingency plan for firefighters should be in place, whereby they can safely retreat until such time as law enforcement personnel arrive at the scene and bring stability to the incident.

- **Uniforms.** Firefighters should be clearly identifiable by the uniforms they wear. It is important to the safety of every firefighter that he or she is recognized as a member of an emergency response organization, and not mistaken by the public or by other responders for anyone else.
- **Risk management.** What do you do if victims start running straight to you? The safety of the firefighter must be first in the mind of every responder at the scene. No responder should be placed in a situation in which the risk outweighs the benefit to be gained by the firefighter's actions. In addition, NFPA 1500 (2007b) states that the organization should provide body armor for all members who operate in areas in which a potential for violence or civil unrest exists.
- **Pubic education programs.** Educating the public in the principles of responder safety as outlined by risk management principles helps keep responders safe. It is essential that the department support public education programs that outline response to violent incidents and ensure each member of the department support.
- **Calling for assistance.** Firefighters should have a special code when they are in trouble or in need of assistance, either by radio signal or by direct communication. It must be a policy or procedure that everyone in the department understands how to implement. (See Figure 11.8.)

Reducing the number of firefighter line-of-duty deaths, especially those resulting from acts of violence, is challenging but not impossible. It is not easy to predict actions that make an incident turn violent or to determine an incident is going to become violent just by the information learned from a call to a dispatch center. Implementation of the preceding recommendations can help reduce the likelihood of responder death and injury as a

FIGURE 11.8 Random acts of violence aimed at firefighters are occurring more frequently. *Courtesy of Martin Grube*

result of response to a violent incident. Always be prepared when responding to an incident that involves a situation that has already resulted in harm to another individual, such as a shooting, stabbing, fight, large crowd gathering, and so on. If the department does not have a policy or procedure in place, such as waiting for the police before approaching the scene, make one or firefighters and emergency responders may find themselves in an uncontrollable situation. Everybody who responds to any type of violent incident must understand how to implement those policies and procedures before they are actually needed.

Nonemergency Response

The response mode of emergency vehicles to an incident continues to be a hot topic of discussion among fire departments. Many departments consider that any call for assistance constitutes an emergency response. Others have policies and procedures stating what type of calls will require an emergency response and what other types of calls will receive something other than an emergency response. Many fire departments have instituted a concept called nonemergency response.

The nonemergency response all but does away with the use of red lights and sirens when responding to incidents that are not considered a true emergency. There is no argument that responding in a nonemergency mode to an alarm for assisting a person back into bed is much safer for all involved.

Departments should do a risk-benefit analysis of their response criteria to establish when it is appropriate to respond in an emergency mode with lights and sirens or to utilize the nonemergency response. It is much safer and wiser to choose to use emergency lights and sirens only in the most severe circumstances.

A **nonemergency response** is an emergency vehicle response to an incident in nonemergency mode (without use of emergency lights and sirens). Such a response should be considered for the following incidents:

nonemergency response ■ An emergency vehicle response to an incident in nonemergency mode (without use of emergency lights and sirens).

- Automatic alarms
- Sprinkler flow alarms
- Outside natural gas leaks
- Wires down
- Requests for additional personnel for staffing relief
- Assisting sick people back in bed
- Assisting medical personnel with additional personnel for a heavy lift
- Carbon monoxide detector alarms without a report of an imminent life hazard
- Rubbish fires
- Open burning not threatening an occupied structure
- Dumpster fires

Opponents of the nonemergency response argue that response times will negatively affect the outcome of the incident: a fire might progress out of control, medical aid to a victim could negatively affect survivability, or a hostage situation could get out of control, creating a standoff that might take hours to resolve. On the other hand, those in favor of the implementation of a nonemergency response policy argue it has significantly reduced the number of serious crashes involving emergency response vehicles.

When an alarm is dispatched as a nonemergency response and call takers/dispatchers receive additional information that life is in danger, persons are injured, or there is a working fire, the call can be upgraded to an emergency response. Then the responding emergency vehicles will activate lights and sirens to continue the response. Most departments that have implemented a nonemergency response policy report a reduction in the number of intersection vehicle crashes and no significant increase in life or property loss (IAFF, 2005, p. 35).

Some departments have implemented a variation of the nonemergency response whereby the closest response vehicle to the incident responds in an emergency mode and the remaining dispatched vehicles respond as to a nonemergency. However, the highest ranking officer responding can have all crews upgrade their response to "emergency" if additional information confirming a true emergency is passed on from either dispatch or a firefighter on the scene.

A policy must be written, understood, and implemented to guide call takers/dispatchers in the proper criteria required to put into action a nonemergency dispatch before such a response mode can be fully operational. Thorough and proper information gathered by call takers/dispatchers can lead to the proper units being dispatched. Utilizing the appropriate response mode will ensure the best use of a responding emergency vehicle is achieved.

The Safe Response

MEANINGFUL SAFETY POLICY

Meaningful safety policy helps ensure an orderly scene where committed resources are designated to life safety, incident stabilization, and property conservation. It promotes safety as the top priority and provides an outline to be used for any type of incident. Once implemented and enforced, a policy can ensure a safe incident scene due to the implementation of an IMS, which provides guidance and improves responder communication and accountability. Policy language may vary among jurisdictions about the where and how, but language regarding safety and safe operating policies and procedures should be universal.

With meaningful safety policy and procedures in place, firefighters can develop a safety-conscious attitude about their own safety as well as that of others. The attitude that line-of-duty death or injury is a badge of honor can no longer exist, because it is not honorable to die or get injured when such a result is avoidable.

All personnel must learn to recognize that a vehicle collision usually is preventable and predictable. Firefighters must be encouraged to do whatever is necessary to prevent such incidents from ever happening and not accept them as an uncontrollable occupational hazard. With more safety consciousness being exhibited by all, it has been learned that the incident scene does not have to be one strewn with occupational hazards.

INCIDENT INVESTIGATION

accident chain ■ The sequence of mistakes and coincidences that lead up to an accident.

Fire departments are encouraged to investigate all emergency vehicle crashes to answer the following questions about the **accident chain**, the sequence of mistakes and coincidences that lead up to an accident:

■ *What was the environment?* Consider the physical surroundings such as weather, surface conditions, access, lighting, and physical barriers.
■ *What human factors were involved?* Consider human and social behaviors, training (or lack of it), fatigue, fitness, and attitudes.
■ *What equipment, if any, was involved?* Consider apparatus, PPE, maintenance and serviceability, proper application, and equipment limitations.
■ *What of the preceding created the event?* Consider the most likely sequence of events and most probable causes.
■ *What injuries or property damages actually were incurred?* Be certain that any and all questions are answered honestly and accurately.
■ *What was learned that can be implemented to create a positive safety attitude?* Consider the practice of good habits, learning from others, and vigilance.

NATIONAL SEAT BELT PLEDGE

Not wearing seat belts while an emergency vehicle is in motion is wrong. Whether it is to oneself or a crew member, correcting this wrong will help reduce the number of firefighter deaths and injuries. In February 2007, retired USFA Assistant Administrator Charlie Dickinson wrote a release regarding the untimely death of a firefighter on April 23, 2005 (Dickinson, 2007). The firefighter was 27-year-old Christopher Brian Hunton, a member of the Amarillo, Texas, Fire Department. Hunton's company was responding to an alarm when he fell out of the fire truck. Unfortunately, Hunton was not wearing a seat belt, much like many other firefighters. Two days later he died from the injuries he received. (See Figure 11.9.)

Some time later, Dr. Burton A. Clark, a program specialist at the National Fire Academy, wrote an article (Clark, 2005) introducing the National Fire Service and EMS Seat Belt Pledge. It is an individual pledge to wear a seat belt whenever one is riding in an emergency vehicle and to further promise to ensure that all others riding in the vehicle wear their seat belts, too. The pledge expects to make it possible that no firefighter will die needlessly from not wearing a seat belt while riding in an emergency vehicle.

TRAFFIC PREEMPTION DEVICES

Traffic preemption devices have proven their worth to communities by helping to make the responses of emergency vehicles timelier and safer by allowing emergency vehicles the right-of-way at intersections. Once installed, the traffic preemption devices should be used on all emergency responses.

Key points to consider once traffic preemption devices are installed include the following:

- When warning lights/sirens are activated, use traffic preemption devices.
- Turn off the traffic preemption emitter and warning devices when ordered to "reduce speed" or upon any order that means there is no longer an emergency.
- The emitter is not to be used during nonemergency functions, such as parades, community functions, store/food runs, and so on.
- All responders qualified to drive emergency vehicles with emitters must be trained on their function and all policies and procedures concerning when to use them.

- Use of the emitter system DOES NOT GUARANTEE or GRANT right of way.
- Install the emitter device so that it will automatically turn off when transmission is in the park position or when the vehicle braking system is applied.
- All emitters must have an on/off switch to allow the unit to be turned off when the vehicle is in nonemergency mode.
- There should be a method for checking the system periodically to ensure it is working properly.

Use of traffic preemption systems allows departments to limit the amount of apparatus responding in emergency mode (lights and siren) on alarms.

COLLISION AVOIDANCE

In safe responses, the vast majority of collisions are avoided by concentrating on the five most common potential hazards:

1. *Driving through intersections.* The largest percentage of major collisions involving emergency vehicles happens at intersections. For a safe response, all occupants of the emergency vehicle must work together during all parts of the response, but especially when going through intersections. Occupants are a driver's second set of eyes. Coming to a complete stop at a **negative right-of-way intersection**—which is an intersection controlled by a stop sign, yield sign, or traffic light—adds only two to three seconds per stop but prevents injury and death. The emergency vehicle driver must ensure that the apparatus has the right of way before proceeding.

2. *Backing operations.* The most common type of emergency vehicle collision involves backing. Those collisions are responsible for a significant percentage of vehicle damage and dollar losses. For a safe response whenever backing the vehicle, NFPA 1500 (2007b) requires at least one guide who has radio contact with the driver. The vehicle also should be equipped with backup safety devices, but those devices are never to be a substitute for a human guide.

3. *Driving speed.* A safe response never involves excessive speed. If you do not arrive on the emergency scene, you cannot make a difference.

4. *Keeping vehicle wheels on the road surface.* Hazards occur when the wheels leave the roadway; when the vehicle sinks in soft ground or otherwise is pulled farther off the road, striking an object or overturning; and when the vehicle shoots across the roadway as a result of overcorrection, which may cause the vehicle to roll over or to strike another vehicle head-on. Therefore, to keep the driver's attention and his or her being in control of the vehicle at all times, the driver should not operate warning devices, read map books or computer monitors, or perform other activities while driving the vehicle, because such activity may result in drifting. Drivers also should never try to pass slowed or stopped vehicles on the right side.

5. *Negotiating curves.* Collisions on curves are typically a result of either excessive speed or failure to keep all wheels on the road surface. Emergency vehicles are heavy and typically have a high center of gravity, which when combined with excessive speed on a curve can result in loss of control or overturning of the vehicle. Excessive speed also may force the vehicle into opposing lanes of traffic, causing a head-on collision. Posted, recommended speeds on road signs approaching curves are based on passenger vehicles on dry roads, not emergency vehicles loaded with various types of equipment. In fact, posted speeds are probably not appropriate for emergency vehicles even in good weather conditions. The best way to avoid collisions on curves is to operate the vehicle at a safe speed, keeping all wheels on the road.

negative right-of-way intersection ■ Any intersection controlled by a stop sign, yield sign, or traffic light.

Other collision avoidance considerations include nonemergency response, call prioritization, equipment innovation, learning from mistakes, and on-scene traffic control.

Nonemergency Response

Emergency vehicles are most likely to be involved in a crash when responding with lights and siren operating. Departments have historically treated every response as an emergency. That view is unnecessary and creates unreasonable danger. Every jurisdiction must look at implementing a policy that requires a nonemergency response on low priority calls. By lessening the number of emergency responses, the collision rate should be reduced. Some jurisdictions have noted a significant drop in emergency vehicle crashes while responding since adopting these types of policies and procedures.

Call Prioritization

Call prioritization remains a critical aspect of the fire and emergency services operation. Standardizing call-taking procedures, call prioritizing, triage, and dispatching ensure requests for fire and emergency services are assigned in order of necessity. Appropriate units are then assigned and respond to each request based on the priority level determined in the call-taking process. Before dispatching can occur, protocols must be written, understood, and related policies and procedures implemented.

Equipment Innovation

In recent years equipment innovation has helped improve the safety of firefighters. It includes the following: a better emergency vehicle design that has produced a safer passenger transport area, noise reduction for better opportunity to hear radio transmissions and for communications between responders, an area in which personnel can be seated and restrained while apparatus is in motion, rollover protection for apparatus, layout and placement of equipment on vehicles for ease of access and removal, and various safety systems such as air bags, open door or compartment warnings, color-coded seat belts, and unbuckled-seat-belt warnings.

Learning from Mistakes

Learning from the mistakes of others is essential in helping to lessen the number of firefighter deaths and injuries. Determining the causes and identifying such patterns and trends can help develop programs to reduce similar deaths and injuries in the future. Studying firefighter close-call and near-miss reports and implementing the lessons learned into local training programs, policies, and procedures can help to cut down on the number of firefighter deaths and injuries. Additionally, the USFA technical report series (USFA, 2008b) contains investigative reports on certain incidents and is designed to share the lessons learned.

On-Scene Traffic Control

The possibility of being struck by another vehicle while operating at the emergency scene is a common hazard. Firefighters operating at the scene of vehicle incidents, fires, medical emergencies, and hazardous materials operations on any public street or highway are in double jeopardy; in addition to being potential targets for passing vehicles, there is inherent danger from the incident itself. Position emergency response vehicles so that they protect firefighters, and always consider moving vehicles that are a threat to safety. (See Figure 11.10.)

The Emergency Responder Safety Institute has been tracking the number of firefighters struck by another vehicle while operating on the emergency scene. The Department of Transportation's Federal Highway Administration and the U.S. Fire Administration, along with the help of the International Fire Service Training Association (IFSTA), have worked together to develop a traffic incident management technical guide and training

FIGURE 11.10 Always position emergency vehicles to protect firefighters on the scene. *Courtesy of Martin Grube*

program for the fire and emergency services (USFA, 2008a). The guide is filled with case studies involving roadway incidents, basic roadway emergency scene safety operations, and incident command for roadway incidents.

Some core elements of a comprehensive Roadway Incident Safety Program that each fire department should have in place include (Sullivan, 2010):

- Roadway incident safety training for all personnel
  - Initial orientation for new recruits before they respond to any emergency
  - Annual in-service training session for all personnel
  - All training in line with SOPs/SOGs and national standards, rules, regulations, and "best practices"
- Roadway incident response procedures
  - Document, authorize, and publish SOPs
  - Comply with NFPA 1500 (2007b)
  - Comply with the *Manual on Uniform Traffic Control Devices* (FHWA, 2009)
- Proper personal protective equipment (PPE) for all personnel
  - Document Occupational Safety and Health Administration–compliant PPE hazard assessment
  - NFPA-compliant turnout gear
  - American National Standards Institute (ANSI)–compliant high-visibility garments
- Multiagency and multijurisdictional cooperation, collaboration, and communication
  - Regular fire department attendance and participation in local and regional traffic incident management committees
  - Ongoing multiagency planning and training on roadway incident response procedures
  - Multiagency review and critique of traffic incidents to improve strategies and tactics at future incidents

- Properly positioned apparatus and traffic control equipment at incidents
  - Park large fire apparatus at an angle upstream of the incident work area
  - Turn front wheels away from the incident scene and properly chock units when parked
  - Properly deploy advance warning devices, including MUTCD-compliant high-visibility signs and cones
  - Position ambulances downstream with the loading area doors angled away from moving traffic, whenever possible
  - Park all emergency equipment on one side of the road
  - Effectively place police cars for traffic control and scene safety
  - Effectively use any available safety service patrol apparatus
- Fire apparatus enhanced visibility design features
  - Emergency warning lights designed for on-scene protection
  - NFPA 1901 (2009b)–compliant high-visibility (reflective and florescent) chevrons on the rear of apparatus, road cones, and PPE

Title 23 of the U.S. Code of Federal Regulations charged the U.S. Department of Transportation (DOT) with developing a manual on uniform traffic control standards, which each state would be required to adopt. DOT developed the *Manual on Uniform Traffic Control Devices* (MUTCD) (FHWA, 2003). In the past, emergency scenes were not explicitly covered by the MUTCD, but the 2003 edition of the MUTCD has a section (6I) on "The Control of Traffic Through Incident Management Areas." The section applies to all incidents on or near the nation's roadways. Because these are federal regulations, not standards, they must be followed.

The goals of emergency traffic control as outlined in the MUTCD are improving responder safety on the incident scene, keeping traffic flowing as smoothly as possible, and preventing the occurrence of secondary crashes. To help meet those goals, there are five main parts of Section 6I of the MUTCD:

- *General.*   These are requirements for interagency coordination, training, visibility, estimating incident scope and length, ETC sign colors, and use of initial control devices, such as road flares and traffic cones.
- *Major traffic incidents.*   These are incidents that exceed 24 hours. For them, full MUTCD work zone requirements must be implemented.
- *Intermediate traffic incidents.*   These incidents range from 30 minutes to two hours in duration. They typically require lane closures. Typical vehicle collisions with injuries fall into this category.
- *Minor traffic incidents.*   These are incidents whose duration is less than 30 minutes. Simple actions, such as the use of initial control devices, will be sufficient to handle the incident. Minor, non-injury collisions and stalled vehicles are examples of minor traffic incidents.
- *Use of emergency vehicle lighting.*   This is direction on the appropriate types of lighting for use at nighttime roadway incidents.

The MUTCD places a significant amount of emphasis on doing a proper size-up of the scope and severity of the incident within 15 minutes of the arrival of the first firefighter. Included in the size-up, and it must be determined by the firefighter what is the magnitude of the incident, are the estimated time duration that the roadway will be blocked or affected, and the expected length of the vehicle backup that will occur as a result of the incident. (For every one minute a lane of traffic is blocked, four minutes of backup are developed.) The information must then be used to set up appropriate measures to handle the incident. (See Figure 11.11.)

The main parts of a traffic incident management area as outlined in the MUTCD are the advance warning area that tells motorists of the situation ahead, the transition area

where lane changes/closures are made, the activity area where responders are operating, and the incident termination area where normal flow of traffic resumes. The distances for the advance warning and transition areas will differ depending on the speed limit in the area of the incident. Higher speed limits require longer advance warning and transition areas. The MUTCD contains charts that detail the appropriate length based on the speed limit in the area.

Emergency Vehicle Driver/Operators

Being the driver/operator of an emergency vehicle places a tremendous amount of responsibility on a responder's shoulders. A driver/operator must consider vehicle size and handling, configuration, and special conditions under which the vehicle is operated. If an emergency vehicle driver/operator does not obey all departmental policies and procedures, the emergency vehicle and crew have less of a chance of arriving at the scene safely.

Emergency vehicle driver/operators are responsible for the safe cooperation of their vehicle. Lack of driver/operator skill, lack of experience, or even overconfidence can be a safety issue. (See Figure 11.12.)

No driver or candidate for driver should be allowed to drive on the road until he or she has had both classroom and practical training in operating vehicles that are the same size and configuration as those that will be driven as a part of his or her job functions. If the candidate does not perform successfully in the training, it should be documented and retraining mandated if necessary. As the candidate performs each section of training, the pass date and instructor signature should accompany the documentation. Because all emergency vehicles handle and operate differently, a separate evaluation should be conducted and a documentation sheet kept for each emergency vehicle that the individual is being trained to drive.

FIGURE 11.12 Arriving on the emergency scene safely is more important than response time. *Courtesy of Travis Ford, Nashville Fire Department*

Driver/operator training should focus on defensive driving. Driver/operators also must understand and operate within the range of their warning devices, mainly sirens that project only 300 feet at 10 mph and at 12 feet or less at 60 mph in front of the emergency response vehicle (De Lorenzo, 1991).

Included in the training, at a minimum, should be a road test of at least 15 miles of driving experience in traffic. The driver/operator candidates should be required to demonstrate proficiency in driving and operating the particular emergency vehicle in the type of driving situations they will experience during a normal shift. Besides the initial training and certification, documented ongoing training and annual recertification should be an integral part of the organization's training program. Operating in a driving simulator, while an effective learning tool, is not considered a suitable substitute for actual driving time (USFA, n.d., p. 48).

Although the operator of an emergency vehicle is responsible for vehicle safety at all times, the crew must take some measures to be safe to ensure everyone "arrives alive" at the scene and returns safely. Firefighters should not dress while the emergency vehicle is in motion. All responders need to ride within a fully enclosed portion of the cab, and those not riding in enclosed seats should wear helmets, hearing protection, and eye protection. The emergency vehicle must not be allowed to move until all responders, including the driver, are seated with seat belts fastened. The emergency vehicle should have seat belts large enough to accommodate a responder in full PPE, and there should be a seated position with a working seat belt for everyone riding in the emergency vehicle. Remember that speed is less important than arriving safely at the destination.

STAFFING AND RESOURCES

Regardless of the incident, one of the greatest impacts on firefighter safety is that of staffing. Nothing can affect safety more than not having adequate staffing at an incident.

FIGURE 11.13 Having the proper resources on the emergency scene impacts firefighter safety.
Courtesy of Martin Grube

This is especially true in a volunteer or paid-on-call department in which members are allowed to respond directly to the scene of an incident. Recently updated standards NFPA 1710, *Standard for the Organization and Deployment of Fire Suppression Operations, Emergency Medical Operations, and Special Operations to the Public by Career Fire Departments* (2010a), and NFPA 1720, *Standard for the Organization and Deployment of Fire Suppression Operations, Emergency Medical Operations, and Special Operations to the Public by Volunteer Fire Departments* (2010b), both contain minimum requirements relating to the organization and the deployment of fire departments. If firefighters arrive on the scene without proper personnel or resources, only the task that can be performed safely should be considered until the other proper resources arrive. (See Figure 11.13.)

Summary

Firefighter deaths and injuries as a result of emergency vehicle response are among the easiest to prevent. The vast majority of them can be eliminated with safe driving practices and responders wearing their seat belts. Cultural changes regarding the safety of emergency vehicles are most needed and most difficult for some firefighters to accept. Departments must eliminate the tendency to operate emergency vehicles in a reckless manner under the guise of the urgency of the response. Rather, they must realize that the fire and emergency services cannot solve someone's problem if responders become part of the problem themselves. Responders must realize that some of the practices adopted over the years in the name of efficiency (such as putting on an SCBA in the cab while en route) are actually placing them in an extremely vulnerable position that is not justified by any corresponding tactical gains.

Additionally, departments must develop, implement, and enforce strict policies and procedures and hold every firefighter accountable. Firefighters must learn to operate safely on the scene of violent incidents and know how the incident management system can help with communication and accountability.

Finally, it is important to understand that NFPA 1710 and NFPA 1720 can assist organizations in utilizing proper resources.

Review Questions

1. Discuss the type of incident for which a non-emergency response is appropriate.
2. Discuss the policies and procedures that fire departments should create and enforce to ensure emergency vehicle response safety.
3. Identify and explain two reasons why firefighters should always be seated and seat belted whenever a vehicle is in motion.
4. List some of the recommendations regarding violent incidents and firefighter safety.
5. Discuss the various collision avoidance strategies.

References

Campbell, K. L. (1999, August). *Traffic Collisions Involving Fire Trucks in the United States.* Prepared for Freightliner Corporation by author, Center for National Truck Statistics, University of Michigan Transportation Research Institute. Report Number UMTRI-99-26. Ann Arbor, MI: University of Michigan. Retrieved November 26, 2009, from http://deepblue.lib.umich.edu/bitstream/2027.42/1292/2/92640.0001.001.pdf

Clark, B. A. (2005, June 21). "Leadership: We Killed Firefighter Brian Hunton." From *Firehouse.com.* Updated June 14, 2007. Retrieved December 13, 2009, from http://www.firehouse.com/node/46085

De Lorenzo, R. A. (1991, December). "Lights and Sirens: A Review of Emergency Vehicles." *Annals of Emergency Medicine, 20*(12), 1331–1335.

Dickinson, C. (2007, February 28). *Chief's Corner Archive: The National Seat Belt Pledge.* Emmitsburg, MD: U.S. Fire Administration.

Drive for Life. (2005, July 7). *Consumer Safety News: Drivers Feel Driving Is Getting Riskier, Yet Most Admit to Dangerous Behaviors.* Retrieved April 8, 2010, from http://safedrivingtest.com/press_releases/7-05driving.pdf

Federal Highway Administration. (2003). *Manual on Uniform Traffic Control Devices (MUTCD). 23 Code of Federal Regulations (CFR), Part 655, Subpart F.* Washington, DC: American Association of State Highways and Transportation Officials. Retrieved October 26, 2009, from http://mutcd.fhwa.dot.gov/

Federal Highway Administration. (2009). *Manual on Uniform Traffic Control Devices (MUTCD): Chapter 6I. Control of Traffic Through Traffic Incident Management Areas: Section 6I.01 General.* Washington, DC: Author. Retrieved April 8, 2010, from http://mutcd.fhwa.dot.gov/HTM/2009/part6/part6i.htm

International Association of Fire Chiefs. (n.d.). *Guide to IAFC Model Policies and Procedures for Emergency*

Vehicle Safety. Fairfax, VA: Author. Retrieved November 26, 2009, from http://www.iafc.org/associations/4685/files/downloads/VEHICLE_SAFETY/VehclSafety_IAFCpolAndProceds.pdf

International Association of Fire Fighters. (2005). *Emergency Vehicle Safety Program: Improving Apparatus Response and Roadway Operations Safety in the Career Fire Service: Guides*. Washington, DC: Author.

Karter, M. J., Jr., and Molis, J. L. (2009). *U.S. Firefighter Injuries—2008*. Quincy, MA: National Fire Protection Association. Retrieved April 7, 2010, from http://www.nfpa.org/assets/files//PDF/OS.firefighterInjuries.pdf

Ludwig, G. (2008, October). "It's the Box, Stupid!" *Firehouse* magazine, p. 40. Retrieved April 15, 2009, from http://www.docstoc.com/docs/41208420/Its-the-Box-Stupid!

National Commission on Fire Prevention and Control. (1973, May 4). *America Burning: The Report of the National Commission on Fire Prevention and Control*. Washington, DC: NIST. Retrieved September 24, 2009, from http://www.fire.nist.gov/bfrlpubs/fire73/PDF/f73004.pdf

National Fallen Firefighters Foundation. (2005–2009). *Everyone Goes Home: 16 Firefighter Life Safety Initiatives*. Emmitsburg, MD: Author. Retrieved November 26, 2009, from http://www.everyonegoeshome.com/initiatives.html

National Fire Protection Association. (2007a). *NFPA 1451, Standard for a Fire Service Vehicle Operations Training Program*. Retrieved December 1, 2009, from http://www.nfpa.org/AboutTheCodes/AboutTheCodes.asp?DocNum=1451

National Fire Protection Association. (2007b). *NFPA 1500, Standard on Fire Department Occupational Safety and Health Program*. Retrieved December 1, 2009, from http://www.nfpa.org/aboutthecodes/AboutTheCodes.asp?DocNum=1500

National Fire Protection Association. (2008). *NFPA 1001, Standard for Fire Fighter Professional Qualifications*. Retrieved December 1, 2009, from http://www.nfpa.org/AboutTheCodes/AboutTheCodes.asp?DocNum=1001

National Fire Protection Association. (2009a). *NFPA 1002, Standard for Fire Apparatus Driver/Operator Professional Qualifications*. Retrieved December 1, 2009, from http://www.nfpa.org/AboutTheCodes/AboutTheCodes.asp?DocNum=1002

National Fire Protection Association. (2009b). *NFPA 1901, Standard for Automotive Fire Apparatus*. Retrieved December 1, 2009, from http://www.nfpa.org/AboutTheCodes/AboutTheCodes.asp?DocNum=1901

National Fire Protection Association. (2010a). *NFPA 1710, Standard for the Organization and Deployment of Fire Suppression Operations, Emergency Medical Operations, and Special Operations to the Public by Career Fire Departments*. Retrieved October 25, 2010, from http://www.nfpa.org/aboutthecodes/AboutTheCodes.asp?DocNum=1710

National Fire Protection Association. (2010b). *NFPA 1720, Standard for the Organization and Deployment of Fire Suppression Operations, Emergency Medical Operations, and Special Operations to the Public by Volunteer Fire Departments*. Retrieved October 25, 2010, from http://www.nfpa.org/aboutthecodes/AboutTheCodes.asp?DocNum=1720

National Volunteer Fire Council. (2009). *Emergency Vehicle Safe Operations for Volunteer and Small Combination Emergency Service Organizations*. Retrieved January 31, 2011, from http://www.nvfc.org/files/documents/EVSO_2009.pdf

Sullivan, J. (2010, February 1). "Protecting Firefighters at Roadway Incidents." *Fire Engineering, 163*(2), 115. Retrieved April 8, 2010, from http://www.fireengineering.com/index/articles/display/9219330283/articles/fire-engineering/volume-163/issue-2/departments/speaking-of_safety/protecting-firefighters.html

U.S. Fire Administration. (n.d.). *Safe Operation of Fire Tankers*. Emmitsburg, MD: Author. Retrieved November 26, 2009, from http://www.usfa.dhs.gov/downloads/pdf/publications/fa-248.pdf

U.S. Fire Administration. (1987). *America Burning Revisited: National Workshop at Tyson's Corner, Virginia*. U.S. Government Printing Office: 1990-724-156/20430. Emmitsburg, MD: Author. Retrieved September 24, 2009, from http://www.usfa.dhs.gov/downloads/pdf/publications/5-0133-508.pdf

U.S. Fire Administration. (2002, April). *Firefighter Fatality Retrospective Study*. National Fire Data Center Report FA-220. Emmitsburg, MD: Author. Retrieved December 25, 2009, from http://www.usfa.dhs.gov/downloads/pdf/publications/fa-220.pdf

U.S. Fire Administration. (2008a, April). *Traffic Incident Management Systems*. Emmitsburg, MD: Author. Retrieved April 8, 2010, from http://www.usfa.dhs.gov/downloads/pdf/publications/tims_0408.pdf

U.S. Fire Administration. (2008b, July 31). *Press Release: USFA Releases After-Action Critiques Technical Report*. Emmitsburg, MD: Author. Retrieved November 26, 2009, from http://www.usfa.dhs.gov/media/press/2008releases/073108.shtm

U.S. Fire Administration. (2009, September). *Firefighter Fatalities in the United States in 2008*. Emmitsburg, MD: Author. Retrieved November 19, 2009, from http://www.usfa.dhs.gov/downloads/pdf/publications/ff_fat08.pdf

Occupational Behavioral Health in the Emergency Services

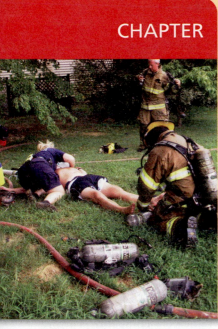

Richard Gist

Vickie Taylor

John F. Neeley

Courtesy of Travis Ford, Nashville Fire Department

OBJECTIVES

After reading this chapter, the student should be able to:

- Identify at least three potentially traumatic events experienced by firefighters.
- Distinguish between *stress, strain,* and *resilience.*
- Describe essential components of a fire and emergency services member assistance program and the Trauma Screening Questionnaire.
- Apply the concept of knowledge translation to counseling and psychological support in the fire and emergency services.
- Distinguish between the principles of critical incident stress management and those of psychological first aid.

Overall Occupational Health Demands of the Fire and Emergency Services

Firefighting and emergency medical services rank among the most stressful and demanding occupations (Gist and Woodall, 1995). The Jobs Rated Almanac Web site (see http://www.archure.net/psychology/stressjobs.html)—using a rating system based on objective elements such as work environment, income, job outlook, physical demands, and job stress—ranked rescue professions near the very bottom. Firefighter ranked 181 and EMT ranked 196 out of 200 occupations (Needleman, 2009). However, when broken down into individual categories, firefighter is rated as the second most stressful job and worst working environment in the United States in both categories, second only to the president of the United States. (See Figure 12.1.)

The conditions can be harsh, the work can be grueling, and its impacts include regular run-ins with aspects of life that most people try to avoid. Pay for a career firefighter could never be called extravagant, and in many communities, those vital and demanding services are provided without pay by citizen volunteers. At the same time, firefighting as an occupation is held in high esteem and is capable of delivering great rewards.

A sidebar to the same Jobs Rated Almanac report cited data from the National Opinion Research Center at the University of Chicago, which ranked firefighter as third highest in job satisfaction. Hiring for career spots is fiercely competitive. More than 1,000 applicants camped on the sidewalk to secure an application for a metropolitan fire and emergency services organization in which only 35 openings existed, and many of those hopefuls spent the entire weekend on line (Associated Press, 2009). Once employed, most career firefighters choose to continue in the field. Uncommonly low rates of turnover have long been reported among career firefighters (U.S. Department of Labor, 2009).

FIGURE 12.1 Firefighters are expected to perform in all types of weather. *Courtesy of Martin Grube*

FIGURE 12.2 Firefighters must deal with frequent encounters involving loss and tragedy. *Courtesy of George Russell, Nashville Fire Department*

Rescue professionals live and work in a world that is an essential paradox. As one firefighter put it after a particularly difficult incident, "[This is] the best job in the world and the only one I ever wanted; I'm the luckiest man in the world to work with these guys. Today, though, it is really tough." Pride and passion are big parts of the job but so are risk and peril. The work not only brings frequent encounters with loss and tragedy but also provides opportunities to intervene in ways that can make a real difference. It can push one from idle boredom to intense activity in a matter of moments. It generates strong emotions, but demands an ability to regulate or suppress emotions in order to do the work and deal with its implications. Those sorts of contrasts and conflicts generate episodes of **stress**, the tension associated with specific events or circumstances; but, even more importantly, they contribute to **strain**, the tension associated with ongoing pressures and difficulties. (See Figure 12.2.)

STRESS, STRAIN, AND RESILIENCE

What is the difference between stress and strain? How does one affect the other? Both play important roles in how fire and emergency services affects providers, both on the job and at home. Think of it like this: if you take a new rubber band and hold it gently between both hands, you can give it a strong tug and it will simply stretch. Once you relax the force pulling on it, it will return to its original condition. The tug you gave it is a *stress*—a temporary but strong force. You can repeat the process over and over again, and the rubber band will return to its original condition. In the fire and emergency services that is called **resilience**, the capacity to maintain healthy functioning and adjustment following exposure to potentially traumatic events.

Yet now take the same rubber band and begin by placing a constant tension on the rubber band, stretching it and holding it tight. That places a *strain* on it. If you then give it the same tug (stress) you gave it before, the strained rubber band is likely to

stress ■ Tension associated with specific events or circumstances.

strain ■ Tension associated with ongoing pressures and difficulties.

resilience ■ Capacity to maintain healthy functioning and adjustment following exposure to potentially traumatic events.

fray or even break. It loses its resilience. Over time, it will wear thin. The problem is not simply a function of the stress placed on the rubber band from the occasional tug; it is strongly affected by the strain on the rubber band when those tugs occur. To preserve and promote resiliency, it is vital to understand and deal with both stress and strain.

BEHAVIORAL HEALTH DEFINED

behavioral health ■
Includes those aspects of health and wellness that involve thoughts, emotions, and behavior, including elements of education, counseling, and therapy intended to promote healthy behavior and ameliorate clinical disorders.

Behavioral health includes those aspects of health and wellness that involve thoughts, emotions, and behavior, including elements of education, counseling, and therapy intended to promote healthy behavior and ameliorate clinical disorders.

Much of the attention originally given to the impact on behavioral health of a fire and emergency services career focused almost exclusively on the role of those stressful incidents that can make some days on the job seem almost unbearable. Line-of-duty death (LODD), mass-casualty incidents, pediatric fatalities, and such are indeed very upsetting and can be emotionally disruptive. Although it may seem on the surface that those incidents would lead directly to psychological problems in providers, the scientific evidence presents a much more complex picture. Four factors in particular help to demonstrate why firefighters' work experiences cannot be so simply explained:

- Nature of the incident
- Post-incident intervention
- Differences in individual response
- Interconnected nature of factors

Nature of the Incident

The nature of the incident itself is not typically the most important element in determining how a particular firefighter handles its impact. Resilience is influenced by a wide range of situational factors, which include the experience and disposition of each particular firefighter, the social and occupational context of a given incident, the actions and attitudes of the individuals surrounding the firefighter, and many other factors that can vary from person to person, from one department to another, and from one incident to the next. Those factors can even vary for the same individual on different days or in different contexts. Just as every passenger involved in a bus accident would not require the same EMS treatment, so every firefighter involved in a rescue would not be treated the same way for the emotional impact he or she experiences.

Post-Incident Intervention

Attempts to reduce feelings of stress after the incident has ended are limited in what they can realistically accomplish. Impacts related to participation in the incident are typically influenced most by how the incident was managed and executed, and even more particularly by the effectiveness of management as seen from the operational level. That is not a clinical issue, but a departmental factor. Regardless, stress can still get to the firefighter. One way to control critical incident stress (CIS) typically involves effective use of the incident command system (ICS). However, even when ICS works well, a crew or multiple crews can have a horrible experience and endure some critical stress, depending upon the size and scope of the incident. ICS has to be addressed well in advance and practiced in the routine work of the department so that it will be effective when applied to extraordinary events that occur on any incident.

FIGURE 12.3 Firefighters initiate advanced life support on the emergency scene to an unresponsive firefighter. *Courtesy of Travis Ford, Nashville Fire Department*

Differences in Individual Response

One firefighter's unshakable incident may be another's routine day on the job. Individual response often depends on particular elements of a particular person's life at a particular point in time. Those impacts may be connected to any number of things—issues with coworkers, problems at home, personal challenges such as drinking or depression, or anything that can affect pretty much anyone at some point or another. Sometimes one can see it coming; other times it seems to come out of the blue. Yet it is not enough to deal with the incident without addressing how it connects to whatever else is involved; and that takes a whole different type of intervention—typically a personal and confidential helping relationship with a qualified professional skilled in the issues involved. Setting that up and making it accessible requires professional helping systems that are constantly ready and available to deal with the entire range of problems firefighters and their families encounter—whether directly related to the job or not. (See Figure 12.3.)

Interconnected Nature of Factors

The preceding factors and others have to be interconnected to have much of an effect on how firefighters, their crews, and their departments weather difficult incidents and maintain their health and well-being. It is impossible to take stressful incidents out of the equation—the fire and emergency services exists to manage stressful incidents. But things can be done to make the experience less stressful and make resilience easier to attain. A truly functional approach to behavioral health provides resources to help at personal, professional, and departmental levels to ensure that firefighters have the necessary tools to work effectively, remain resilient, and enjoy the rewards of a job known for its dangerous and demanding nature.

Critical Incident Stress Debriefing and Critical Incident Stress Management

The terms *critical incident stress debriefing* and *critical incident stress management* have sometimes been interchangeable. However, there is a difference in the way both specifically look at stressful events.

CRITICAL INCIDENT STRESS DEBRIEFING

The early efforts of the fire and emergency services to manage the impact of stressful events on the job paved the way for much of the research and information now available regarding occupational exposure to traumatic incidents. **Critical incident stress debriefing (CISD)** is a seven-stage process of recounting the nature and details of a traumatic experience, discussing thoughts and feelings connected with the experience, and presenting information about stress and coping. Although imprecise, the delivery motive was to help all firefighters deal with the difficult incidents they encountered from time to time. It has typically been presented as an early intervention to be conducted very shortly following the experience and is intended to follow a carefully scripted process argued to be critical to its impact (Mitchell, 1983).

CISD became somewhat of a social movement in fire and emergency services, championed by an enthusiastic cadre of "peer providers" trained to deliver the intervention in weekend workshops. The movement provided a niche for individuals who sought a role in the aftermath of major events. Many persons saw a need for some means of assisting firefighters with the difficult events they must sometimes face, and many were attracted to the roles CISD prescribed for peers. The rubric was to train local CISD teams to deliver the intervention, and a rapidly increasing industry grew to support those teams with workshops and other programming. (See Figure 12.4.)

critical incident stress debriefing (CISD) ■ A seven-stage process of recounting the nature and details of a traumatic experience, discussing thoughts and feelings connected with the experience, and presenting information about stress and coping.

FIGURE 12.4 Fire and emergency services personnel being provided with CISD training.

Based on its simplicity, the ease of acquiring skills, and its seeming consistency with then-current beliefs about traumatic stress and early intervention, CISD quickly grew footholds in other areas as well. Early articles and workshops promoted the intervention as a powerful preventative for post-traumatic stress disorder (PTSD) and similar negative outcomes (Mitchell, 1983). There were also claims regarding the effectiveness and benefit of intervention in reducing absenteeism, preventing burnout and turnover, and saving lives.

CRITICAL INCIDENT STRESS MANAGEMENT

Before long, CISD was expanded to include other interventions such as *defusing* (short sessions immediately following the event or even at its scene, before a full-scale debriefing would be mounted), *demobilization* (sessions similar to defusings held as a part of scaling down large or long-lasting events such as disaster or mass-casualty incidents), and a variety of other education and intervention ideas. The package became known as **critical incident stress management (CISM)**, a collection of prescriptive interventions focused on presumed impacts of certain occupational events and purported to prevent post-traumatic stress disorder and other psychological impacts, centered around a specific "signature" intervention (critical incident stress debriefing). CISM was promoted through the same mechanisms that had propelled CISD. Standards such as NFPA 1500, *Standard on Fire Department Occupational Safety and Health Program* (2007a), came to include CISM as a requirement for fire departments. There are seven core components of CISM:

critical incident stress management (CISM) ■ A collection of prescriptive interventions focused on presumed impacts of certain occupational events and purported to prevent post-traumatic stress disorder and other psychological impacts, centered around a specific "signature" intervention (critical incident stress debriefing).

1. Precrisis preparation
2. One-on-one crisis intervention/counseling
3. Family crisis intervention
4. Disaster or large-scale incident support programs
5. Defusing
6. Critical incident stress debriefing
7. Follow-up and referral

CISM seemed an easily achievable, inherently logical approach to helping firefighters deal with occupational stress. It seemed generally consistent with accepted beliefs about dealing with traumatic events. Many firefighters reported that having an opportunity to talk about how they felt seemed to be helpful.

Its widespread use, however, also led to more rigorous investigation of its impacts by independent researchers in psychology, psychiatry, and public health. Surprisingly to many, the results did not reveal an appreciable preventive effect and in fact suggested that some might end up having more difficulty resolving their experience as a result of their participation in debriefing (see McNally, Bryant, and Ehlers, 2003, for a definitive overview). Some of the research pointed to negative outcomes for individuals who received debriefing. However, those negative outcomes did not come directly from active fire and emergency services professionals, but from civilians involved in one-on-one applications of the debriefing sessions. Still, some authoritative guidelines for treatment of persons exposed to traumatic events now recommend against routine debriefing.

On the other hand, the stress those sorts of incidents can create for firefighters is very real. Those kinds of experiences happen with unfortunate regularity—they are essential parts of what the fire and emergency services does for the communities it serves. The larger and more active the department in which one works, the more traumatic incidents one is apt to see and the more often one is apt to see them. It is important to be prepared to deal with their impacts and that the department provides personnel with access to resources that can make a difference. (See Figure 12.5.)

FIGURE 12.5 Fire departments should have a process in place to deal with the stress of an incident involving one of their own. *Courtesy of Buddy Byers, District Chief, Nashville Fire Department*

Behavioral Health Needs of the Emergency Service

Just a generation earlier, there was considerable resistance to the idea of behavioral health services for firefighters. Before the early 1980s, very little attention regarding the impact of the work or approach to prevention, mitigation, or treatment appeared in the fire and emergency services. Firefighters were expected to stand tough and "suck it up." That much has changed, and the need for behavioral health assistance for fire and EMS personnel is now widely accepted (Gist and Woodall, 1995).

MEMBER ASSISTANCE PROGRAMS

It is well recognized that personnel and their families must have the resources to deal with the various complications that those occupations bring to their lives. Help also needs to be available so they can deal with the common problems of living that everyone sometimes faces. Health and safety standards such as NFPA 1500 require the availability of **member assistance programs** as a resource providing access to confidential counseling and assistance to firefighters and families for basic behavioral health concerns. Member assistance programs (often called employee assistance programs, or EAPs) are usually self-referral, and EAPs are basically considered to help firefighters deal with issues that might adversely affect their overall job performance. Fire and emergency services member assistance programs typically include access to a range of basic behavioral health services, including counseling for various problems in living, access to alcohol and addiction services, and assistance in gaining treatment for more serious clinical conditions. Many programs are also open to families of fire and emergency services personnel, with marriage counseling and family therapy among the services routinely provided. The services of member assistance programs are strictly confidential and that confidentiality is protected

member assistance program ▪ A resource providing access to confidential counseling and assistance to firefighters and families for basic behavioral health concerns.

by law. It is essential that EAP services be available to all firefighters and their families when needed.

Research has raised concerns about intervention and treatment approaches that have been commonly used with firefighters. Hence, serious questions have arisen: Which form should those services take? Who should provide them? What standards should apply for various interventions? How should the services be evaluated? There are many ways to approach treating those conditions and, as seen with CISD, even treatments that seem reasonable and have been widely accepted may not represent the best care according to objective standards. Ensuring that firefighters receive care that meets current standards of evidence-informed practice (see the later section on evidence-informed practice) requires that fire departments become much more informed and much more prescriptive in how they establish their member assistance programs.

Human responses to life's many challenges are anything but simple, and a firefighter's occupation can make things even more complex. When considering PTSD, for example, firefighters present a very different picture than do persons from other backgrounds. When surveyed in various studies, firefighters report more signs of exposure to distressing events than do many others, but such behavior should be expected, given the nature of their work. Just when those reactions should be considered symptomatic and in need of professional help presents a complicated set of questions.

A good member assistance program should be an easily accessible and approachable resource to help firefighters and their families sort through their questions and experiences. Whenever personal problems, whatever their genesis, begin to interfere with the ability to get along with others, do the things important to self and family, enjoy those aspects of life that have typically held personal meaning, or be productive in the workplace, it is time to seek a confidential visit with the member assistance program to find out what help may be available. If a firefighter sees changes in a coworker that cause concern, it is important to encourage the coworker to make that contact as well.

SIDE EFFECTS OF INTERVENTION

Whereas it is important to have help available when needed and desired, it is also important not to force help on others when it may not be needed. One might think that any effort to help should be at least mildly beneficial, but that is not necessarily the case. Just as there can be unwanted side effects associated with taking medications, there can be unwanted side effects from other forms of intervention. The act of intervention itself can have unanticipated impacts on the individual's view of his or her reactions and can communicate notions about its course and severity that may prove counterproductive. The statement, "I didn't know I was sick until you sent for a doctor," captures the essence of a sometimes subtle but often far-reaching impact that can result from advocating professional (or professional-appearing) intervention in place of the types of comfort and support typically offered by coworkers and friends.

Once people start to look at their reactions as symptoms of illness or injury, their actions and expectations change to match those they hold for dealing with sickness. Such a change is not necessarily healthy. It is absolutely important to respond promptly and affirmatively where illness or injury exist because, for many conditions, the sooner treatment can be rendered, the better the outcome is likely to be. But it is also a medical maxim to avoid providing treatment until a condition in need of treatment is identified. Even then, it is typically wiser to be conservative about matching the treatment given to the nature and extent of the patient's condition.

The bottom line is that every diagnostic test, every effort at prevention, and every treatment or intervention carries a risk as well as a benefit. Weighing the risk imposed against the benefit derived can be a complicated matter and should not be taken lightly.

Indeed, the effort of fire departments to deal with occupational exposures to **potentially traumatic events** (PTEs)—the range of events that could qualify as Criterion A exposures in the diagnostic rubric for post-traumatic stress disorder and hold the potential to adversely affect vulnerable individuals—has become one of the best studied examples of those difficulties with respect to behavioral health. Although it has become common practice to offer or even compel immediate interventions such as debriefing for virtually anyone exposed to many types of PTE, by far the greater majority of persons exposed will not experience clinical-level disruptions and, even among those who do, many regain their perspective over time without intervention. Given that well-meaning interventions (such as immediate debriefing) have proven ineffective overall and may even cause increased difficulty for some, it is clearly time to rethink this approach. Again, some of the ineffectiveness of debriefing is a result of a team operating outside its realm and not following proper debriefing procedures.

Evidence-Informed Practice

Decision makers in fire departments do not typically encounter or engage information sources such as the medical journals and technical reports commonly associated with knowledge translation in health care. **Knowledge translation** means systematic efforts to translate research information into treatment approaches that are usable and accessible by clinical practitioners and health care consumers. Information flow in a firefighter's world is driven by trade conferences, trade publications, and regional training offerings. Such information flow is in contrast to **evidence-informed practice**, defined as the application of procedures that reflect available evidence, but not at the level of specific tests of specific treatments covered by evidence-based health care. **Evidence-based health care** is the application of systematic assessments of research evidence to enhance treatment selection and promote efficacious care.

Preparing fire departments to make fully informed decisions about occupational behavioral health care for their personnel takes deliberate efforts to bring scientists and "consumers" together. The National Fallen Firefighters Foundation (NFFF), therefore, initiated a systematic approach to knowledge translation as its central objective for **occupational behavioral health**, which consists of applications of behavioral health research and practice specifically focused on the occupational setting.

The strategic plan employed a consensus process much like that used in developing standards in both medicine and safety. It began by bringing together researchers whose programs deal with areas important to the occupational behavioral health needs of the fire and emergency services. Researchers worked in conjunction with a similar number of key representatives from constituency organizations and standards bodies of the fire and emergency services to act on several critical topics:

- Occupational exposure to potentially traumatic events (PTEs)
- Evidence-based approaches to assessment and treatment of behavioral conditions
- Standards and systems to support delivery of behavioral health services
- Promotion of health-enhancing behavior and reduction in behavioral aspects of risk

The first of those meetings took place in December 2008. Researchers whose work has centered on prevention and early intervention programs related to traumatic stress joined with health and safety representatives selected by six constituency organizations of the fire and emergency services. The purpose was to recommend ways to enhance the capacity of fire and emergency services to care for its personnel (the first objective). A second meeting, held in April 2009, examined standards and recommendations for addressing other behavioral health issues (the second and third objectives). A third session explored the role of peer support systems in promoting health, wellness, and safety, and in facilitating effective utilization of behavioral health resources.

The products of those efforts are being translated into new and enhanced approaches to delivering evidence-informed behavioral health care to the fire and emergency services. Included are protocols for support of firefighters and intervention for occupational exposure to PTEs, recommended standards for member assistance programs, Web-based training programs to help mental health professionals serving firefighters and their families to learn and employ evidence-based treatments for clinical conditions, and interactive Web-based approaches to self-help for learning effective ways to cope with various difficulties and develop behaviors that promote health, wellness, and safety. Those are ultimately intended to become elements of industry standards and prevailing practices.

AN EVIDENCE-INFORMED PROTOCOL FOR TREATING PTES

The first area to be addressed was occupational exposure to traumatic stress. Both academic and industry participants were acquainted with research regarding the limitations of debriefing and related approaches. A suggested protocol had been drafted by NFFF consultants as a component of recent revisions to the National Association of Emergency Medical Services Physicians medical direction text (Gist and Taylor, 2009) and had also been published in a major trade journal for employee assistance providers (Gist and Taylor, 2008). That protocol was derived from current best practices guidelines, consensus, and systematic reviews. The following principles guided its development:

- Matching organizational response to personnel needs and preferences
- Early, reliable, and nonintrusive assessment
- Stepped care
- Evidence-based treatment of clinical conditions

Matching Organizational Response to Personnel Needs and Preferences

College landscape architects long ago learned to hold off pouring concrete for the sidewalks of a new quadrangle until the students had worn pathways in the grass. The wise ones pave those pathways, creating sidewalks that lie where the students actually prefer to walk. Too many helping approaches have instead begun by designing help delivery systems, expecting people to adapt their help-seeking patterns to fit those systems. Systems designed in such a way are notoriously underutilized.

Halpern, Gurevicj, Schwartz, and Brazeau (2009) conducted a systematic study of help-seeking preferences among Canadian EMS providers. Medics, supervisors, and dispatchers were questioned in detail about coping strategies, resource preferences, and preferred patterns for seeking help. Systematic interviews in group and individual sessions were transcribed, coded, and analyzed for common themes.

Most suggestions from medics concentrated on very practical workplace steps, such as a half-hour to one-hour "time-out" period to be spent alone or, as another option, with coworkers of their choosing. Medics also valued direct expressions of support and interest from supervisors. Providers who indicated that they might like more specific or detailed discussions said they would choose to do so at their own pace and in their own contexts in the days and weeks following an incident.

Approaches such as that of Halpern et al. (2009) first look to identify how and where help is typically sought. Then, like paving those pathways in the campus grass, they work to strengthen the routes people choose to take. Supportive contacts utilizing principles of **psychological first aid** (Brymer et al., 2006) represent the current standard for evidence-informed best practice with respect to immediate assistance and can easily be incorporated into those first contacts preferred by street-level personnel. Psychological first aid may be useful in calming and supporting persons exposed to potentially traumatic events. Core elements of psychological first aid and examples are shown in Table 12.1.

Extraordinary events requiring more focused support often benefit from more systematic visits involving "upward contacts" (Taylor and Lobel, 1989). That is best accomplished

psychological first aid ■ An evolving set of evidence-informed activities that may be useful in calming and supporting persons exposed to potentially traumatic events.

| TABLE 12.1 | Psychological First Aid |

ELEMENT	EXAMPLE
Contact and engagement	Identify crew experiencing a PTE; offer assistance if desired.
Safety and comfort	Allow for time-out in quarters or other comfortable location away from incident scene.
Stabilization (if needed)	Determine whether immediate care or support is needed to calm or re-equilibrate; relieve from duty if indicated or requested.
Information gathering: current needs and concerns	"Hot wash" the incident to gather information in a systematic and nonintrusive way; follow up queries as indicated.
Practical assistance	Provide any assistance indicated to address immediate concerns.
Connection with social supports	Offer opportunity to interact with peers of own choosing as desired; may offer choice of leaving duty to be with family or significant others if indicated.
Information on coping	May come in form of carefully constructed handouts, peer discussion, or at conclusion of hot wash.
Linkage with collaborative services	Reminder should be given regarding availability of confidential member assistance program and how to access.

Source: National Child Traumatic Stress Network and National Center for PTSD, *Psychological First Aid: Field Operations Guide,* September, 2005.

by peer-level firefighters seen as having experienced and mastered similar challenges. It can also be structured around the principles of psychological first aid. Such an approach differs from the "peer counselor" concept common to CISM and similar intervention programs in a subtle but very important way.

The contacts suggested here are conceived as informal supportive visits rather than as interventions with a formal preventive intent. They are more about providing presence than about performing any sort of structured intervention. Such contacts are typically well received and provide much needed solace and solidarity. They can also help personnel to feel more comfortable in seeking further assistance if and when they feel it is needed.

Early, Reliable, and Nonintrusive Assessment

Early, reliable, and nonintrusive assessment is essential to effective treatment. Many firefighters will experience some level of distress following difficult duty, but the greater majority will not see that distress rise to levels that require clinical treatment. Most persons, firefighters or not, respond to even deeply unsettling experiences with resilience rather than requiring recovery (Bonanno, 2004). It cannot be easily determined in the beginning just who will rebound or who will need extra assistance to regain their footing.

The Trauma Screening Questionnaire (TSQ), illustrated in Figure 12.6, is a simple, straightforward, and nonintrusive self-report that can help distinguish those for whom resolution is progressing well and suggest who may require full assessment for clinical treatment of PTSD (Brewin et al., 2002). Screening is the broad use of structured instruments or techniques as the first step in identifying those who may be in need of clinical treatment for specific conditions.

The screening becomes effective about three to four weeks following the traumatic event. The scale consists of 10 yes or no questions regarding whether a particular symptom has been experienced more than twice in the preceding week. Scoring is done simply by counting the number of affirmative responses—six or more answered "yes" suggests that further assessment is needed. The screening can also be combined with a short depression screen (e.g., Henkel et al., 2004).

Trauma Screening Questionnaire

	YES, AT LEAST TWICE IN THE PAST WEEK	NO
1. Upsetting thoughts or memories about the event that have come into your mind against your will		
2. Upsetting dreams about the event		
3. Acting or feeling as though the event were happening again		
4. Feeling upset by reminders of the event		
5. Bodily reactions (such as fast heartbeat, stomach churning, sweatiness, and/or dizziness) when reminded of the event		
6. Difficulty falling or staying asleep		
7. Irritability or outbursts of anger		
8. Difficulty concentrating		
9. Heightened awareness of potential dangers to yourself and others		
10. Being jumpy or being startled at something unexpected		

FIGURE 12.6 Trauma Screening Questionnaire. Brewin, C.R., Rose, S., Andrews, B., Green, J., Tata, P., McEvedy, C., Turner, S., and Foa, E.B. (2002). Brief screening instrument for post-traumatic stress disorder. *British Journal of Psychiatry, 181:*158–162.

Stepped Care

Stepped care, providing progressively more intensive clinical care based upon demonstrated clinical need, is now recommended in place of sweeping early intervention models. Studies of cardiac patients following major coronary events have found that a significant minority actually do better if not enrolled in seemingly benign interventions such as symptom education and supportive visits (Frasure-Smith et al., 2002). Some studies of early interventions based on debriefing techniques also showed those kinds of paradoxical impacts.

Cardiac patients who typically used repressive coping strategies fared as well as or better than others if left to their own devices but deteriorated when involved in well-meaning interventions that challenged their normal patterns of coping. It has often been claimed that firefighters need intervention because they tend to utilize repressive coping, and those who appear to use repressive coping are often targeted for those interventions. Such approaches must clearly receive very serious reconsideration.

Discomfort that persists or becomes troublesome, even in those who do not screen positively for probable PTSD, should be addressed by referral to employee assistance providers or other counseling resources. Self-help programs with demonstrated efficacy in bolstering symptom management skills may also help, especially for those reluctant to accept referral. If those approaches fail to help or discomfort seems too great, referral should be made to a competent specialist.

Evidence-Based Treatment of Clinical Conditions

Evidence-based treatment of clinical conditions by competent and credentialed specialty providers should be the standard of care for serious cases of PTSD, depression, or other disorders. Evidence-based treatments consistent with current authoritative guidelines should form the basis for clinical intervention. Current treatment guidelines favor use of cognitive behavior therapy (CBT) utilizing graded exposure. The approach has been shown to be effective in a range of applications, including treatment of PTSD in World Trade Center survivors (Levitt et al., 2007). On the other hand, many treatments typically

stepped care ■ Provision of progressively more intensive clinical care based upon demonstrated clinical need.

employed in routine therapy have been found to be relatively ineffective in treating conditions such as PTSD.

It is not necessarily a simple matter to locate and identify properly trained specialists. A number of certifications and credentials that claim to indicate specialty in traumatic stress have appeared, but none of those so far could be said to be comparable to the boards that certify medical specialists through extensive postgraduate training, supervised practice, and rigorous examinations. Methods to provide reliable information to help fire departments evaluate resources for professional assistance are also being explored as a part of NFFF's knowledge translation process.

DRAFT PROTOCOL FOR RESPONSE TO PTE EXPOSURE

A draft protocol was produced to guide fire departments through a series of steps to deal with occupational exposure to PTEs. It begins with providing practical assistance and support as requested, followed by evidence-based screening if indicated later. Clinically significant cases are referred for specialist treatment using evidence-based interventions, and access to employee assistance services is encouraged to address additional issues aggravated or exacerbated by the exposure. The protocol is outlined below, with a process diagram presented in Figure 12.7. Steps in the protocol include:

- Experience of a potentially traumatic event (PTE)
- Time out/hot wash
- TSQ screening
- Complete assessment
- Treatment by specialty clinician

Experience of a Potentially Traumatic Event (PTE)

A trauma for one responder may be a routine experience to another. The reaction is in large measure subjective, driven by individual experiences, sensibilities, and personal situations. The first question should be, "Does the firefighter desire assistance?" If so, the responder may request that the protocol be initiated; if not, expression of concern and the availability of help if needed may be all that is immediately required.

Time Out/Hot Wash

The hot wash is an element of the military after-action review (AAR) process that, especially if carried out using principles of psychological first aid, can make the time-out advocated by Halpern et al. (2009) useful, helpful, and nonintrusive. It is widely used in military circles and recognized for its capacity to help mitigate stressful impacts as well as its importance in preserving lessons learned and improving combat performance (Marine Corps Community Services, 2009). It centers around a few simple, open-ended questions about the experience.

Typical questions in a hot wash session include:

- What was the nature of the call?
- What aspects were particularly successful?
- Were there parts that could have gone better?
- What lessons could be learned from the experience?
- Who should be told about what has been learned?

If that appears sufficient, the process may be complete. However, if serious issues are obvious, referral for complete assessment is prudent. If still unsure, a quick and nonintrusive screen such as the TSQ may be employed at three to four weeks.

TSQ Screening

As noted earlier, the TSQ is simple, straightforward, and easily scored. If fewer than six items score positive, resolution can generally be expected. If six or more items receive positive

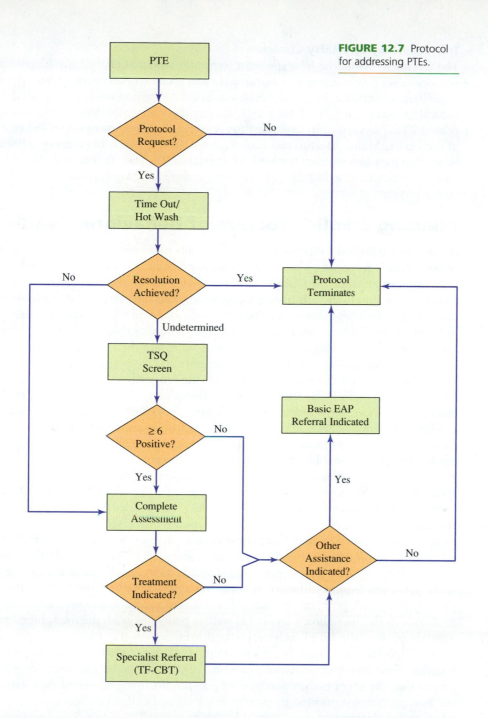

FIGURE 12.7 Protocol for addressing PTEs.

responses, referral for a more complete assessment is indicated. Even if a firefighter screens negative for PTSD, he or she may still need assistance with symptom regulation or compounding life issues. If so, appropriate referral for basic EAP assistance should be considered.

Complete Assessment

A complete assessment can typically be accomplished by a qualified EAP counselor or other mental health provider using appropriately validated instruments. Specialty treatment may not be indicated, but the EAP can help with symptom management or external stressors that might be complicating the firefighter's usual capacity to deal with the circumstance. For those instances that indicate the need for clinical treatment, referral should be made to a competent specialist fully qualified in appropriate evidence-based techniques.

Treatment by Specialty Clinician

The treatment should be by a specialist (typically a board-certified psychiatrist; a licensed, doctoral-level psychologist; or a certified clinical social worker) with advanced training and supervised experience in specific evidence-based treatment models supported by current clinical guidelines (e.g., CBT for PTSD, anxiety disorders, and depression). Occupationally related PTSD has typically responded to relatively short treatment cycles (12 to 25 sessions in the Levitt, Malta, Martin, and Cloitre, 2007, study), but is often accompanied by other issues that may benefit from further EAP assistance (e.g., family impacts). Accordingly, evaluation for other needed or desired assistance completes the protocol.

Building a Solid Program of Behavioral Health

A solid and complete behavioral health program must cover considerably more than the organization's response to potentially traumatic occupational incidents. Firefighters and their families must have access to assistance with a wide range of behavioral health issues. Those issues should include not only counseling and intervention with problems in living but also positive behavioral steps toward wellness, fitness, and safety. Many efforts are now underway to help fire departments, their behavioral health service providers, their unions and employee organizations, and personnel and their families learn to develop and use those various components wherever support and assistance can improve the relationship between life on the job and life outside the workplace.

Cardiac events are the leading cause of line-of-duty deaths and injuries in the fire and emergency services. Most of the controllable aspects of cardiac risk are behavioral in nature—smoking, nutrition, obesity, fitness. A complete behavioral health program should address those with evidence-supported interventions developed to help firefighters change risk inducing behaviors and establish healthy patterns of behavior and activity. The second leading cause of line-of-duty death—emergency vehicle accidents—also involves behavior. Preventing those accidents requires effective approaches to behavior change. Those behavioral issues include attitudes and habits that are typically deeply embedded, not just in particular firefighters but in the culture and environment that surrounds them. Much research has been reported in other enterprises that may be of considerable value to the fire and emergency services if adapted for use. That effort is also a part of ongoing NFFF behavioral health initiatives.

Problems at home lead to problems at work, as frequently as or even more often than work problems lead to difficulty at home. An effective behavioral health program has to provide help with problems outside the work setting and must especially offer help with problems in the firefighter's family (as well as problems *with* the firefighter's family). Yet another part of the NFFF behavioral health initiatives has focused on providing more detailed specification regarding what components departments should require in putting together their member assistance programs, and in providing guidance regarding how to ensure that the services they purchase or provide for firefighters and their families reflect the best practices supported by current research and clinical evidence.

Critical elements of a behavioral health program include:

- Counseling for problems in living and life crises, including marriage, family, and child-rearing issues
- Assessment for clinical conditions requiring specialty care (e.g., alcohol and addiction, clinical depression, PTSD, and other anxiety disorders)
- Access to specialty services meeting accepted standards for evidence-based treatment when indicated; and advocacy and assistance in securing needed treatment (including proper utilization of health benefits and identification of other resources when needed). Many member assistance programs also provide direct assistance or access to other resources for things as varied as legal problems, financial assistance, child and elder care, or other practical needs.

Future of Health and Safety Programs

Not much more than a decade ago, it was rare to find a fire department that required an annual physical examination for firefighters, much less one that had a designated fire department physician who supervised a prescribed program of occupational health care. Any physical examination was likely to be done by a primary care physician according to whatever standards he or she used in daily practice. What determined fitness for duty was typically left up to the physician's individual judgment.

Attention to the health of firefighters has been changing in major ways. A recent survey by the International Association of Fire Chiefs (Fischler, 2006) indicated that nearly 70% of all career departments and almost half of volunteer organizations required annual physicals by the time of the survey. Two-thirds of career departments and more than half of fully volunteer organizations used NFPA 1582, *Standard on Comprehensive Occupational Medical Program for Fire Departments* (2007b), as the basis for their physical examinations. NFPA 1582 prescribes a rigorous protocol for evaluation of physical health, but it is pertinent to note that only 12% of career departments and less than 10% of volunteer agencies included a mental health component in their annual examinations.

Behavioral health has yet to approach that level of sophistication. NFPA 1500, the industry standard for fire department health and safety programs, requires that all departments provide a member assistance program, but it does not specify what services should be provided by what level of provider, nor does it offer any guidance with respect to protocols for assessment or evidence-based standards of treatment. The current version of the standard prescribes a CISM program, despite a lack of consensus in terms of research evidence to suggest that CISM is as effective as originally thought. Although research evidence is not conclusive either way, some researchers consider CISM ineffective and perhaps even counterproductive.

Currently, all but two states in the United States have local assistance strike teams to assist in line-of-duty deaths sponsored by National Fallen Fighters Foundation. Yet, at present there are no good data to indicate what firefighters are actually receiving in the way of behavioral health care, much less how effective those programs and services are, based on objective study of outcomes. The outcomes from the NFFF Everyone Goes Home initiative's consensus process are intended to provide a major step forward in bringing behavioral health into parity with standards and practices taking firm root in other areas of occupational health for the firefighter.

Summary

Firefighters cannot escape being involved in some type of tragic incident. Although, firefighters' need for behavioral health assistance is challenging and complex, it is no longer in question. Programs to help address the impacts of the career on firefighters and their families have become an accepted part of the industry. Early programs gained strong followings, but certain approaches central to most were later shown in a variety of research studies to be ineffective overall and even counterproductive to some. However, many fire departments still use debriefing and other related interventions. Research in areas related to occupational behavioral health in the fire and emergency services has grown much more sophisticated. New techniques with strong evidence behind them have become available, and new technologies make it possible to put those techniques into the hands of those who assist personnel in need.

The focus for researchers, fire departments, and the behavioral health professionals who serve them is now on employing knowledge translation activities designed to identify the best evidence, adapt or refine techniques that use evidence to help firefighters, and develop mechanisms that put those techniques to best use. But at the end of the day, the fire and emergency services must still depend on firefighters to do the things required to keep themselves and their loved ones healthy and resilient while remaining committed to an important and demanding line of service.

Member assistance programs provide many resources to support firefighters and their families with the issues and demands of daily living. Firefighters and their families can use member assistance programs to help identify resources, solve problems before they get out of hand, and help keep family communicating and connected. Responders are encouraged to take advantage of such assistance when incidents on the job leave them rattled, rather than waiting until resulting behavior gets in the way of the things they value and enjoy. Responders are also encouraged to help their coworkers realize when they might consider assistance—it is much more likely to help save a life in this way than to rescue a colleague on the emergency scene.

Review Questions

1. List three major components that might fall under the broad topic of behavioral health as defined in this chapter.
2. List three factors that might contribute to resilience.
3. Discuss the contributions and the limitations of the critical incident stress management movement in the U.S. fire service.
4. Discuss how the NFFF protocol for response to potentially traumatic events differs from earlier approaches to intervention.
5. Discuss why it is important to encourage proactive, voluntary use of member assistance programs.

References

Associated Press. (2009, February 2). "Hundreds Turn Out to Apply for Miami Fire Department Job." Retrieved February 5, 2009, from http://www.justnews.com/news/18623606/detail.html

Bonanno, G. A. (2004). "Loss, Trauma, and Human Resilience—Have We Underestimated the Human Capacity to Thrive After Extremely Aversive Events?" *American Psychologist, 59,* 20–28.

Brewin, C. R., Rose, S., Andrews, B., Green, J., Tata, P., McEvedy, C., et al. (2002). "Brief Screening Instrument for Post-Traumatic Stress Disorder." *British Journal of Psychiatry, 181,* 158–162.

Brymer, M., Jacobs, A., Layne, C., Pynoos, R., Ruzek, J., Steinberg, A., et al. (2006). *Psychological First Aid: Field Operations Guide* (2nd ed.). Washington, DC: National Center for PTSD.

Fischler, D. H. (2006, May 1). "How's Your Physical? Results from the IAFC's Survey on Annual Physicals." *IAFC On Scene, 20*(8), 4.

Frasure-Smith, N., Lespérance, F., Gravel, G., Masson, A., Juneau, M., and Bourassa, M. G. (2002). "Long-Term Survival Differences Among Low-Anxious, High-Anxious and Repressive Copers Enrolled in the Montreal Heart Attack Readjustment Trial." *Psychosomatic Medicine, 64*, 571–579.

Gist, R., and Taylor, V. H. (2008). "Occupational and Organizational Issues in Emergency Medical Services Behavioral Health." *Journal of Workplace Behavioral Health, 23*, 309–330.

Gist, R., and Taylor, V. H. (2009). "Prevention and Intervention for Psychologically Stressful Events." In R. Bass, J. H. Brice, T. R. Delbridge, and M. R. Gunderson (Eds.), *Medical Oversight of EMS* (Vol. 2, pp. 386–396). Dubuque, IA: Kendall/Hunt Publishing.

Gist, R., and Woodall, S. J. (1995). "Occupational Stress in Contemporary Fire Service." *Occupational Medicine: State of the Art Review, 10*, 763–787.

Halpern, J., Gurevicj, M., Schwartz, B., and Brazeau, P. (2009). "Interventions for Critical Incident Stress in Emergency Medical Services: A Qualitative Study." *Stress and Health, 25*(2), 139–149. [First published online on October 9, 2008, by John Wiley & Sons, Ltd. © 2008.]

Henkel, V., Mergl, R., Coyne, J. C., Kohnen, R., Möller, H. J., and Hegerl, U. (2004). "Screening for Depression in Primary Care: Will One or Two Items Suffice?" *European Archives of Psychiatry & Clinical Neurosciences, 254*, 215–223.

Levitt, J. T., Malta, L. S., Martin, A., and Cloitre, M. (2007). "The Flexible Application of a Manualized Treatment for PTSD Symptoms and Functional Impairment Related to the 9/11 World Trade Center Attack." *Behavior Research and Therapy, 45*, 1419–1433.

Marine Corps Community Services. (2009). *Leaders Guide for Managing Marines in Distress: Combat and Operational Stress*. Retrieved July 15, 2009, from http://www.usmc-mccs.org/LeadersGuide/Deployments/CombatOpsStress.

McNally, R. J., Bryant, R. A., and Ehlers, A. (2003). "Does Early Psychological Intervention Promote Recovery from Posttraumatic Stress?" *Psychological Science in the Public Interest, 4*(2).

Mitchell, J. T. (1983). "When Disaster Strikes. . . ." *Journal of Emergency Medical Services, 8*(1), 36–39.

National Cancer Institute. (2009). *Fact Sheet: Prostate-Specific Antigen (PSA) Test*. Retrieved May 5, 2009, from http://www.cancer.gov/cancertopics/factsheet/detection/PSA

National Fire Protection Association. (2007a). *NFPA 1500, Standard on Fire Department Occupational Safety and Health Program*. Retrieved March 1, 2009, from http://www.nfpa.org/aboutthecodes/AboutTheCodes.asp?DocNum=1500

National Fire Protection Association. (2007b). NFPA 1582, Standard on Comprehensive Medical Program for Fire Departments. Retrieved March 1, 2009, from http://www.nfpa.org/aboutthecodes/AboutTheCodes.asp?DocNum=1582

Needleman, S. E. (2009, January 26). "Doing the Math to Find the Good Jobs." *Wall Street Journal*, p. D2.

Taylor, S. E., and Lobel, M. (1989). "Social Comparison Activity Under Threat: Downward Evaluation and Upward Contacts." *Psychological Review, 96*, 569–575.

U.S. Department of Labor. (2009). "Firefighting Occupations." *Occupational Outlook Handbook* (2008–2009 ed.). Washington, DC: U.S. Government Printing Office.

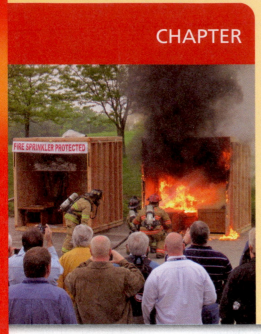

Courtesy of Vickie Pritchett

13
Public Education, Code Enforcement, and Residential Sprinklers

Beverley E. Walker

KEY TERMS

community risk analysis, *p. 315*

community risk reduction, *p. 300*

economic incentives, *p. 301*

education initiatives, *p. 301*

enforcement initiatives, *p. 300*

engineering initiatives, *p. 300*

intervention strategy, *p. 317*

OBJECTIVES

After reading this chapter, the student should be able to:

■ Explain the history of fire prevention and public fire education.

■ Explain the five categories of risk reduction initiatives.

■ Identify the risk reduction requirements for the different levels of professional qualifications.

■ Explain the public's misconceptions about residential sprinklers and the facts that overcome them.

■ Recognize the five steps in the risk reduction process.

■ Demonstrate the importance of code enforcement and prefire planning in the risk reduction process.

Fire Prevention's Role in Safety and Survival

Fire prevention plays a major role in the safety of firefighters in all types of situations, because it reduces the number of incidents in which dangerous emergency response activities must take place. Everyone Goes Home (NFFF, 2005–2009) is one program that lays out ways firefighters can perform their jobs in a safer manner through public education, enforcement of codes, and installation of home fire sprinklers.

It seems obvious that by supporting fire prevention and public safety education and enforcement of codes, the number of fire and emergency responses can be reduced, thereby lowering the number of instances in which firefighters can be killed or injured. Even so, some complain that a reduction in the number of fires will cause a reduction in staffing in fire departments, the closing of fire stations, or other changes that affect job security or budget allotments. This is not true. Fire departments will always be needed as public safety entities in all communities.

What will result from increased prevention is that both communities and firefighters will be safer. Saving lives and protecting property is the mission of the fire service. What better way to accomplish that than to prevent fires and other emergencies from occurring and educate individuals on the proper actions to take when unwanted fires or other types of incidents occur? (See Figure 13.1.)

Consider the following: On March 28, 1994, three FDNY firefighters died as a result of an apartment fire on Watts Street. They were Jimmy Young, Christopher Siedenberg, and John Drennan. That fire occurred when a bag of trash was left sitting on a stove too near a pilot light. The trash ignited, and due to a recent coat of polyurethane on the stairs, the crew was caught in the flashover. Firefighter Young died instantly. Firefighter Siedenberg died the next day. Firefighter Drennan lived for 40 days, with severe burns

FIGURE 13.1 Increased public fire education can decrease the number of structure fires. *Courtesy of Martin Grube*

over 65% of his body. Three firefighters died because of a careless act, an act that could have been prevented through education.

John Drennan's wife, Vina, has spoken on a number of occasions in support of education and firefighter safety. In her speech at the 7th Annual National Fire and Emergency Services Dinner on April 26, 1995, Mrs. Drennan stated:

> You didn't become a firefighter to talk about fire safety, but let's elevate this important task to the prominent place it deserves. . . . You honor the firefighters who died at the Watts Street fire. Their death certificates read, "Cause of death: multiple burns." Yet we in this room know that the real cause of their deaths was nothing more than carelessness. . . . Learn fire prevention strategies and take seriously the fire safety lessons that every city has in place already. (Bukowski, 1995; Drennan, 1995)

The importance of public education, code enforcement, and residential sprinklers plays an important role in firefighter safety and survival.

What Is Prevention?

There are a variety of terms used to describe the concept of prevention: *public education, fire and life safety education,* and the process called *community risk reduction.* **Community risk reduction** can be defined as those programs, initiatives, and services that eliminate or mitigate the risk of and effects from fire, injuries, natural disasters, hazardous materials incidents, and other emergency events. But whatever prevention is called, it is a proactive approach to fighting fires, reducing injuries, and protecting citizens.

Community risk reduction can be divided into three main areas of intervention—engineering, enforcement, and education, historically referred to as the three Es of prevention: (See Table 13.1.)

■ *Engineering.* **Engineering initiatives,** or environmental interventions, are measures that involve changes in the physical environment through the design, development, and manufacture of safety products. Examples include fire sprinklers, smoke alarms, helmets, and vehicle air bags.

■ *Enforcement.* **Enforcement initiatives** involve passing, strengthening, or enforcing laws, regulations, and standards that are used with penalties to influence the actions

community risk reduction ■ Those programs, initiatives, and services that eliminate or mitigate the risk of and effects from fire, injuries, natural disasters, hazardous materials incidents, and other emergency events.

engineering initiatives ■ Measures that involve changes in the physical environment through the design, development, and manufacture of safety products.

enforcement initiatives ■ Activities that involve passing, strengthening, or enforcing laws, regulations, and standards that are used with penalties to influence the actions of people and businesses.

TABLE 13.1	The Es of Prevention
	THE THREE ES OF PREVENTION
Engineering	Using technology to create safer products or modifying the environment to make it safer
Enforcement	Imposing rules and regulations that require the use of safety initiatives
Education	Providing information (facts) about risk and prevention to promote a behavioral change
	PLUS TWO MORE
Economic Incentive	Using monetary measures to discourage unsafe behaviors or encourage safe behaviors
Effective Emergency Response	Mitigating the risk promptly and effectively when it occurs

FIGURE 13.2 Community risk reduction and the five Es that affect its success.

of people and businesses. Examples include enforcement of life safety codes, exit widths, occupancy capacities, and product safety standards.

■ *Education.* **Education initiatives** provide information about risk reduction and prevention to a population, with the end result being a voluntary behavioral change that increases the ability of individuals to be responsible for their own safety. Education can raise awareness of a problem, provide additional information and knowledge, and ultimately produce a desired behavior. However, it is effective only if people do what they are supposed to do, such as wear their seat belts, maintain their smoke alarms, and install fire sprinklers.

Economic incentives, a fourth E, has been added only recently. Such measures can be used to influence behavior. Some are negative and include such things as fines, tickets, or citations, which discourage people from choosing unsafe behaviors. Alternatively, an economic incentive may be positive and include benefits such as a property tax and insurance rate reduction for homeowners who install residential sprinklers.

There is also a fifth E, effective emergency response, which increases the response capability of firefighters. These initiatives involve the ability of firefighters to mitigate or intervene during emergencies, thereby reducing the risk to those involved. Mitigation can come in the form of a strategically placed firehouse/station, the use of automatic aid agreements, up-to-date training of fire department personnel, and the use of citizen groups such as community emergency response teams (CERTs). CERTs are basically citizens helping out the community during emergencies when the local fire department resources are overwhelmed.

Each of the Es can contribute to the development of realistic, comprehensive, and effective interventions to prevent or mitigate risks in the community. They are more effective used collectively than individually. Education, however, must be an integral part of each one. And each of them prevents or mitigates the risks that are presented to firefighters each and every day. (See Figure 13.2.)

Some Prevention History

Historically, fire departments in the United States have been reactionary. That is, they responded to fires and other emergencies in a prompt and effective manner. The earliest departments even promoted the use of community fire wardens, whose job it was to ensure that when fires were needed, they were guarded and safe. However, other than in a few cities such as Boston that specifically prohibited wooden chimneys in homes, there was little active work to prevent fires from occurring.

education initiatives ■ Activities that provide information about risk reduction and prevention to a population, with the end result being a voluntary behavioral change that increases the ability of individuals to be responsible for their own safety.

economic incentives ■ Monetary measures that are used to influence behavior either positively or negatively.

A FOCUS ON PREVENTION

In 1947, President Harry S. Truman commissioned The President's Conference on Fire Prevention. That group was brought together to review the fire problem in the United States and to identify deficiencies in the U.S. fire service. Its purpose was to determine ways to increase the level of awareness about the fire problem and to increase the "work of fire safety in every community" (USFA, 2002c).

The educational recommendations that resulted from the conference were the implementation of fire prevention and safety in school classrooms, teaching a variety of life safety measures, as well as the development of safe attitudes and behaviors. It was suggested that fire safety education be incorporated into a school's educational curriculum. The report recommended better building design, the use of technology for prevention, as well as an increased focus on prevention education. (See Figure 13.3.)

The use of fire protection equipment that provided for the prompt discovery and extinguishment of fires in all buildings was recommended. The equipment included hand-extinguishing equipment, hoses, standpipes, automatic sprinklers, and alarm systems. The current concept of the three Es of prevention, mentioned earlier, was developed as a result of the conference report.

In 1972, President Richard Nixon convened the National Commission on Fire Prevention and Control. The commission produced the famous report entitled *America Burning* (NCFPC, 1973), which has had a great impact on the fire and emergency services, fire prevention, and fire protection in the United States. Overall recommendations of *America*

FIGURE 13.3 Fire departments are still encouraged to educate children about fire prevention. *Courtesy of Gerri Penney, Community Education Coordinator, Palm Beach County Fire Rescue*

Burning included the improvement of fire protection features of buildings, improvement in the safety of living environments through design and materials, and improvement in fire and emergency services training and education programs. It identified the need for the education of the American public about the dangers of fire and appropriate fire safety measures.

The Commission made six recommendations regarding fire safety education in schools, teacher training programs, and media campaigns:

- The Department of Health, Education and Welfare should include fire safety education in the schools throughout the school year in accreditation standards.
- The proposed U.S. Fire Administration should sponsor fire safety education courses for educators to provide a teacher cadre for fire safety education.
- States should include fire safety in programs educating future teachers and require knowledge of fire safety as a prerequisite for teaching certification.
- The proposed U.S. Fire Administration should develop a program, with adequate funding, to assist, augment, and evaluate existing public and private fire safety education efforts.
- The proposed U.S. Fire Administration, in conjunction with the Advertising Council and the NFPA, should sponsor an all-media campaign of public service advertising that is designed to promote public awareness of fire safety.
- The proposed U.S. Fire Administration should develop packets of educational materials appropriate to each occupational category that has special needs or opportunities in promoting fire safety. (NCFPC, 1973, pp. 169–170)

Specifically, the *America Burning* report affirmed the need for a greater emphasis on fire prevention in fire and emergency services organizations. It identified the need for fire departments to expend more effort to educate children on fire safety, to educate adults through residential inspections, and to enforce fire safety codes. The report recommended that economic incentives be provided for built-in fire protection. "Perhaps most important, Americans need to be encouraged to install early-warning fire detectors in their homes where most fire deaths occur" (NCFPC, 1973, p. xi). The report stated that public education about fire was the single activity that could have the greatest potential for reducing losses due to fire. It recommended the implementation of residential fire inspections by local fire departments, and education on how to act and react when fires start in the home.

The impact of the publication has been enormous in the areas of fire prevention, fire protection, fire service training, and public education. Based upon the recommendations of the committee, the U.S. Fire Administration (USFA) and the National Fire Academy (NFA) were created. The report confirmed that U.S. citizens must be educated about fire safety.

In subsequent reports—*America Burning Revisited* (USFA, 1987) and *America at Risk* (USFA, 2002b)—it was found that many of the original recommendations of *America Burning* still needed to be implemented or enforced, including the efforts of local fire departments in the education of young children about fire and fire safety. *America Burning Revisited* identified two primary problem areas: first, that public awareness remained an obstacle that still could be overcome by way of a national education campaign; and second, the lack of information needed to analyze the fire problem and evaluate programs. The 2002 *America at Risk* report recommended community-based programs to encourage or require the use of residential fire sprinklers and smoke alarms.

Another report is *Solutions 2000* (USFA, 1999) of the North American Coalition for Fire and Life Safety Education, an association comprised of fire and emergency services professionals and advocates for at-risk groups of people (the very young, the very old, and those with disabilities) who cannot take lifesaving actions in a timely manner. Most importantly, the report identified the high-risk audiences that need to be the focus of fire and life safety in all communities. In addition, the report's solutions addressed each of the three Es of prevention.

In January 2002, the North American Coalition for Fire and Life Safety Education reconvened to examine the recommendations of the *Solutions 2000* report. Interestingly, in its report, entitled *Beyond Solutions 2000* (USFA, 2002a), the group reported there had not been much progress in the recommendations of the original *Solutions 2000* report. Both of those documents are still consulted by fire safety educators and fire prevention specialists when working with these high-risk audiences.

FIRES THAT INSPIRED CHANGE

Many catastrophic fires throughout history have provided the basis for the development of fire codes and other prevention interventions. Many of the events prompted changes in codes and regulations concerning the design, installation, and maintenance of engineering and prevention measures. The following devastating events are benchmarks in the history of fire prevention. Most fire codes and building design requirements came about as a result of fires such as these:

- *Iroquois Theater Fire, Chicago, Illinois, 1903.* That fire claimed the lives of 602 people. The canvas scenery in the playhouse was painted in highly flammable oil-based paints. A hot stage light ignited a velvet curtain, causing the backdrops to go up in flames. There were no automatic sprinklers for the stage, and the stage's fire curtain did not close properly. There was no emergency lighting, the smoke and heat vents for the stage were not functional, and many exit doors either were locked or did not open in the proper direction.
- *Triangle Shirtwaist Factory, New York City, 1911.* The company was located in a high-rise building, occupying the eighth, ninth, and tenth floors. An employee tossed a cigarette into a bin containing scrap material, causing it to ignite immediately. While employees on the eighth and tenth floors were able to escape due to early discovery and notification, the 250 employees on the ninth floor were not as fortunate. They were not informed of the fire and the dangerous situation it presented. The fire occurred at 4:45 P.M., around the time when the business was shutting down for the day. As was the custom to deter theft, security guards had already locked one of the two exit doors on the ninth floor. Some individuals were able to escape from the unlocked exit, but it was quickly made inaccessible by the fire. There were 145 lives lost as a result of that fire.
- *Coconut Grove Nightclub, Boston, Massachusetts, 1942.* A fire in the nightclub claimed the lives of 492 individuals. The club's walls were covered with paper palm tree decorations that ignited when someone struck a match. The nightclub was filled to twice its capacity, having approximately one thousand occupants at the time the fire erupted. As with the Iroquois Theater fire, there were no automatic sprinklers, the exit doors did not swing in the proper direction, and many doors and windows were sealed shut. The primary exit was a revolving door, greatly slowing the evacuation process. People panicked and got caught in the doorway, causing it to jam.
- *Winecoff Hotel, Atlanta, Georgia, 1946.* That event was the deadliest hotel fire in U.S. history, causing 119 fatalities. The building had only one exit stairway, which became impassable early in the fire's development. Doors had been propped open, and there was no fire suppression system in the building. The building was not equipped with a fire alarm system, so there was no way to notify the hotel's occupants when the fire started in the early morning hours.
- *Our Lady of Angels School, Chicago, Illinois, 1958.* This is the deadliest school fire in U.S. history. It started in a basement trash can in a stairwell, which quickly

engulfed the wooden staircase. The location of the fire blocked the escape route for the occupants of the second floor. The exit corridors had combustible walls and ceilings, and once again there were no automatic fire sprinklers. In addition, there was no automatic fire alarm system, the stairwell was not enclosed, and there was a delay in notification of the fire department. The fire resulted in the death of 92 schoolchildren and three nuns.

- *Beverly Hills Supper Club, Southgate, Kentucky, 1977.* It was Memorial Day weekend in 1977, and the building was packed with over three thousand patrons. There was inadequate egress and exit identification, no fire suppression systems, and no alarm system. The fire started in an unoccupied room and was believed to be electrical in nature. It burned for some time before being discovered and quickly spread to other parts of the building. The fire took the lives of 165 individuals, many of whom were enjoying a show at the opposite end of the building. It was noted that employees attempted to fight the fire prior to notifying the fire department or the building's occupants.

- *MGM Grand Hotel, Las Vegas, Nevada, 1980.* The fire caused 87 fatalities and almost 700 injuries. Although the hotel portion of the building was protected by a sprinkler system, there were no sprinklers in the casino area. The hotel contained many unprotected vertical shafts and had openings that allowed smoke to enter and fill the exit stairwells. Once the occupants entered the stairwells, they were unable to return to their floor or room due to automatically locking doors.

- *Station Nightclub, West Warwick, Rhode Island, 2003.* One recent fire in the United States took the lives of one hundred patrons. Pyrotechnics used during a band performance ignited soundproofing foam at the back of the stage. There were no automatic sprinklers. Individuals panicked and failed to use secondary exit routes. There was a rush for the front door, which became jammed, and most of the victims died trying to use that exit.

NATIONAL FIRE PROTECTION ASSOCIATION

No history of fire prevention would be complete without discussing the National Fire Protection Association (NFPA). The NFPA was organized in 1896 and is still a leader in fire prevention today. Its mission is to "reduce the worldwide burden of fire and other hazards on the quality of life by providing and advocating consensus codes and standards, research, training and education" (NFPA, 2009a). The NFPA publishes over three hundred codes and standards relating to fire prevention, building design, professional standards, equipment standards, and many other areas. Many consider that nonprofit organization to be a world leader in fire prevention and advocacy for public safety.

In 1920, President Woodrow Wilson issued the first National Fire Prevention Day proclamation, and since 1922, Fire Prevention Week has been observed in the Sunday through Saturday period in which October 9 falls. Sponsored by the NFPA, Fire Prevention Week was originally established to commemorate the Great Chicago Fire of 1871, the tragic conflagration that killed more than 250 people, left 100,000 homeless, destroyed more than 17,400 structures, and burned more than 2,000 acres (Bennie, 2008). The fire began on October 8, but continued into and did most of its damage on October 9, 1871.

According to the National Archives and Records Administration's Archives Library Information Center, Fire Prevention Week is the longest running public health and safety observance on record (NFPA, n.d., *About Fire Prevention*).

It is during this time that many fire departments conduct their only fire prevention and public education events.

HISTORY OF RESIDENTIAL FIRE SPRINKLERS

The history of fire sprinklers, specifically residential fire sprinklers, is not as long running as the history of prevention. Fire sprinklers were first used in textile mills in New England over one hundred years ago after a series of devastating fires claimed many lives and destroyed entire businesses. Since that time schools, office buildings, factories, and other commercial buildings have benefited from fire protection sprinkler systems. Automatic sprinklers are part of building and fire codes for commercial structures in communities today.

Fire sprinklers in residences have not been as readily accepted as in commercial occupancies. Until recent years, few one- and two-family homes were equipped with fire sprinklers, and almost no ordinances required them. But in 1986, the City of Scottsdale, Arizona, passed a comprehensive sprinkler ordinance stating that all new single-family, multifamily, and commercial structures for which building permits were issued would require sprinklers. This was one of the first municipalities in the United States to implement such a drastic ordinance.

Studies by the USFA have indicated that the installation of residential fire sprinkler systems could save thousands of lives, prevent a large percentage of injuries, and eliminate millions of dollars in property loss. Yet homebuilders, politicians, and even a few fire departments have failed to support such ordinances and requirements in communities. (See Figure 13.4.)

However, recent action at the International Code Council (ICC) annual meeting in 2008 in Minneapolis, Minnesota, adopted new language in its 2009 edition of the International Residential Code requiring the installation of residential fire sprinklers in newly constructed townhomes (to become effective January 1, 2010) and newly constructed single-family homes and duplexes (effective January 1, 2011). Those actions were reaffirmed by the ICC during its meeting in Baltimore, Maryland, in 2009. As a result of those actions, all nationally recognized model codes now call for mandated installation of residential fire sprinklers in new construction.

FIGURE 13.4 The installation of residential fire sprinkler systems could save thousands of lives.
Courtesy of Vickie Pritchett

Risk Reduction in Today's Fire Service

Community risk reduction activities in today's fire and emergency services are varied and sometimes unclear. Some fire departments today are still reactive in nature; that is, they see their main function as responding to fires and other emergencies. They operate on the theory that they will be able to arrive soon enough to have a positive effect on the emergency incident. Certain fire departments are more proactive, with prevention and education taking a more aggressive role. Those departments believe that proven technology and built-in protection, effective code enforcement, and aggressive public education programs are a real part of their mission in the community. For others, prevention is a function that is usually performed in the form of annual code inspections to ensure that businesses and other commercial occupancies are adhering to the local codes and ordinances concerning fire and building safety. In still others, fire and life safety education activities occur only during Fire Prevention Week in October.

In a survey conducted by the Home Safety Council in 2006 (Home Safety Council, 2007), it was reported that 86% of the fire departments surveyed conducted some type of fire and life safety education, most frequently using uniformed personnel who have multiple responsibilities. Those instances of a department having a dedicated fire and life safety educator on staff were only 12%. The survey found that 85% of the fire departments spent less than 10 hours per week on fire and life safety education activities.

Many fire departments have smoke alarm programs that involve providing alarms for low-income or elderly residents of a community. In the Home Safety Council's survey (2007), 51% of the fire departments responding reported having some kind of smoke alarm installation program. Some may be installation programs, whereas others may simply be giveaway programs. It was undetermined how many departments had any kind of follow-up program to evaluate the effectiveness of this activity. Nor was it clear how many departments were providing education about the alarm's maintenance, and what actions should be taken when the alarm activates. (See Figure 13.5.)

FIGURE 13.5 Providing and installing smoke alarms for the community is an important part of reducing death and injuries. *Courtesy of Vickie Pritchett*

Many fire departments have absorbed emergency medical services into their operations, so they work to prevent unintentional injuries as well as to prevent fires. That includes child passenger safety programs, with many fire stations serving as child passenger seat inspection stations. Other departments are involved in drowning prevention programs, bike and other wheeled sports safety, and poison prevention.

Code enforcement and prefire inspections are more often a part of a fire department's prevention activities. This is due to local regulation or ordinance requirements, and the benefit to a community's Insurance Services Office (ISO) rating. Those departments conducting annual code enforcement inspections receive a higher rating than those that do not, which directly affects the community's property insurance rates.

RESIDENTIAL SPRINKLER INITIATIVE

Recently, California, New Hampshire, and Pennsylvania have officially adopted the International Residential Code (IRC) that mandates fire sprinklers be installed in all residential structures beginning January 1, 2011. In addition, at present, over 700 communities in the United States have an ordinance that requires all residential structures to be sprinklered. Many others have some form of an ordinance with varying restrictions. This number is continually changing, as communities and fire departments deal with the politics of this legislation.

In 2009, the USFA published the following statement concerning residential sprinklers (USFA, 2009b):

> The U.S. Fire Administration has promoted research, development, testing, and demonstrations of residential fire sprinkler systems for more than 30 years. The research regarding residential fire sprinkler systems has indisputably demonstrated the following:
> - Residential fire sprinklers can save the lives of building occupants.
> - Residential fire sprinklers can save the lives of firefighters called to respond to a home fire.
> - Residential fire sprinklers can significantly offset the risk of premature building collapse posed to firefighters by lightweight construction components when they are involved in a fire.
> - Residential fire sprinklers can substantially reduce property loss caused by a fire.
> The time has come to use this affordable, simple, and effective technology to save lives and property where it matters most—in our homes. . . . The U.S. Fire Administration supports . . . the recently adopted changes to the International Residential Code that require residential fire sprinklers in all new residential construction. (USFA, 2009b)

Many national organizations, coalitions, and advocacy groups are working to promote the use of residential sprinklers and the adoption of related local ordinances. However, there is still a lot of opposition from homebuilders and politicians to such requirements.

Much of the opposition arises due to misconceptions about residential fire sprinklers. Many individuals still believe that when a fire occurs all the sprinkler heads will activate. Of course, that simply is not true. Because sprinklers are activated by heat, only the sprinkler in the immediate fire area will activate. Studies have shown that residential fires are usually controlled by one sprinkler head. Homebuilders assert that the added cost of sprinklers will make housing unaffordable to first-time buyers. In actuality, a sprinkler system costs 1% to 2% of the total construction costs. It is comparable to what most people pay for a carpet upgrade. (See Figure 13.6.)

Another misconception is that a smoke alarm provides adequate protection, when in fact it can do nothing to extinguish a fire. Some people believe that a smoke alarm will cause the fire sprinkler to activate. In reality, smoke alarms work off different technology, and fire sprinklers are activated by heat only.

It is up to fire department personnel to educate the public, including homebuilders, insurance representatives, and politicians, on the realities of sprinkler systems and their

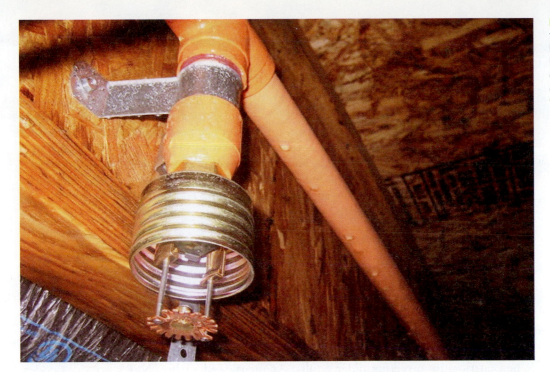

lifesaving benefits, both to the residents and to fire departments. The Home Fire Sprinkler Coalition is the primary resource for fire departments to educate the community about home fire sprinklers.

NATIONAL SUPPORT

National organizations that are working to lead fire departments toward risk reduction are the USFA, the National Fire Academy (NFA), the NFPA, and the Federal Emergency Management Agency (FEMA). Each is a front-runner in the prevention and mitigation of fires and other injuries. Their work at the national level is being enhanced by work done at the local level, and all of these agencies are focused on providing assistance to local fire departments in their efforts for prevention and risk reduction activities.

U.S. Fire Administration

The USFA was created as a result of recommendations from the original *America Burning* report (NFPC, 1973). Presently, its mission is to provide national leadership to local fire and emergency services for prevention, preparedness, and response. Its goals are (USFA, 2009c):

- Reduce risk at the local level through prevention and mitigation.
- Improve local planning and preparedness.
- Improve the capability of fire and emergency services for response to and recovery from all hazards.
- Improve the professional status of fire and emergency services.
- Lead the nation's fire and emergency services by establishing and sustaining USFA as a dynamic organization.

The USFA, through its National Fire Data Center, collects data on various fire problems throughout the nation and promotes national marketing campaigns. Many of the campaigns are available to those at the local level. One well-known program is the Sesame

Street Fire Safety Station program, which targets preschool children. There are additional educational and marketing programs available that target older adults, those who are economically challenged, and those with disabilities.

National Fire Academy

The National Fire Academy, established as a result of the *America Burning* report (NFPC, 1973), falls under the direction of the USFA. The National Fire Academy has been a leader in providing training and education for fire service personnel in the areas of public education, code enforcement, engineering, and investigations, among others. Presently, the National Fire Academy provides training to individuals at the local level to properly assess community risk, conduct risk reduction programs, and evaluate their effectiveness. Other classes revolve around code enforcement, applying codes to plans review, and fire suppression and detection systems.

National Fire Protection Association

Whereas both the USFA and the National Fire Academy are government agencies or institutions, the NFPA is a national nonprofit organization that has supported fire prevention since its inception in 1896. The prevention process began in the early 1900s with an NFPA employee, Franklin H. Wentworth, who distributed fire prevention bulletins. Among the many achievements of the NFPA is the creation of a mascot, Sparky the Fire Dog®, the development of the Learn Not to Burn® curriculum for preschool and elementary age children, the Risk Watch® all hazards curriculum, and the Remembering When curriculum targeting older adults. As mentioned earlier, NFPA provides codes, standards, and performance guidelines for many aspects of the fire and emergency services, and is a major contributor to fire prevention and safety in the United States. (See Figure 13.7.)

Federal Emergency Management Agency

In recent years, FEMA has focused on the prevention aspect of disaster preparedness. Although many terrorist incidents and natural disasters cannot be prevented, they can

FIGURE 13.7 According to NFPA 1021, firefighters should be able to deliver a public education program. *Courtesy of Gerri Penney, Community Education Coordinator, Palm Beach County Fire Rescue. The name and figure of Sparky® are registered trademarks of the NFPA, Quincy, MA 02169*

be anticipated or predicted. And individuals can prepare themselves for such disasters to mitigate damages that might occur. FEMA is responsible for the coordination of the Citizens Corps, the community emergency response teams (CERTs), and the Fire Corps volunteer cadre being trained and organized in many communities in the United States today.

PROFESSIONAL QUALIFICATION STANDARDS

One of the reasons for a lack of community risk reduction activities in many departments is that many members feel that it is not what they were hired to do. The culture of the fire service promotes this type of thinking, which is the offspring of the reactive nature the fire service has historically demonstrated. However, departments are embracing the concept of risk reduction and the associated activities. It is, however, a slow change process. (See Figure 13.8.)

For those fire service personnel who feel prevention is not one of their job responsibilities, a look at the professional qualification standards will show differently. NFPA 1001, *Standard for Fire Fighter Professional Qualifications,* requires that new firefighters have the skills to deliver a public education presentation as well as perform a home safety inspection. Qualifications for Firefighter II state the following:

Section 6.5.1. Perform a life safety survey in a private dwelling . . . so that fire and life safety hazards are identified, recommendations . . . are made to the occupant, and unresolved issues are referred to the proper authority.

Section 6.5.2. Present fire safety information to station visitors or small groups . . . so that all information is presented, the information is accurate, and questions are answered or referred.

Section 6.5.3. Prepare a pre-incident survey . . . so that all required occupancy information is recorded, items noted, and accurate sketches or diagrams are prepared.

More importantly, in looking at NFPA 1021, *Standard for Fire Officer Professional Qualifications,* specifications are found regarding the ability to deliver presentations and

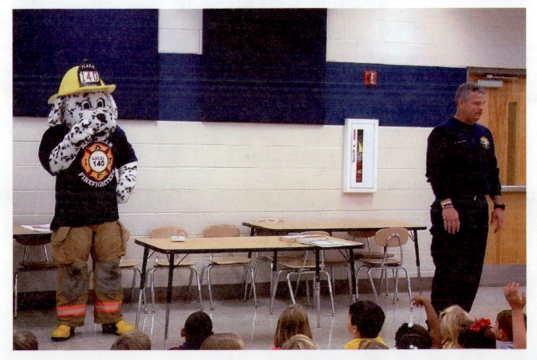

FIGURE 13.8 Delivering a public education presentation is one of the responsibilities of firefighters. *Courtesy of Dave Brasells*

develop departmental education programs. The following qualifications are found for Fire Officer I:

Section 4.3.1. Initiate action on a community need so that the need is addressed.

Section 4.3.2. Initiate action on a citizen's concern . . . so that the concern is answered or referred to the correct individual for action.

Section 4.3.4. Deliver a public education program, given the target audience and topic, so that the message is conveyed clearly.

Fire Officer III standards require much the same, but are more involved. According to the standard, it is the responsibility of the Fire Officer III to prepare "community awareness programs to enhance the quality of life by developing non-traditional services that provide for increased safety, injury prevention and convenient public services" (NFPA 1021, Section 6.3.1).

The Fire Officer IV standard requires the individual to project a positive image of the fire and emergency services organizations in the community by attending, participating in, and assuming a leadership role in community events. In addition, the individual in this position should monitor local, state, and federal legislative activities to enhance fire and emergency services effectiveness (NFPA 1021, Section 7.3 et seq).

Additionally, there are professional qualification standards specifically for the fire and life safety educator and inspector: NFPA 1031, *Standard for Professional Qualifications for Fire Inspector and Plan Examiner,* and NFPA 1035, *Standard for Professional Qualifications for Public Fire and Life Safety Educator.* The intent of NFPA 1035 is to develop clear and concise job performance requirements, including knowledge, skills, and abilities, for an individual to perform as a public fire and life safety educator. Similarly, NFPA 1031 provides job performance requirements for those individuals wishing to serve as competent fire inspectors. Both standards are applicable to fire and life safety educators. They were designed to assist fire departments and other professionals in the task of preventing or mitigating situations that can endanger lives, property, and the environment. (See Figure 13.9.)

FIGURE 13.9 Plan examiners help in preventing situations that can endanger lives, property, and the environment. *Courtesy of Travis Ford, Nashville Fire Department*

When looking at all of these standards, it is clear that prevention and education should be an integral part of any fire department. By trying to prevent fires and other emergency situations, firefighters are working to keep themselves safe and prevent those incidents that bring harm to themselves and their coworkers.

Risk reduction is everyone's job. Although there may be individuals whose primary assignments relate directly to fire prevention, such as inspectors and educators, it is still incumbent upon each and every fire department member to promote risk reduction. Increasing fire prevention and public education programs increases firefighter safety. It allows fire service members to uphold the oath of protecting the lives and property of citizens.

So What Is Needed?

Fire prevention is a win-win activity, a positive for both firefighters and the community members that fire personnel are sworn to protect. The best way to save lives and protect property is to prevent the events that take them. Fire prevention is certainly a positive for fire departments. Not only does it enable individuals to do their jobs more efficiently and effectively, it removes some of the danger of the job and decreases the chances for firefighter death or injury. So, what is needed to increase the effectiveness of fire prevention and public education in fire departments today?

Several important elements have been found in successful risk reduction efforts at the local level. These include:

- Departmental commitment, from the chief to the newest recruit
- Training for new recruits and ongoing monthly training
- Targeted and focused community educational programs

In addition, there needs to be a national focus within the fire and emergency services and its leaders on prevention and risk reduction measures.

DEPARTMENTAL COMMITMENT

In departments that successfully implement prevention activities within their communities, there is a commitment of the department to the process—not just an individual commitment but also a collective commitment from the entire department.

At the center of successful public education and risk reduction programs is a catalyst—an individual within the department who is a motivated, enthusiastic visionary with great departmental skills. Such individuals must have the appropriate leadership from their fire and emergency services departments to maintain their motivation and develop their skills.

The department leader's commitment to risk reduction needs to filter down throughout the entire department by supporting programs that are clearly part of the department's mission and values. That means that risk reduction is a focus that should be included in the department's mission statement. It should be a process that is included and supported through the department's strategic plan and the planning process. When this happens, community risk reduction will become institutionalized in the department; as a departmental priority, it will become a value within the department's culture.

Departmental culture drives the behavior of fire department personnel and affects the entire direction of the department. In order for risk reduction efforts to be successful, there has to be a departmental culture committed to risk reduction activities and efforts in the community. Such a department's management and personnel demonstrate willing support of risk reduction efforts, and ensure that all personnel, regardless of rank, clearly understand their responsibilities and duties for reducing community risk and thereby reducing the risk to department members.

The departmental leader is responsible for changing the department's culture. As mentioned previously, the culture of the department can be a hindrance in community risk reduction initiatives. A department's culture often can seem to be difficult to identify, but it comes from the senior officials. The leader must connect the new values or cultural change to activities that are long term and that show a benefit to the department's members. In addition, the leader must always be a visible advocate for community risk reduction.

Commitment to risk reduction is demonstrated throughout the department when risk reduction activities receive priority equal to training and emergency response activities. And commitment becomes concrete when resources are dedicated to risk reduction activities. Resources can be in the form of people, time, funding, and equipment. A committed department will seek funding for risk reduction resources as a regular part of its budget process.

The committed department will ensure that risk reduction activities are a written part of job descriptions for all personnel, and that performance evaluations take into consideration involvement in such activities.

TRAINING

In order for risk reduction to be an organized and effective part of a fire department's operations, individuals must receive the proper training. Prevention education is more than simply entertainment. It is the changing of behaviors through instruction and has to be conducted as such. And in order for that to happen, all personnel must receive the proper training in risk reduction.

In many departments, station tours and other public education details are relegated to the newest recruit. Although this is not a bad practice, it sometimes indicates the lack of importance placed on such activities. It is seriously doubtful that the same recruit would be sent into a fire without the proper instruction and training in firefighting skills. Nor should this recruit be allowed to conduct education activities without the benefit of the same training and guidance. (See Figure 13.10.)

FIGURE 13.10 Many fire departments still give station tours on a regular basis. *Courtesy of Travis Ford, Nashville Fire Department*

As stated earlier, NFPA 1001, *Standard for Fire Fighter Professional Qualifications,* requires that individuals in a Firefighter II position should be able to perform firehouse/station tours and other public education activities accurately and effectively. In many states, recruit training courses allow for instruction in public education activities and information. However, many times that requirement is only a couple of hours, or maybe an eight-hour block at the most. The amount of time allocated may depend upon the total length of the training course. However, fire and life safety education should be a course requirement for all new firefighters, and some departments incorporate the entire Educator I certification class into their initial recruit training. All new graduates should understand the importance of risk reduction in their community, in their department, and the vital role it plays in their own safety.

Risk reduction must be included in monthly in-service training for all firefighters and departmental members. It is not enough for firefighters to be given training in risk reduction as a part of their initial training; they should receive updates and new information continually. Just as firefighters review and update their emergency scene capabilities and skills, so should they review and update their risk reduction skills.

In addition, prevention and risk reduction should be a part of the promotional or career advancement requirements for fire officers. Those seeking advancement to officer and higher should be required to achieve certain levels of training or complete certain risk reduction classes as part of their qualifications for promotion. They must demonstrate a commitment to risk reduction through their initiative and participation in community risk reduction activities. Again, standards for fire officers require increasing levels of community involvement and must be considered when promotions are made.

TARGETED AND FOCUSED EDUCATIONAL PROGRAMS

For risk reduction programs to be effective, they must follow an organized process. Although it is easy to schedule presentations at schools, pass out brochures, and conduct firehouse/station tours, those activities may not be targeting the community's worst problems or reaching the audience most at risk. Those activities are important, but they are most effective as part of an organized and planned community-wide program.

Successful programs that reduce fires and prevent injuries are always the result of effective planning. Many notable risk reduction programs around the country prove this to be true. The USFA, in its publication *Public Fire Education Planning* (2008), presents a five-step process for designing and implementing successful risk reduction programs:

Step 1:   Conduct a community analysis.
Step 2:   Develop community partnerships.
Step 3:   Create an intervention strategy.
Step 4:   Implement the strategy.
Step 5:   Evaluate the results.

Conducting a Community Analysis

A **community risk analysis** is a process that identifies fire and life safety problems and the demographic characteristics of those at risk in a community. The analysis allows the worst problems to be identified, so that they will be the targets of the risk reduction program.

Many times, risk reduction programs are based on a perceived need or problem that is inaccurate or wrong. It is hard to get people to participate in a program that is unnecessary or not addressing the real problem. If the problem has not been identified, defined accurately, or targeted appropriately, all that is accomplished is loss of money, time, and resources. This frustrates those conducting the programs, as well as those who fund them.

community risk analysis ■ A process that identifies fire and life safety problems and the demographic characteristics of those at risk in a community.

FIGURE 13.11
Geographic information
systems can help narrow
down where specific
events are occurring.
*Courtesy of Don Oliver,
Fire/Rescue Chief, Wilson
Fire/Rescue Services*

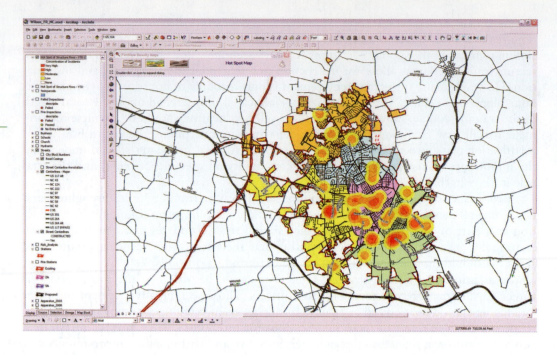

A community risk analysis provides a factual overview of the community's risk issues. Whenever possible, it should be based on local data, not national events. Data to be examined should include local fire reports from the National Fire Incident Reporting System (NFIRS). Geographic information systems (GIS) can help identify and map those areas with vacant buildings, vehicle crashes, incendiary fires, and the like. The state fire marshal's office may have compilation data to see how a community ranks within the state. However, if only a very limited amount of local data is available, then information from a state or perhaps even at a national level may be used, but with caution, because each community is different. (See Figure 13.11.)

The information obtained from local reports should include what type of incidents are occurring and the causes; the geographic location of the events (perhaps indicating a cluster, or specific neighborhood); time of day and day of week; and how often they are occurring. Demographic information needs to be examined for the area(s) where the events are occurring. If the data cannot be narrowed down to specific areas, then overall community demographic information may be used. A problem statement, developed from the information gathered and analyzed, will then provide the focus for the risk reduction program. (See Figure 13.12.)

The process of conducting a community assessment to determine the what, who, where, and why of a community's problem is the foundation for an effective risk reduction program in any community.

Developing Community Partnerships

It is important for the risk reduction process always to be centered on teamwork. It is impossible for one individual to meet the needs of larger communities, or even smaller communities, when volunteers provide the fire and emergency services protection. Partnering with other agencies means sharing the time, resources, and equipment, and it makes risk reduction programs more effective and credible.

A community partner can come in a variety of shapes and sizes. It might be an individual that is looking for a community cause to support. It could be a group or organization that is already working with the target population or in the overall area of risk

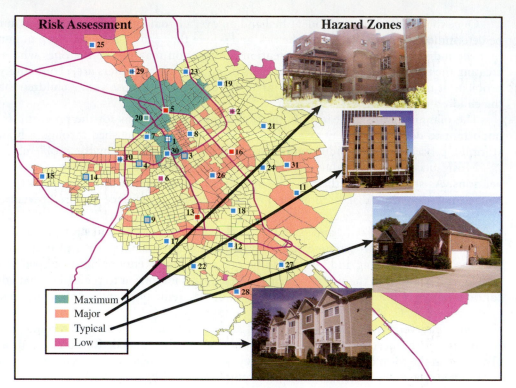

FIGURE 13.12
Geographic information systems can help a fire department identify high-risk areas for public information and code enforcement activities. *Courtesy of Don Oliver, Fire/Rescue Chief, Wilson Fire/Rescue Services*

reduction. It might be a group that is impacted financially by the problem, such as insurance agencies or those providing disaster services. And certainly partners in the risk reduction process should include members of the population that are affected by the risk.

Once partners have been identified, a planning team should be developed. It is important to identify those partners who can best deliver the risk reduction message or those who have the resources needed to address the problem. Partners who have leadership, skills, credibility, influence, and resources are those who should be brought together to work on this important project.

Creating an Intervention Strategy

Development of an **intervention strategy** will help to gain long-term success for risk reduction. It is a detailed plan that outlines the work necessary for a successful fire or life risk reduction program. It is similar in function to the emergency planning for a community building or industry and includes what will be done, where it will occur, how implementation will occur, and who will conduct the program once it is developed.

Intervention strategies are based upon the examination of the community risk analysis and the work of the planning team. The team should develop a specific statement about the community's identified risk problem and then set a goal for the risk reduction program based upon that problem. A goal is a broad statement about the problem and what the team sees as the new condition it would like to create. An example of a problem statement and goal would be:

intervention strategy ■
A detailed plan that outlines the work necessary for a successful fire or life risk reduction program; includes what will be done, where it will occur, how implementation will occur, and who will conduct the program once it is developed.

Problem.   A high number of cooking fires are occurring within the homes of older adults residing in the community's public housing apartment project.

Goal.   Decrease the number of cooking fires occurring within the homes of older adults who live in public housing apartments by 25% in the next two years.

Once the problem has been identified and an overall goal set for the program, it must be determined which groups affect the risk and where they can be reached. Keep in mind that the individuals needing the intervention information may not be the ones who are creating the risk. For example, if preschoolers using matches or lighters are a community's problem, it may be more important to target the caregivers of those young children than the children themselves.

The planning team must identify the prevention interventions for the program. It is important to develop as many as possible, which can be accomplished through a brainstorming process and by conducting research among those affected by the problem.

Most successful risk reduction efforts use a combination of the prevention interventions. As stated earlier, the three Es of prevention—engineering, enforcement, and education—that were discussed in the 1947 President's Conference on Fire Prevention are still effective today.

The planning team must approach the development of intervention strategies and the whole risk reduction process in a scientific manner. Team members need to understand why the risk is occurring and how it can be prevented. Interventions must be specific in providing a solution to the risk issue being targeted. It is important to remember that education alone may not produce the desired level of risk reduction, and all strategies must be examined.

Once interventions have been identified and described, it is time to determine what resources will be needed and how the program will be evaluated. Resources may not always be in the form of funding. Instead, they may be equipment and tools to make modifications to the environment, such as the improvement of lighting near the stove or the installation of smoke alarms. Certain interventions may require the support of community or political leaders, such as the implementation of a residential sprinkler requirement would need.

Evaluation, though a separate step, must be introduced during this stage of program development. Objectives should be developed that define what is going to be done, who is going to do what, when it should be done, and how effectiveness will be determined. Developing objectives can be tedious and time-consuming, but it is crucial to find out whether the program goal has been achieved and whether the risk reduction process has been successful.

When intervention strategies are specific and based on fact, an accurate evaluation process is possible. Intervention strategies must be realistic and within the capabilities of the target audience. Each strategy must be community based and include the fire department planning team, target audience, and other stakeholders—all as a part of the problem-solving process. Strategies must be implemented over time, and must be continually monitored and evaluated so that adjustments can be made, if needed.

Implementing the Strategy

At the implementation stage, the interventions are tested and the plan is put into action. An action plan is developed that includes the step-by-step work necessary to meet the objectives. There may be a temptation to put the program into action without an implementation plan, but this only leads to a lack of accountability and disorganization. A small pilot program should be conducted to identify and resolve any problems encountered with the program's design. Many times a program looks good on paper, but when it is actually implemented, it does not work or there are aspects that have not been considered by the planning team. The pilot program offers an opportunity to evaluate the program on a small scale and make adjustments where necessary.

Once the pilot program has been conducted and modifications made as necessary, it is time to market the program, initiate the activities, and monitor progress. Marketing can be done through a variety of sources, such as the local media, newsletters, direct mailings, or community meetings. Whatever medium is used, it should be one that the target audience uses and that will be effective in communicating the message to them.

The program should continually be monitored to ensure that the goal and the objectives of the intervention plan are being met. It is important to report successes to encourage a continuation of the program and to analyze problems so that revisions can be made as necessary.

Evaluating the Results

Because the primary goals of a risk reduction program are to reach a target audience, impact the problem, and reduce loss from fire, it makes sense that the primary goal of the evaluation process is to determine whether that has occurred. Without evaluation, the planning team and the community at large can only guess whether risk reduction efforts are successful. There is too much at stake to simply guess!

Evaluation is often an overlooked part of a risk reduction process. Many times there is a fear of working with numbers or simply a lack of knowledge of how to perform program evaluations. Many times people are afraid of the results of an evaluation.

In order to effectively evaluate a program, there must be a means for collecting data. This might be in the form of a survey or questionnaire sent to the target audience. Once collected, the data must be compared with the baseline data gathered at the beginning of the program development.

If the evaluation results were not as anticipated, the planning team should reexamine the intervention strategies and the implementation process. Team members will be able to assess potential areas for change and note possible adjustments and modifications. Questions to ask include:

- Was the correct risk issue identified in the community analysis?
- Was the correct target audience identified?
- Did the interventions reach the target audience?
- Were the interventions the right ones for the identified risk issue?
- Were the interventions within the capabilities of the target audience?
- Were there outside forces that affected the outcome?

The results of the evaluation process should be shared with the planning team, the fire department, and the community. If a revision of the program is desired, it is important that everyone understands the reason for change. If the desired results have been achieved, this does not mean that the program ends. There needs to be some type of maintenance program to ensure that safe behaviors are maintained, as well as an enforcement aspect to ensure that the engineering initiatives are maintained.

The five-step process for designing and implementing a successful risk reduction program is a never-ending cycle of data collection and examination, and of devising intervention strategies to address identified issues.

It also must be remembered that the process is meant to reduce the risk to firefighters. When individuals change their behaviors and employ engineering initiatives to prevent fires or provide notice to occupants of a hazardous condition, it has lessened the risk to fire service members.

NATIONAL INVOLVEMENT

One aspect of success is fire and emergency services involvement at the national level in support of community risk reduction. There must be a greater organizational focus on prevention, not only locally but at the national level as well. Since 1947 the Hartford's Junior Fire Marshal Program has been providing safety education materials. The program is recognized as one of the earliest efforts to teach families about the basics of fire safety. Chiefs, fire marshals, public educators, and all fire and emergency services leaders must be advocates for risk reduction initiatives and national leaders in fire and injury prevention efforts.

Certain agencies at the national level help promote risk reduction throughout the country. Created in 1972, the U.S. Consumer Product Safety Commission helps protect the public from unreasonable risk of death and injury caused by numerous products that pose a fire, electrical, chemical, or mechanical hazard. It has created partnerships with the USFA and Home Safety Council in developing prevention materials. This partnership has created a comprehensive Internet site (see www.firesafety.gov) that provides fire prevention resources and information for citizens, the fire and emergency services, children, the media, and public information officers on eliminating residential fire deaths. Safe Kids Worldwide is probably one of the most widely recognized organizations. Originating in 1988, its main focus is children under the age of 14, and the prevention of fires and burns is just one aspect of its work.

The Home Safety Council, another nationally recognized organization, is working for risk reduction by developing home safety tips and checklists. Its focus is on safety in the home, dealing with all types of injury issues for all ages. Again, fire prevention is only a part of its mission. It is working to promote fire safety and risk reduction at all levels, and to provide cost-effective materials as well as materials and programs that target the high-risk audiences.

Since 1991 the People's Burn Foundation has produced and distributed educational and training programs funded by the U.S. Department of Homeland Security on the reality of burn injuries and how to prevent them. Those programs include firefighter burn prevention, situational awareness, interactive training simulation, community awareness, juvenile fire setters, college fire survival, and cyanide poisoning (PBF, n.d.).

The NFPA Center for High-Risk Outreach is another organization that works nationally and abroad to reduce risks related to fire. The Center for High-Risk Outreach was created in 1995 with a mission to reduce deaths and injuries from fires and burns among the high-risk populations: the very young, older adults, and the impoverished. The Center's advisory committee provides expert advice on technical and educational issues related to high-risk audiences. The NFPA, with the help of the USFA, has recently conducted research in behavioral issues caused by cooking and smoking, along with how to reduce the death rate from fire in rural areas.

Beginning in 2004, Fire Team USA has brought key stakeholders from communities together to learn about the benefits of fire sprinklers. Its free workshops provide attendees with the latest resources on fire prevention methods and alternatives that affect quality of life. Common Voices, begun in 2007 as a coalition of fire safety advocates, involves individuals who have been directly affected by fire. The advocates turn the tragedy of losing a family member or being a burn survivor themselves in getting involved in policy decisions and becoming a voice that makes the fire problem real.

Vision 20/20 is a recent committee that is working for fire prevention at the national level. With funding from the U.S. Department of Homeland Security Assistance to Firefighters Grant (AFG) and Fire Prevention and Safety Grants (FP&S) program, the Institution of Fire Engineers established a steering committee comprised of noted fire and emergency services and related agency leaders to guide a national strategic planning process for the fire loss prevention that should result in a national plan to coordinate activities and fire prevention efforts. In October 2008, the Vision 20/20 committee—intended for action, not just recommendations—reported five different strategies for national fire prevention efforts. These are:

- *Strategy 1.* Increase advocacy for fire prevention.
- *Strategy 2.* Conduct a national fire safety education/social marketing campaign.
- *Strategy 3.* Raise the importance of fire prevention within the fire service.
- *Strategy 4.* Promote technology to enhance fire and life safety.
- *Strategy 5.* Refine and improve the application of codes and standards that enhance public and firefighter safety and preserve community assets. (Institution of Fire Engineers, US Branch, 2008)

There are other ways that risk reduction can be promoted and supported in the fire and emergency services at the national level. Numerous fire and emergency services professional journals and publications exist, yet none are specifically dedicated to risk reduction. Some might offer a page or an occasional article, but fire prevention and public education activities are not a major focus on a regular basis. Professional journals should dedicate more space to articles regarding risk reduction initiatives and invest more time soliciting information and submissions from those involved with public education and risk reduction activities. This would assist fire departments in their risk reduction efforts, as well as give risk reduction an equal footing with other fire and emergency services topics.

Having a means by which to share best practices, validated programs, and evaluation tools among organization leaders and prevention specialists will go a long way in institutionalizing fire prevention in the fire and emergency services as an industry, not just organization by organization.

RISK REDUCTION THROUGH ENFORCEMENT

Enforcement is part of all community risk reduction initiatives as the force behind the educational and engineering interventions. Fire and emergency services leaders must become involved in the enforcement process, because it involves public policy, public policy takes place in the political arena, and fire and emergency services leaders are, by proxy, politicians at the local level. And although it is a role that other fire and emergency services members can support, it is up to officers and leaders to be advocates in the political arena.

The fire and emergency services must bring legislation to the attention of local leaders. Chiefs and chief officers need to communicate the importance of fire prevention and risk reduction to local community leaders. Local leaders must see that fire protection has economic benefits for the community and contributes to the community's quality of life. Fire and emergency services leaders need to promote the adoption and enforcement of current model building and fire codes, including those that require fire sprinklers in one- and two-family dwellings. (See Figure 13.13.)

In some areas, local political leaders try to get involved in the enforcement process. They want to assist their constituents by approving variations to code requirements or by ignoring code violations. The situation must not be tolerated by fire department leaders, because to do so only puts firefighters at risk. It is an opportunity for the fire service to educate local and state political leaders in appropriate fire safety measures for the community and the risk reduction strategies being designed and implemented.

Fire inspections and planning surveys must be a regular part of the risk reduction activities of the fire departments. When a department becomes familiar with the buildings and occupancies in its response area during nonemergency conditions, it will be able to identify potential problems or hazards that can occur during emergency situations. Such an activity should be a fire department's training agenda.

RESIDENTIAL SPRINKLERS

According to the USFA, there has never been any multiple loss of life in a fully sprinklered building. Property losses are 85% less in residences with fire sprinklers compared to those without sprinklers. The combination of automatic sprinklers and early warning systems in all buildings and residences could reduce overall injuries, loss of life, and property damage due to fire by at least 50% (USFA, 2009c). This figure includes injuries and loss of life of firefighters as well. According to the NFPA's Fire Sprinkler Initiative, 90% of residential fireground firefighter deaths occur in one- and two-family homes (NFPA, 2009c).

FIGURE 13.13 Public information officers play an important role in providing the community with fire prevention and risk reduction information. *Courtesy of Martin Grube*

At the present time, all model safety codes include the use of home fire sprinklers in new one- and two-family homes. Despite this, anti-sprinkler legislation has been filed in a number of states. Sprinkler opponents are pushing state legislation that would restrict a community's ability to make its own decision about model safety codes for new construction and would prevent them from implementing new sprinkler mandates in one- and two-family homes.

Fire departments must band together and advocate for voluntary installation of residential sprinklers and the enactment of local sprinkler legislation. Evidence shows the value of local fire service leaders speaking up at the local level, and banding together at the state level, to lobby for sprinkler legislation. For example, in January 2009, fire and emergency services leaders in Anne Arundel County, Maryland, successfully lobbied for a sprinkler ordinance in one- and two-family homes. Even though there was existing legislation relating to townhomes and condominiums, until then leaders had not been successful in extending the reach to one- and two-family homes. This time they were, to the point of including first-owner mobile homes and even some renovations (Sutherland, 2009). (See Figure 13.14.)

As a part of community risk reduction efforts, fire service personnel must educate the public about residential fire sprinklers and the benefit of installation. Many individuals believe that sprinklers are too expensive and are not cost effective, yet studies actually show that they are affordable and pay off in a variety of ways (Butry, Brown, and Fuller, 2007). In areas that have required sprinkler systems for years, the cost

FIGURE 13.14 Side-by-side burn demonstrations are an effective tool to demonstrate the impact of residential fire sprinklers. *Courtesy of Vickie Pritchett*

is substantially lowered by market competition. According to the Fire Protection Research Foundation, an affiliate of NFPA, the cost of installing sprinkler systems averages $1.61 per sprinklered square foot (NFPA, 2008, pp. iii, 15, and 21). The cost includes all costs to the builder associated with the system including design, installation, and other costs such as permits, additional equipment, and increased tap and water meter fees, if applicable.

Summary

Community risk reduction plays as much of a part in firefighter safety and survival as fitness for duty and the use of seat belts. Oftentimes the subject is overlooked as a part of risk reduction for firefighters, but its impact is huge. Risk reduction must be valued by the entire department and receive the support it is due.

When individuals understand the role that risk reduction, code enforcement, and residential sprinklers play in overall safety, then a difference will be made in the statistics relating to both civilian and firefighter deaths and injuries related to fire.

Review Questions

1. Discuss the steps in the risk reduction process and explain why each is important to the success of the overall process.
2. Explain the five categories of risk reduction initiatives and how they work together to reduce community risk.
3. State the misconceptions that exist about residential sprinklers and why they are not true.
4. Compare community risk reduction activities today with those of one hundred years ago. What has changed? What has not?
5. Explain the role of risk reduction in firefighter safety and survival. How important is it?

References

Bennie, P. (2008). *The Great Chicago Fire of 1871*. New York: Infobase.

Bukowski, R. W. (1995, November/December). "Modeling a Backdraft: The Fire at 62 Watts Street." *NFPA Journal*, 89(6), 85–89. Retrieved November 6, 2009, from http://fire.nist.gov/bfrlpubs/fire95/PDF/f95090.pdf

Butry, D. T., Brown, M. H., and Fuller, S. K. (2007, September). *Benefit-Cost Analysis of Residential Fire Sprinkler Systems*. NISTIR 7451. Gaithersburg, MD: National Institute of Standards and Technology. Retrieved November 25, 2009, from http://www.fireteamusa.com/Documents/NISTIR_7451_Oct07.pdf

Drennan, V. (1995, April 26). Transcript from speech at the 7th Annual National Fire and Emergency Services Dinner, Washington, DC.

Home Safety Council. (2007, April). *The Home Safety Council and Johns Hopkins Surveyed Fire Chiefs to Identify Fire and Life Safety Public Education Practices and Barriers*. Washington, DC: Home Safety Council. Key findings retrieved November 25, 2009, from http://homesafetycouncil.org/AboutUs/Research/re_FLSE_w001.asp

Institution of Fire Engineers, US Branch. (2008, October 6). *Vision 20/20: National Strategies for Fire Loss Prevention Final Report*. Alexandria, VA: Author.

Retrieved November 6, 2009, from http://www.strategicfire.org/08report.pdf

National Commission on Fire Prevention and Control. (1973, May 4). *America Burning: The Report of the National Commission on Fire Prevention and Control*. FA-264. U.S. Government Printing Office #1973-O-495-792. Retrieved April 15, 2009, from http://www.usfa.dhs.gov/downloads/pdf/publications/fa-264.pdf

National Fallen Firefighters Foundation. (2005–2009). *Everyone Goes Home: 16 Firefighter Life Safety Initiatives*. Emmitsburg, MD: Author. Retrieved November 26, 2009, from http://www.everyonegoeshome.com/initiatives.html

National Fire Academy. (2008). *Strategies for Community Risk Reduction*. Emmitsburg, MD: Author.

National Fire Protection Association. (n.d.). *The Fire Sprinkler Initiative*. Quincy, MA: Author. Retrieved November 6, 2009, from http://www.firesprinklerinitiative.org/

National Fire Protection Association. (n.d.). *About Fire Prevention Week*. Quincy, MA: Author. Retrieved March 24, 2010, from http://www.nfpa.org/itemDetail.asp?categoryID=375&itemID=17365&URL=Learning/Public%20Education/Fire%20Prevention%20Week/About%20Fire%20Prevention%20Week&cookie%5Ftest=1

National Fire Protection Association. (2008, September). *Fire Research: Home Fire Sprinkler Cost Assessment Research Project.* Quincy, MA: Fire Protection Research Foundation. Retrieved November 25, 2009, from http://www.nfpa.org/assets/files//PDF/Research/FireSprinklerCostAssessment.pdf

National Fire Protection Association. (2009a). *About NFPA: Mission Statement.* Retrieved November 4, 2009, from http://www.nfpa.org/categoryList.asp?categoryID=143&URL=About%20NFPA

National Fire Protection Association. (2009b). Standards for Professional Qualifications: 1001, 1021, 1031, and 1035. Retrieved November 3, 2009, from http://www.nfpa.org/aboutthecodes/list_of_codes_and_standards.asp

National Fire Protection Association. (2009c, May). "Fire Departments Stand Together." In *Fire Sprinkler Initiative Update: Bringing Safety Home.* Retrieved November 25, 2009, from http://f.chtah.com/i/47/272412627/051509FSI.html

National Fire Protection Association. (2010). *Center for High-Risk Outreach.* Retrieved March 23, 2010, from http://www.nfpa.org/categoryList.asp?categoryID=201&URL=Learning/Public%20Education/Center%20for%20High-Risk%20Outreach

People's Burn Foundation. (n.d.). *To Hell and Back: An Educational Program on the Reality of Burn Injury.* Indianapolis, IN: Author. Retrieved March 24, 2010, from http://www.pbfeducation.org/home

Sutherland, S. (2009, March/April). "The Case for Home Fire Sprinklers." *NFPA Journal.* Quincy, MA: National Fire Protection Association. Retrieved November 25, 2009, from http://www.nfpa.org/publicJournalDetail.asp?categoryID=&itemID=42313&src=NFPAJournal

U.S. Fire Administration. (1987). *America Burning Revisited.* U.S. Government Printing Office: 1990-724-156/20430. Retrieved April 15, 2009, from http://www.usfa.dhs.gov/downloads/pdf/publications/5-0133-508.pdf

U.S. Fire Administration. (1999). *Solutions 2000: Advocating Shared Responsibilities for Improved Fire Protection.* A Report of the North American Coalition for Fire and Life Safety Education. Retrieved November 6, 2009, from http://www.usfa.dhs.gov/downloads/pdf/solutions2000.pdf

U.S. Fire Administration. (2002a). *Beyond Solutions 2000: Advocating Shared Responsibilities for Improved Fire Protection.* A Report of the North American Coalition for Fire and Life Safety Education. Retrieved November 6, 2009, from http://www.usfa.dhs.gov/downloads/pdf/beyondsolutions2000.pdf

U.S. Fire Administration. (2002b, June). *America at Risk: America Burning Recommissioned.* FA-223. Retrieved April 15, 2009, from http://www.usfa.dhs.gov/downloads/pdf/publications/fa-223-508.pdf

U.S. Fire Administration. (2002c, July). *1947 Fire Prevention Conference.* Retrieved November 6, 2009, from http://www.usfa.dhs.gov/about/47report.shtm

U.S. Fire Administration. (2008, June). *Public Fire Education Planning: A Five Step Process.* FA-219. Retrieved November 6, 2009, from http://www.usfa.dhs.gov/downloads/pdf/publications/fa-219.pdf

U.S. Fire Administration. (2009a). *Residential Sprinkler Myths and Facts.* Retrieved November 6, 2009, from http://www.usfa.dhs.gov/citizens/all_citizens/home_fire_prev/sprinklers/facts.shtm

U.S. Fire Administration. (2009b). *Chief's Corner: International Code Council Votes to Retain Residential Fire Sprinkler Requirement.* Retrieved March 24, 2010, from http://www.usfa.dhs.gov/about/chiefs-corner/103009.shtm

U.S. Fire Administration. (2009c, April 15). *Residential Sprinkler Myths and Facts: The Arguments Against Sprinklers.* Emmitsburg, MD: Author. Retrieved March 29, 2010, from http://www.snohomishfire.org/Community%20Information/Residentail%20Fire%20Sprinklers/Basic%20Information/Myths%20and%20Facts%20FEMA.pdf

U.S. Fire Administration. (2009d, April 15). *Strategic Plan.* Retrieved November 6, 2009, from http://www.usfa.dhs.gov/about/strategic/

14

Emergency Response Vehicles and Equipment

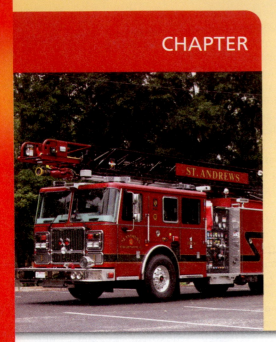

A. K. Rosenhan

Courtesy of A. K. Rosenhan

KEY TERMS

aerial devices, *p. 331*

emergency response vehicle, *p. 327*

fire equipment, *p. 327*

fire hose, *p. 340*

personal alert safety system (PASS), *p. 344*

personal protective ensemble (PPE), *p. 340*

pump capacities, *p. 328*

Quint apparatus, *p. 332*

self-contained breathing apparatus (SCBA), *p. 343*

Storz coupling, *p. 342*

triple combination pumper, *p. 331*

OBJECTIVES

After reading this chapter, the student should be able to:

- Explain the basic safety features in the design of emergency response vehicles and equipment.
- Recognize safety aspects of the operation of emergency response vehicles and equipment.
- Demonstrate a basic knowledge of the functions of emergency response vehicles and equipment.
- Identify the basic types and configuration of emergency response vehicles and equipment.

Types of Emergency Response Vehicles and Fire Equipment

An **emergency response vehicle** is a motorized fire apparatus that is equipped with emergency light and siren. **Fire equipment** includes tools, fittings, nozzles, and loose equipment carried on emergency response vehicles. This chapter identifies the various emergency response vehicles and equipment, with particular attention to their safety features. A basic history of the development of emergency response vehicles, firefighting tools, and safe operating considerations also is discussed.

An injured, incapacitated, untrained, unsafe, or ineffectual firefighter endangers self, crew, and public. To get the job done, under a variety of emergency scenarios, requires the utmost vigilance on the part of each and every firefighter, the supervision and administration, and the supporting departments.

The mission of the firefighter has become even more complex due to the proliferation of new building and interior finish materials, and construction techniques; the addition of emergency medical service (EMS) operations, hazardous materials, technical rescue, and the like; and the lack of resources to ensure an adequate arsenal for the firefighters to attack any problem. As a result, emergency response vehicles and equipment have become quite sophisticated. And firefighters are required to have the knowledge and skills necessary to use them.

A firefighter relies on two basic elements when doing the job:

- Proper vehicles and equipment
- Knowledge of how to use them safely, effectively, and efficiently

The first element means the hardware must be capable of doing the job in a reliable and safe manner for both operators and the public. The second element is the product of training, practice, and the ability of the user to pick and correctly use the proper tool and equipment. And certainly the dedication of the firefighter is of paramount importance to bring about a good result. (See Figure 14.1.)

emergency response vehicle ■ Motorized fire apparatus that is equipped with emergency light and siren.

fire equipment ■ Tools, fittings, nozzles, and loose equipment carried on emergency response vehicles.

FIGURE 14.1 Firefighters must have the basic working knowledge of their emergency response vehicles and equipment in order to be safe, effective, and efficient.
Courtesy of Martin Grube

TABLE 14.1	Internet Sources for Information
WEB SITE	**ACTIVITY**
EVTcc.org	Emergency Vehicle Technician Certification Program
fama.org	Fire Apparatus Manufacturers' Association
fdsoa.org	Fire Department Safety Officers Association
FEMSA.org	Fire and Emergency Manufacturers and Services Association
ul.com	Underwriters Laboratories set standards for mobile fire equipment and emergency response vehicles, building materials, and fixed fire protection hardware
http://www.femalifesafety.org/about.html	Fire Equipment Manufacturers' Association

ADVOCATES FOR EMERGENCY RESPONSE VEHICLE AND EQUIPMENT SAFETY

Fire and emergency services organizations such as the National Fallen Firefighters Foundation, Firefighter Close Calls, the Secret List, and various state organizations promote vehicle, personnel, and operational safety. Table 14.1 lists the Web sites for several of those types of organizations. Those and other groups have sponsored various projects and published the results on their Web site or in print.

ROLE OF NFPA AND OTHER STANDARDS

There is a myriad of organizations that develop standards related to the safe operation and use of emergency response vehicles and equipment. Those organizations include the American Society of Mechanical Engineers, the National Fire Protection Association (NFPA), the National Highway Traffic Safety Administration (which develops the Federal Motor Vehicle Safety Standards, known as FMVSS), the Occupational Safety and Health Administration (OSHA), the Society of Automotive Engineers, developers of various fire and building codes, and developers of a host of various industry standards (such as chassis builders). Those groups set the standards for specifying pump capacity, hose specifications, and industry benchmarks for the fire and emergency services. **Pumping capacities** are ratings in increments of 250 gallons per minute (950 L/min) (minimum capacity 250 gpm/950 L/min) at a discharge pressure of 150 psi (10.2 bar) from draft through 20 feet (6.1 m) of suction hose for emergency response vehicles with permanently mounted fire pumps. NFPA 1901, *Standard for Automotive Fire Apparatus,* references these other materials.

Whereas the Federal Motor Vehicle Safety Standards and OSHA standards are legally the law, others are put into place only after adoption by the authority having jurisdiction (AHJ). Even without the AHJ adopting various NFPA and other standards, it should recognize the standards considered to be the "standard of care," and deviation from or ignoring them can be a problem from a legal standpoint.

All of the requirements within NFPA and other applicable standards are to ensure the safest available emergency response vehicles, equipment, and operations possible. Users of the vehicles and equipment have the responsibility to ensure that all vehicles and equipment are properly maintained (i.e., the inspection of vehicle components for wear, malfunction, damage) and serviced per manufacturer's recommendations for oil changes, filter service, brake adjustments, and the like. All personnel, from the chief to any purchasing and safety committee members, should be familiar with any safety-related issues of every emergency response vehicle and item of equipment in their inventory. Many states have additional requirements and training standards or have adopted the OSHA standard

pump capacities ■ A rating in increments of 250 gallons per minute (950 L/min) (minimum capacity 250 gpm/950 L/min) at a discharge pressure of 150 psi (10.2 bar) from draft through 20 feet (6.1 m) of suction hose for emergency response vehicles with permanently mounted fire pumps.

on personal protective equipment (OSHA, 2010), which may supercede NFPA standards and other adopted practices.

The NFPA standards are continually being updated with new editions published approximately every four years. Comparison of the various editions shows a greatly increased emphasis on personnel safety, change to performance standards in lieu of design standards, and recognition of the complexity of the mission of the fire and emergency services in general. The NFPA procedures for modifying a standard are well documented and provide for a Tentative Interim Amendment (TIA) for changes considered of an emergency nature to be incorporated into a standard prior to the issuance of a new edition. Therefore, TIAs are amendments to NFPA documents that have not gone through the entire code and standards-making process and yet have become effective between editions of a document. A recent example is the requirement to equip pre-connect hose lines with a cover to prevent inadvertent deployment of the nozzle and hose.

Any acquisition of emergency response vehicles and equipment must ensure that requirements of current editions of all applicable standards are met.

TRAINING AND INFORMATION SEEKING

Training

Training, in a classroom environment as well as during physical operations, is the basis for safe and effective emergency scene operations. A formal training regime is important in any fire department, and constant attention to new and appropriate fire and emergency services developments is necessary. Training is particularly vital in the use and limitations of emergency response vehicles and equipment. (See Figure 14.2.)

Even a well-designed vehicle will not prevent all incidents: the driver/operator is the ultimate controller. Continued and consistent training is most important as it relates to driver/operators of emergency response vehicles. Various NFPA standards such as NFPA 1002, *Standard for Fire Apparatus Driver/Operator Professional Qualifications,*

FIGURE 14.2 Training helps firefighters understand the use and limitations of emergency response vehicles. *Courtesy of Martin Grube*

FIGURE 14.3 This live wire fell across the emergency vehicle on scene, making it hazardous and unusable. Because anything can happen on the emergency scene, emergency vehicle placement is crucial during a working structure fire. *Courtesy of Martin Grube*

are relevant to firefighter safety through emergency response vehicles and equipment. NFPA 1071, *Standard for Emergency Vehicle Technician Professional Qualifications*, deals with professional qualifications required of personnel engaged in the diagnosis, maintenance, repair, and testing of emergency vehicles. NFPA 1915, *Standard for Fire Apparatus Preventive Maintenance Program*, contains the requirements for systems and components that are unique to emergency response vehicles. (See Figure 14.3.)

The very nature of the business of a fire department poses numerous challenges, because every run provides the opportunity for a problem to develop. The proper licensing of drivers, in-service training, and review are very important. Although many areas across the United States have access to a statewide academy for fire and emergency services training, there may be difficulty in scheduling such classes due to financial and time limitations. Volunteer firefighters, who comprise the majority of U.S. fire departments and, unfortunately, are involved in a majority of emergency response vehicle accidents, have limited ability to attend such classes. Several companies that provide insurance to the fire and emergency services, such as the Emergency Services Insurance Program (ESIP) and VFIS, serving volunteers, provide free driving training courses and other material to their customers. Larger fire departments and statewide agencies may have the resources to use driving simulators. There are computer-based training courses via CD and DVD as well as the traditional printed course work.

Fire departments should have a mandatory set of requirements for driver/operators that includes periodic review of driving license status, in-service training, and management review. In addition, insurance companies that specialize in fire department coverage and other organizations have recommended or required training and certification requirements for driver/operators of emergency response vehicle operations.

Information Seeking

The firefighter is responsible to be knowledgeable about the profession. The knowledge may be gained by subscribing to and reading various fire and emergency services publications (e.g., see *Fire Apparatus & Emergency* magazine at www.fireapparatusmagazine.com).

Some of these publications are free and all provide valuable and interesting material. A wealth of textbooks, online courses, and manufacturer's material relates to the use of various emergency response vehicles and equipment.

Emergency Response Vehicles

TYPES OF EMERGENCY RESPONSE VEHICLES

The emergency response vehicle is the basic tool that provides transportation for firefighters and equipment to an emergency scene. Such vehicles are outfitted for various functions; they can carry a water supply for initial fire attack, have a fire pump and hose, have aerial capabilities, have rescue capabilities, and provide various support functions. The combination of vehicular engineering coupled with the required emergency scene operations should result in a system that provides an elevated level of safety at the emergency scene. As a result, the need for significant planning, training, and constant vigilance related to safe and effective operation is present.

The seven types of emergency response vehicles as defined in NFPA standards follow:

1. The long-standard **triple combination pumper** has a permanently mounted fire pump with a minimum capacity of 750 gallons per minute (2,839 L/min), a minimum complement of hose, and a minimum size water tank. Its primary purpose is to combat structural and associated fires.

2. **Aerial devices** are aerial ladders, elevating platforms, aerial ladder platforms, or water towers that are designed to position personnel, handle materials, provide continuous egress, or discharge water. Certain devices have pumping capacity, and there are various types of designs to elevate and maneuver firefighters and nozzles. (See Figure 14.4.) Aerial devices may be of a "straight" or a simple telescoping

triple combination pumper ■ An emergency response vehicle with a permanently mounted fire pump with a minimum capacity of 750 gallons per minute (2,839 L/min), a minimum complement of hose, and a minimum size water tank whose primary purpose is to combat structural and associated fires.

aerial devices ■ Aerial ladders, elevating platforms, aerial ladder platforms, or water towers that are designed to position personnel, handle materials, provide continuous egress, or discharge water.

FIGURE 14.4 There are various types of emergency response aerial vehicle designs. *Courtesy of A. K. Rosenhan*

(a)   (b)

FIGURE 14.5 Various designs of emergency response vehicle tenders. *(a) Courtesy of A. K. Rosenhan, (b) Courtesy of Phil Fraser, Cedar Hill Fire Department*

Quint apparatus ■ A combination emergency response vehicle with a permanently mounted fire pump, a water tank, a hose storage area, a power aerial ladder or elevating platform with a permanently mounted waterway, and a complement of ground ladders.

section design or having articulated sections or a combination thereof. Additionally, certain aerial device configurations may or may not have ladder or personnel capacities but are for generating elevated master streams only.

3. Mobile water supply vehicles, also called tenders, have water tanks of 2,000 gallon (7,570 L) or greater capacity. (See Figure 14.5.)

4. Combination emergency response vehicles called **Quint apparatus**, or Quints, have a permanently mounted fire pump, a water tank, a hose storage area, a power aerial ladder or elevating platform with a permanently mounted waterway, and a complement of ground ladders. (See Figure 14.6.)

5. Special service emergency response vehicles are multipurpose vehicles that primarily provide support services at emergency scenes.

FIGURE 14.6 Quints are considered combination emergency response vehicles. *Courtesy of A. K. Rosenhan*

6. Initial attack vehicles are equipped with a permanently mounted fire pump with a minimum capacity of 250 gallons per minute (950 L/min), a water tank, and a hose body. Its primary purpose is to initiate a fire suppression attack on structural, vehicular, or vegetation fires, and to support associated fire suppression operations.

7. Mobile foam emergency response vehicles are equipped with a permanently mounted fire pump, foam proportioning system, and foam concentrate tank(s). Their primary purpose is for use in the control and extinguishment of flammable and combustible liquid fires in storage tanks and other flammable liquid spills.

There are additional combinations of emergency response vehicles:

■ The Quad configuration preferred by some fire and emergency services organizations is equipped with a fire pump, hose, water, and ground ladders.
■ Wildland units are especially useful in off-road operations. (See Figure 14.7.)

There are no separate categories for rescue vehicles, which are included under special service emergency response vehicles.

There are two basic chassis classifications for emergency response vehicles:

■ A custom chassis built specifically for the fire and emergency services
■ A commercial chassis being a standard truck chassis adapted for fire and emergency services use (See Figure 14.8.)

In addition to the requirements for the emergency response vehicle chassis itself, there are numerous considerations for the configuration and amount of suppression hardware to be included. Various materials are available for vehicle bodies, including steel (both Galvaneal and stainless), aluminum, and plastic. Water tanks may be of the same materials or a combination thereof.

All emergency response vehicles have been designed and constructed to provide specific combinations of functions and equipment for a specific set of jobs and localities.

FEATURES OF EMERGENCY RESPONSE VEHICLES

Although many of the automotive aspects of emergency response vehicles have simply been by-products and adaptations from the commercial truck market, other features have been developed as a result of specific requirements and needs of the fire and emergency services. Input from firefighter unions, fire and emergency services organizations, insurance companies, and the legal profession has added additional design/performance requirements and properties to both emergency response vehicles and equipment.

FIGURE 14.8 Truck chassis are adapted for the fire and emergency services emergency response vehicles. *Courtesy of A. K. Rosenhan*

Features of emergency response vehicles include:

- Body
- Hose storage configurations
- Emergency response vehicles for EMS use

Body
The body itself will usually consist of an operator position; a hose bed storage area; various configurations of inlet and outlet fittings as well as master stream devices; a water tank; compartments for various equipment items; and external equipment such as ladders, suction hose, and tools. Figures 14.9 and 14.10 show a side-mounted and a top-mounted control panel, respectively. There are also rear-mounted control panels as well as hybrid configurations, some completely enclosed for weather conditions.

Hose Storage Configurations
Hose storage configurations vary greatly with the typical pre-connected attack hose lines coming off the side of the body and larger diameter supply and discharge lines coming off the rear of the vehicle. Both front and rear suction inlets are common, as are "trash lines" on the front bumper or side-mounted lines and rear attack lines.

Emergency Response Vehicles for EMS Use
Over the years, as the fire and emergency services has taken on the added responsibility of providing emergency medical services (EMS), the design of emergency response vehicles has changed. Additional equipment such as defibrillators, medical supplies, and other medical equipment storage requirements is taking up valuable storage volume on the emergency response vehicle. Some emergency response vehicles have even been configured for patient transport.

PURCHASING CONSIDERATIONS
Emergency response vehicles come in a variety of automotive designs, colors, configurations, and equipment inventories. Particularly in the U.S. fire and emergency services,

FIGURE 14.9 Side-mounted driver/operator pump control panel.
Courtesy of A. K. Rosenhan

FIGURE 14.10 Top-mounted driver/operator pump control panel.
Courtesy of A. K. Rosenhan

there are seldom two emergency response vehicles that are identical, even in the same department. Such variation is expensive, as each vehicle is more or less custom built. Some fire departments try to pool their purchases with other fire departments or make multiple purchases. In addition to the expense aspect of custom vehicles, there is the problem of uniformity as it relates to training and operations. When a fire department's vehicle is different, even from the same manufacturer, the differences can cause confusion for firefighters trying to find tools and equipment on similar vehicles. Likewise, the inventory of repair parts, maintenance capabilities, and other operational considerations are important from both an economic and emergency response vehicle availability standpoint. The need for a fire department to standardize emergency response vehicles and equipment becomes obvious. The effects of economics, training, and operational safety significantly vary from department to department. Some purchasers are constrained to a low bid, but good specification writing and a thorough review of submitted bids will eliminate non-standard or significantly varied purchases and ensure the purchaser gets what is desired.

ROLE OF NFPA 1901

Although there are many combinations of design considerations in building the emergency response vehicle, the foundation of such activities is contained in NFPA 1901, *Standard for Automotive Fire Apparatus* (2009), and other standards referenced by NFPA 1901. NFPA 1901 covers all facets of design, performance, and function of various types and configurations of the emergency response vehicle. In fact, that standard is the basis for all types of emergency response vehicles. A review of the latest edition of NFPA 1901 is necessary to understand the emergency response vehicle (including older vehicles) and allied equipment.

The members of the NFPA 1901 Technical Committee—which is composed of representatives from the fire and emergency services, emergency response vehicle manufacturers, and regulatory agencies—are always interested in defining problems and determining solutions.

Over the years NFPA 1901 has basically changed from a design standard to a performance standard that now incorporates the following: specific seat-belt requirements, a data recorder ("black box"), specific retro-reflective stripping requirements, emergency and other lighting requirements, cab occupant noise limits, vehicle vertical center of gravity (VCG) limits, and other very specific functions and features related to firefighter safety such as speed limits. In addition, the industry has provided supplemental restraint systems (air bags), disc brakes (with anti-lock braking system, or ABS), cab and body crash testing per European Union Standard EC29R (a legal instrument of the member nations of the European Union), various suspension configurations, electronic controls for pump and aerial device operations, and tire-pressure monitoring systems. Specific performance standards relating to human factors, environmental issues, and even the mounting of equipment in the cab area are now included in the standard.

EVOLUTION OF NFPA 1901

In the 1980s, statistics indicated an unacceptable number of firefighter deaths and injuries due to falling off or getting knocked off a moving vehicle. Although it was obvious that getting firefighters off the tailboard and exposed riding positions was a great step forward for personnel safety, even the open-type jump seats that became prevalent did not offer sufficient protection. In 1991, NFPA 1901 was updated to include the requirements that all personnel have an enclosed seating arrangement and that seat belts be provided for all occupants. Emergency response vehicle design has come a long way since the days of open cabs and exposed riding positions. (See Figure 14.11.)

As a result, design/configuration requirements of the typical four-door cab came into vogue, for both commercial and custom chassis. Subsequently the three-point safety belt

(a)   (b)

FIGURE 14.11 Emergency response vehicle design has changed throughout the years. *(a) Courtesy of A. K. Rosenhan, (b) Courtesy of George Russell, Nashville Fire Department*

became standard, along with air bags and other restraint devices. In addition, there were requirements for the anchoring of any equipment to be carried inside the passenger compartment to eliminate flying objects in the event of a crash.

New requirements for improved emergency vehicle visibility and warning devices, along with secondary braking devices for emergency response vehicles with more than a 36,000-pound gross vehicle weight rating, occurred in 1996. Other improvements in 1999 included anti-lock braking systems, handrails, and diamond-plate slip-resistant stepping surfaces, and manufacturers were required to conduct predelivery testing.

In the 1990s, it became evident that problems existed in the frequency, severity, and type of accidents involving emergency response vehicles. NFPA data showed that some 25% of all firefighter deaths and injuries occurred during travel to and from the emergency scene. Many of those resulted from rollover incidents in which the emergency response vehicle ultimately rolled over after various initial maneuvers, which was due in part to a high vertical center of gravity as well as excessive speed and road conditions. The sloshing of water and shifting loads contributed to the instability of the vehicle. Subsequently, the NFPA recognized some specific problems and incorporated additional design and performance standards into its publications. The 2003 edition of NFPA 1901 required hose bed restraints after several civilians were killed and injured during emergency vehicle response, and made other advancements toward safety.

The 2009 edition of NFPA 1901 incorporated over one hundred additions related to safety. Some of the new requirements for emergency response vehicles include the following:

- Seat-belt warning device with a 60-inch minimum lap belt and 110-inch shoulder harness.
- European-style retro-reflective stripping that is in a chevron pattern at a 45-degree angle covering 50% of the rear.
- Vehicle data recorder that can collect data on the most recent hundred engine hours. The information can be downloaded in the event of an accident.
- Speed restriction for any emergency response vehicle that has a 26,000-pound gross vehicle weight rating will be limited to 68 miles per hour, whereas those with more than 50,000 pounds having 1,250 gallons of water or combined water/foam will be limited to 60 miles per hour.
- Diesel particulate filter that can be activated by the ignition of the engine or manually.
- Location for helmet storage.

Some other requirements that must come with each new emergency response vehicle delivery include traffic vest, traffic cones, illuminated warning devices, automatic external defibrillator (AED), and stepladder or multipurpose ladder.

The most significant change in the 2009 edition dealt with the rollover issue. The standard provided that the stability function may be accomplished by one of three means:

- Electronic stability control (ESC)
- Calculation of the vertical center of gravity to meet specified criteria stated in NFPA 1901
- Testing of the completely loaded vehicle on a tilt table

The first method is an expensive one although it is becoming readily available on commercial truck chassis. There are driver/operators that do not like a computer taking away the control of the vehicle (by manipulating throttle and individual wheel brakes). The second method, although relatively simple to calculate, involves a great deal of detail work in order to gather the necessary data for each part and assembly on a vehicle and provides a somewhat limited analysis.

The third method, whereby the completed and loaded vehicle is tilted 26.5 degrees, is the simplest to do and the most reliable. However, it does require the manufacturer to make a considerable outlay for the equipment. (See Figure 14.12.) The angle value corresponds to a lateral acceleration that is presumed to be the practical limit for safety of an emergency response vehicle. The tilt table method takes into consideration tire and spring squat, as well as body movement, which the other two methods do not.

OLDER EMERGENCY RESPONSE VEHICLES

The NFPA standards have addressed rehabbed or rebuilt emergency response vehicles. Typically fire departments that possess older vehicles may wish to keep using them, depending on what various outside organizations (such as fire insurance rating bureaus and

FIGURE 14.12
Emergency response vehicles' stability testing is now required. *Courtesy of A. K. Rosenhan*

FIGURE 14.13 Service life for emergency vehicles varies from state to state and department to department. *Courtesy of Wayne Haley*

NFPA) recommend. As a result, NFPA 1901 now includes an Annex D for rehabilitated emergency response vehicles, and NFPA 1912, *Standard for Fire Apparatus Refurbishing* (2011), contains additional requirements. Typically, some rating organizations will put a service life of 15 years on a commercial chassis vehicle and 20 years on a custom chassis vehicle, with aerial ladder vehicles having a 25-year rated service life. Those lifetimes vary from state to state, department to department, and in some cases are even based on labor contracts in career fire departments. Obviously, some emergency response vehicles are worn out long before these limits are reached, whereas others are in pristine condition after many years of service. Such is a result of a combination of frequency of use, type of use, and certainly the maintenance and care given the emergency response vehicles. Some rating organizations have policies regarding a Service Life Extension Program (SLEP) that provides for the modernization of older emergency response vehicles. (See Figure 14.13.)

Fire Equipment

Whereas the emergency response vehicle is well regulated, the equipment and tools utilized by firefighters are not always as carefully analyzed. Certainly the various NFPA standards, as well as those of other standards-making organization, have rules and recommendations pertaining to the design and use of tools. Much attention must be given to the use of both hand and powered tools by the firefighter.

SAFE USE OF EQUIPMENT

The use of power saws, hydraulic rescue tools (such as the Jaws of Life), cutters, axes, and other tools provides a number of pinch points, sharp edges, unexpected forces, and other physical phenomena that generate significant injury potential. Hydraulic rescue tools may be powered with internal combustion engines or by electricity. The usual safety considerations for engines, such as hot components and fuel, are obvious, whereas certain battery-operated safety items are of low DC voltage only. Hydraulic rescue tools can operate at pressures in excess of 15,000 psi, and in the case of a leak or failure are a significant hazard to personnel. Certainly the forces generated by such rescue tools are formidable, and operators must always be extremely careful regarding pinch points, sudden movement of the tool due to various means, and the effect of the tool on the material/structure being worked on.

Equipment and tools of particular interest include:

- Personal protective ensemble
- Extinguishing equipment
- Ground ladders
- Self-contained breathing apparatus
- Personal alert safety system

Personal Protective Ensemble

personal protective ensemble (PPE) ■ Collection of protective gear consisting of helmet, coat, pants, boots, gloves, hood, and personal alert safety system (PASS) device, which may be incorporated into a self-contained breathing apparatus.

A basic tenet of firefighter safety is to be equipped with and use personal protective gear, also called **PPE** for **personal protective ensemble**, appropriate for the activity involved. Personal protective ensemble is covered by NFPA 1971, *Standard on Protective Ensembles for Structural Fire Fighting and Proximity Fire Fighting*. Personal protective ensemble is the collection of protective gear consisting of helmet, coat, pants, boots, gloves, hood, and personal alert safety system (PASS) device, which may be incorporated into a self-contained breathing apparatus. The NFPA 1971 Technical Committee has primary responsibility for documents on the design, performance, testing, and certification of protective clothing and protective equipment manufactured for fire departments and personnel. It reviews the selection, care, and maintenance of such protective clothing and equipment that protects hand, foot, torso, limb, head, and interface protection (except respiratory protection). Related NFPA standards include:

- NFPA 1976, *Standard on Protective Ensemble for Proximity Fire Fighting*
- NFPA 1977, *Standard on Protective Clothing and Equipment for Wildland Fire Fighting*
- NFPA 1991, *Standard on Vapor-Protective Ensembles for Hazardous Materials Emergencies*
- NFPA 1992, *Standard on Liquid Splash-Protective Ensembles and Clothing for Hazardous Materials Emergencies*
- NFPA 1994, *Standard on Protective Ensembles for First Responders to CBRN Terrorism Incidents*
- NFPA 1999, *Standard on Protective Clothing for Emergency Medical Operations*

Significant improvements in the design and materials of PPE have been made in the last several years. Components are much lighter, more flexible, and they offer a greater degree of protection than previous designs and materials. However, the increased protection has come at a price, and a significant investment is necessary to fully equip a firefighter. (See Figure 14.14.)

In addition, the exposure to chemicals, hazardous materials of various types, an unexpected fire during rescue operations, and the impact due to collapse or shifting of items all require adequate personal protective gear to prevent or at least mitigate injury. Any operators of rescue tools should be intimately familiar with their capabilities and operations as well as be adequately supervised during their operation.

Extinguishing Equipment

fire hose ■ A hose that can be attached to a hydrant, standpipe, or similar outlet to supply water to extinguish a fire.

Being able to put water on a fire is the basic method of extinguishing a fire, although specific types of fires, such as electrical, chemical, and petroleum, require other extinguishing agents. Since the introduction of **fire hose** (a hose that can be attached to a hydrant, standpipe, or similar outlet to supply water to extinguish a fire) by the Dutch in about 1672, there have been many improvements in delivering water to a fire. The ability to deliver water to the fire requires lightweight hose that does not easily kink, is resistant to chemicals and heat, is not easily cut or damaged, and will continually provide the firefighter with an adequate firefighting capability, thus increasing safety.

The first water mains were simply hollowed-out logs. (See Figure 14.15.) The first fire hose, made of leather or heavy sailcloth, was a great improvement over the previous method of simply squirting water through a nozzle attached to a hand-powered pump. (See Figure 14.16.) Later developments involved a woven cotton jacket with a rubber lining with various sizes being supplied. With fire hose it was possible to get away from the pump and water source and actually enter a fire building. That capability brought other factors into consideration, such as personal protective ensemble, communication between firefighters, and the hazards involved in breathing smoke and products of combustion.

As pumps became more powerful and were ultimately powered by steam and internal combustion engines, the fire hose was greatly improved. Although the standard hose

FIGURE 14.15 The first water mains were made out of hollowed-out logs. *Courtesy of A. K. Rosenhan*

FIGURE 14.16 An example of one of the first fire hoses. *Courtesy of A. K. Rosenhan*

started out being 2.5 inches (63.5 mm) in internal diameter, today's hose is routinely up to 6 inches (152 mm) in diameter, as well as being constructed of synthetic materials that resist mildew, deterioration, and physical damage.

For many years the 2.5-inch (63.5-mm) hose was universally used with larger 3-inch (76.2-mm) and 3.5-inch (89-mm) hose for large flows from fireboats and high-pressure water systems in larger cities. Screw threads and coupling design varied from city to city, and even today hundreds of fire hose threads are used in the United States. (See Figure 14.17.) The 1.5-inch (38-mm) hose universally used to pre-connect hose lines is standard throughout the United States, as are threads used for 1-inch (25.4-mm) reel lines. With the advent of the larger 4-inch (102-mm) hose, the use of **Storz couplings**—non-threaded couplings for fire hose, consisting of interlocking hooks and flanges, that are actuated by a quarter turn—originally imported from Europe, is now commonplace. (See Figure 14.18.) NFPA 1961, *Standard on Fire Hose,* and NFPA 1963, *Standard for Fire Hose Connections,* cover these items.

Ground Ladders

Another basic tool for the firefighter is the ground ladder. Safe, stable, and portable ladders are important to both firefighter safety and emergency scene operations. Ground ladders range in height from 10 feet (3 m) to 55 feet (16.75 m) and may be composed of multiple, extending sections. NFPA 1931, *Standard for Manufacturer's Design of Fire Department Ground Ladders,* and NFPA 1932, *Standard on Use, Maintenance, and Service Testing of In-Service Fire Department Ground Ladders,* list the requirements for such ladders and in-service testing.

Storz coupling ■ A non-threaded coupling for fire hose, consisting of interlocking hooks and flanges, that is actuated by a quarter turn; used mostly for large-diameter hose (LDH) of 4 inches (102 mm) or more; sometimes called a "sexless" or hermaphroditic connection as both ends of a hose have identical fittings.

FIGURE 14.17 There are still various fire hose threads and coupling designs being used. *Courtesy of A. K. Rosenhan*

FIGURE 14.18 Storz coupling.
Courtesy of A. K. Rosenhan

Self-Contained Breathing Apparatus

The use of a **self-contained breathing apparatus (SCBA)**, a wearable respirator that is carried by an individual and supplies breathable air, is now standard for any sort of firefighting. Whether for a simple car fire, a serious structure fire, or even an outdoor fire, the use of SCBA must be mandated due to the high probability of toxic products of combustion. The days of "smoke eaters" are gone; and OSHA, NFPA, and other regulatory bodies require the proper equipping and use of respiratory protection. NFPA 1981, *Standard on Open-Circuit Self-Contained Breathing Apparatus (SCBA) for Emergency Services*, governs the function and use of SCBA for the fire and emergency services. (See Figure 14.19.)

SCBA started out as a simple tank of compressed air, a regulator, and a face piece. Over the years various types of low-air/low-pressure alarms have been added, as have heads-up displays regarding the operating conditions of the overall system, built-in radio communication, and personal alert safety system (PASS) devices, as discussed next.

self-contained breathing apparatus (SCBA) ■ Wearable respirator that is carried by an individual and supplies breathable air.

FIGURE 14.19 There should be no more excuses when it comes to properly wearing your PPE and SCBA. *Courtesy of Travis Ford, Nashville Fire Department*

There are constant developments in SCBA hardware with innovations relating to weight, size of units, and information systems. Given the complexity of modern SCBA apparatus and the obvious importance to life safety while using SCBA, the user should be quite familiar with the normal and abnormal operations of the apparatus.

Personal Alert Safety System

The **personal alert safety system (PASS)** is a battery-operated device that senses movement or lack of movement of a person, and automatically or manually activates an audible alarm signal to alert and assist others in locating a firefighter who is in danger. PASS may be separate devices or incorporated into the SCBA system, and it is specified in NFPA 1982, *Standard on Personal Alert Safety Systems (PASS)*. However, most PASS devices are now integrated into and activated with the use of the SCBA.

Maintenance and Testing

A good maintenance program for emergency response vehicles and equipment is important to ensure all systems are functioning properly. Knowing that the emergency vehicle and equipment will meet the challenge when needed to perform in an emergency situation should be a primary concern. A clean and well-maintained emergency vehicle, along with equipment, is very important to overall safety. However, it is vital to first become familiar with the emergency vehicle and equipment so that one understands what the capabilities and limitations are. A daily preventive maintenance program that includes routine maintenance checks should be conducted every shift. It will help extend the life of the emergency response vehicle and equipment by identifying mechanical problems when they are small.

Some states have rigorous inspection practices. In fact, certain states have emergency response vehicle and equipment regulations and then OSHA steps in occasionally. In addition, some states have U.S. Department of Transportation (DOT) inspection and service requirements to include emergency response vehicles. Other states have nothing. There are a lot of considerations other than type and purpose of emergency response, response run time, and age of emergency response vehicle; because of the demands placed on emergency response vehicles, they have a limited service life.

Because every emergency response vehicle has a limited service life, the fire department should have scheduled maintenance routines, arrange frequent visual and operational inspections by trained personnel, and keep records of such. Proper documentation of vehicle records is important for a variety of reasons. As in the medical field, "if it isn't written down, it didn't happen." Proper documentation can be used to show that maintenance issues are taken care of in a timely manner.

In 2007, NFPA incorporated three previous standards into one. Now, NFPA 1911, *Standard for the Inspection, Maintenance, Testing, and Retirement of In-Service Automotive Fire Apparatus,* contains information from the previous NFPA 1911 standard for service test of fire pump systems on fire apparatus along with NFPA 1914 and NFPA 1915.

NFPA 1915, *Standard for Fire Apparatus Preventive Maintenance Program,* contained material specific to the care and maintenance of all types of emergency response vehicles. The standard identified the minimum recommended scheduled maintenance of emergency response vehicles. The former NFPA 1914, *Standard for Testing Fire Department Aerial Devices,* prescribed testing to be done on a specified schedule. However, two of the main recommendations from the new NFPA 1911 standard are that pre-1991 apparatus be refurbished and placed in reserve and that apparatus more than 25 years old be retired.

NFPA 1071, *Standard for Emergency Vehicle Technician Professional Qualifications,* identifies the training and education needed to qualify as an emergency vehicle technician. It is reported that only 26% of mechanics are actually certified emergency vehicle technicians.

Summary

Emergency response vehicles and equipment serve not only to make emergency scene operations more efficient but also to make those operations safer for the firefighter. For the vehicles and equipment to serve during emergencies, it is imperative that the firefighter be trained to operate them safely, perform daily inspections, and adhere to maintenance programs.

This chapter has reviewed the types and features of emergency response vehicles, with particular attention to safety considerations, and explained the role and evolution of NFPA 1901. In addition, information has been presented on equipment and tools, including discussion on their safe use and the critical role of personal protective ensemble, extinguishing equipment, ground ladders, self-contained breathing apparatus, and personal alert safety systems.

Every emergency response vehicle and equipment has a limited service life. Proper service and repair is an important part of maintaining good operational equipment so that it is usable when the department needs it to be.

Review Questions

1. Discuss the role of emergency response vehicles and equipment in firefighter safety.
2. Identify the seven types of emergency response vehicles identified in NFPA standards.
3. Discuss the requirements for refurbishing older emergency response vehicles.
4. Describe the difference between custom chassis and commercial chassis emergency response vehicles.
5. Discuss the role of scheduled maintenance and testing of emergency response vehicles and equipment in ensuring firefighter safety.

References

National Fire Protection Association. (2007a). *NFPA 1911, Standard for the Inspection, Maintenance, Testing, and Retirement of In-Service Automotive Fire Apparatus.* Retrieved November 5, 2010, from http://www.nfpa.org/aboutthecodes/AboutTheCodes.asp?DocNum=1911

National Fire Protection Association. (2007b). *NFPA 1971, Standard on Protective Ensembles for Structural Fire Fighting and Proximity Fire Fighting.* Retrieved November 5, 2010, from http://www.nfpa.org/aboutthecodes/AboutTheCodes.asp?DocNum=1971

National Fire Protection Association. (2009). *NFPA 1901, Standard for Automotive Fire Apparatus.* Retrieved November 5, 2010, from http://www.nfpa.org/aboutthecodes/AboutTheCodes.asp?DocNum=1901

National Fire Protection Association. (2011). *NFPA 1912, Standard for Fire Apparatus Refurbishing.* Retrieved November 5, 2010, from http://www.nfpa.org/aboutthecodes/AboutTheCodes.asp?DocNum=1912

Occupational Safety and Health Administration. (2010). *29 Code of Federal Regulations Part 1910: Subpart I.* Retrieved November 5, 2010, from http://www.osha.gov/pls/oshaweb/owadisp.show_document?p_table=STANDARDS&p_id=10118

INDEX